FIREFIGHTER EMOTIONAL WELLNESS

JADA HUDSON

FIREFIGHTER EMOTIONAL WELLNESS

HOW TO RECONNECT WITH YOURSELF AND OTHERS

Fire Engineering®
BOOKS & VIDEOS

> **Disclaimer**
> The recommendations, advice, descriptions, and methods in this book are presented solely for educational purposes. The author and publisher assume no liability whatsoever for any loss or damage that results from the use of any of the material in this book. Use of the material in this book is solely at the risk of the user.

Copyright © 2022 by
Fire Engineering Books & Videos
110 S. Hartford Ave., Suite 200
Tulsa, Oklahoma 74120 USA

800.752.9764
+1.918.831.9421
info@fireengineeringbooks.com
www.FireEngineeringBooks.com

Senior Vice President: Eric Schlett
Vice President: Amanda Champion
Operations Manager: Holly Fournier
Sales Manager: Joshua Neal
Managing Editor: Mark Haugh
Production Manager: Tony Quinn
Developmental Editor: Chris Barton
Cover Designer: Trent Farar
Book Designer: Robert Kern, TIPS Publishing Services, Carrboro, NC

Library of Congress Cataloging-in-Publication Data
Names: Hudson, Jada, 1969- author.
Title: Firefighter emotional wellness : how to reconnect with yourself and others / Jada Hudson.
Description: Tulsa, Oklahoma : Fire Engineering Books and Videos, [2022] | Includes bibliographical references and index. | Summary: "An evidence-based self-help book for first responders with interviews with first responders and how they adapt"-- Provided by publisher.
Identifiers: LCCN 2021048194 | ISBN 9781593705725 (paperback)
Subjects: LCSH: Fire fighters--Job stress. | Fire fighters--Mental health. | Fire extinction--Psychological aspects.
Classification: LCC HD8039.F5 H83 2022 | DDC 363.37023--dc23
LC record available at https://lccn.loc.gov/2021048194

All rights reserved. No part of this book may be reproduced, stored in a retrieval system, or transcribed in any form or by any means, electronic or mechanical, including photocopying and recording, without the prior written permission of the publisher.

Printed in the United States of America

1 2 3 4 5   26 25 24 23 22

In loving memory of my parents,
James B. Hudson and Odelia Hudson

My father gave me my spark and my determination. He was a man who loved his family to his core. His tremendous energy and passion for life made him so fun to be around. I carry many of the important life lessons he taught me, and I'm so very grateful. My mother had such a beautiful, quiet presence. She was equally as strong as my father in different ways; her strength was subtle but unwavering. She loved her family with a fierce loyalty that could be counted on no matter what. The confidence that she would be supportive and that her love was unconditional, no matter what horrible place I found myself in, always helped me get through any situation. When I think of my two parents together, I simply received everything I needed. Their belief in me gave me a belief in myself, and it's that foundation that has made this book possible.

Contents

Foreword by Robert S. Hoff..xv
Introduction xvii
Acknowledgments xxi

1 Reconnecting: What It Means to Be Emotionally Well...................................1
   What Is *Connected*?...2
   Actions of Balanced People 3
   The Firefighter Support Web 4
   Looking at Ourselves from Seven Dimensions of Wellness.............................4
   Reflection Questions 9

2 Rescuer's Depression 11
   The Unique Kind of Depression First Responders Develop 13
   Avoiding Emotion in First Responder Jobs 14
   Firefighters: Heroes and Humans 15
   Instrumental Personality Type 15
   Self-Medicating the Feelings Away..16
   The Difference Between Clinical Depression and Hidden Depression 18
   Depression and Anger 19
   Depression and Sleep Disturbances..19
   What Happens When First Responders Stay Busy 20
   The Downside of Being Busy...21
   What If I Do Not Want to Sort Through My Thoughts? 22
   Staying Busy and Neglecting Yourself 22
   How to Stay Intentionally "Un-Busy"..23
   The Overwhelming Experience of Facing Trauma On and Off the Job 23
   How Can I Begin Healing from Depression?...24
   Reflection Questions 25

3 Needing To Be Needed and Loneliness..29
   Mark's Story...29
   The Rescuer Identity...31
   Where Is the Line Between Wanting to Help and Needing to Be Needed?..............32

What Causes First Responders to Need to Be Needed? 33
The Unshakable Loneliness of an Emotionally Neglected Child. 34
What Is the Upside? 34
Becoming Amazing Firefighters. 36
The Difference Between Needing to Be Needed and Compassion 37
Overcoming the Need to Be Needed 38
How Mark Overcame His Need to Be Needed 39
The Loneliness Epidemic. 39
The Physical, Mental, and Behavioral Impacts of Loneliness. 40
Joiner's Theory on Loneliness 40
The Link Between Broken Attachments in Childhood and Loneliness. 41
When Loneliness Shows Up at Work 42
Your View of Yourself. 43
Reflection Questions 44

4 Substance Abuse and Addiction 47
Brett's Story. 47
What Is Addiction? 47
What Does Addiction Do for the Individual?. 48
The Development of an Addiction 49
The Biology of Addiction 50
Risk Factors for Addiction. 51
Childhood Emotional Neglect. 52
What Scientists Discovered About Rat Parks and Addiction. 53
How to Identify an Addiction 54
Connection: What People with Addictions Actually Crave 54
Self-Care: What People with Addictions Need to Prioritize. 57
Relapse and Triggers. 57
Signs of Relapse 58
What Does True Recovery Look Like? 61
Supporting Someone with an Addiction 61
10 Ways Family, Friends, and Peers Can Help Someone in Recovery 62
Treatment for Addictions 64
Reflection Questions 64

5 Suicidal Ideation 67
Ryan Elwood's Story. 67
Joiner's Theory on Suicide 69

When Suicide Hit My Own Family... 71
Situations That Contribute to the Desire to Commit Suicide 73
Practicing Self-Harm... 75
First Responder Personality Type and Risk 75
Why Does It Feel Like I Hear About One Suicide After Another? 76
Point Cluster Suicide in Fire Departments: Chief Hojek's Story 77
Shneidman's Theory on Suicide 79
Acute Suicidal Affective Disturbance (ASAD) 80
Retirees' Risk of Suicide... 81
The Truth for Those Who Consider Suicide 82
Are Addiction, Depression, and Suicide Linked?..................................... 83
Where to Turn for Help 83
Reflection Questions 84

6 Pediatric Death.. 87
Why Pediatric Death Is So Hard to Process 88
Jack's Story.. 88
Factors That Make a Pediatric Death Harder to Process.......................... 89
Mike's Story 89
Processing Pediatric Death... 91
When Pediatric Death Becomes More Than You Can Bear 91
Rick's Story.. 92
Reflection Questions 93

7 Stress and Trauma... 95
What Is Stress?.. 95
Chronic Stress... 96
What Is Trauma?.. 97
Reactions to Trauma.. 98
Trauma and the Brain 99
Mirror Neurons... 101
The Freeze Response 101
What Is the Result of the Freeze Response? 104
Mind-Body Intervention 104
Disconnecting After Trauma 105
Trauma and the Nervous System 105
First Responders Misdiagnosed with "Bipolar Disorder"........................ 107
Reflection Questions 110

8 Post-Traumatic Stress Disorder (PTSD) 113
What Is Post-Traumatic Stress Disorder? 114
Symptoms of PTSD 114
Ben's Story 115
PTSD Assessment 116
Complex PTSD 116
Moral Injury 117
Reflection Questions 119

9 Resiliency 121
What Is Resiliency? 121
What Is the Difference between Coping and Resiliency? 122
Politics and Resiliency—Blake's Story 122
If It Does Not Kill You, Does It Make You Stronger? 124
Daily Thinking: Optimism Versus Pessimism 125
How to Stay on Top of Your Thoughts 127
The Seven Resiliencies 128
Marks of Resilient People 130
Recharging Your Batteries 133
What Can You Control? 133
Resiliency Through Transition 134
What to Do When You Are in Transition 134
Assistance Can Help You Hold Steady 135
Adjustment Disorder in First Responders 136
Searching for More of Yourself 136
Maybe You Need to Slow Down 137
Reflection Questions 137

10 Post-Traumatic Growth 141
Growing from the Pain 143
A New Perspective on Life 144
Reflection Questions 145

11 First Responder Self-Care 147
Anthony's Story 147
What Is Self-Care? 149
Where Does Self-Care Start? 150
The Duty to Care for Yourself 150
Self-Care Practices for First Responders 151

Exercise for Mental Health 152
Exercise in Overcoming Addictions 156
Self-Care for Anger 156
Nutrition for Mental Health 157
Nature for Mental Health 160
Intentional Actions for Happiness 160
Self-Care Actions That Are Also Treatments 161
Science Supporting Yoga and Meditation for Trauma and PTSD 164
What Occurs in the Body During Yoga and Meditation? 166
What Occurs in the Brain During Yoga and Meditation? 167
Reflection Questions 168

12 How I Treat Trauma in My Practice 173
Accelerated Resolution Therapy 173
How Does ART Work? 174
Is It Effective? 174
Mike's Story Using ART 175
Narrative Therapy: Writing Your Way to Healing 176
Prolonged Exposure/Imaginal Exposure 178
How Does Prolonged/Imaginal Exposure Work? 179
What Happens During Prolonged/Imaginal Exposure? 179
Reflection Questions 180

13 Your Place in the Brotherhood 183
Making Personal Breakthroughs Using Firefighting Tactics 184
What the Recruits Walked Away With 184
Using Your Unique Strengths in the Brotherhood 185
Introverted Leadership 186
Peer Support 187
Brotherhood in the First Responder Careers 188
Core Values Assessment 189
Reflection Questions 190

14 Paternalistic/Maternalistic Leadership, Training, and the Importance of Alliances 193
Brad's Story 193
Trajectories for People Facing a PTE 194
Challenge Appraisal Versus Threat Appraisal 194
Other Factors Influencing Challenge Appraisal Versus Threat Appraisal 194

- Leadership That Prevents PTSD ... 196
- Training Subordinates for Physiological Toughness ... 196
- Training Implications for Leaders ... 198
- Modeling Vulnerability and Coping Flexibility ... 200
- Preparing Subordinates for PTEs ... 200
- How to Lead Before a PTE ... 201
- What Is Paternalistic/Maternalistic Leadership? ... 202
- How to Mitigate Anxiety as a Leader ... 202
- How to Lead During a PTE ... 204
- How to Lead After a PTE ... 205
- Are Critical Incident Stress Debriefings Effective? ... 206
- Using Fire and Rescue Alliances ... 207
- Leaders Gearing Up for Stratospheric Success ... 209
- Reflection Questions ... 209

15 Healthy Relationships—Premarital and Marriage ... 213
- The First Responder Family ... 213
- Children ... 214
- First Responders Need Time to Recover: 24-On/48-Off Shift Patterns ... 214
- Communication ... 214
- Paramilitary Communication Style in First Responders ... 214
- How First Responder Stress Can Impact Intimacy ... 215
- When First Responders Have an Overactive SNS ... 216
- The Unique Way First Responders Experience and Respond to Conflict ... 216
- Basics of Listening Well ... 217
- Finances ... 217
- Differing "Money Scripts" ... 218
- Your Unconscious Beliefs About Money ... 219
- Money Scripts Examples ... 220
- The Emotions of Money ... 221
- Illogical Money Beliefs ... 222
- The Importance of a Budget in Reducing Marital Financial Stress ... 223
- Setting a Money Agreement in Your Marriage ... 225
- The Danger of Overworking ... 226
- Sex and Intimacy ... 226
- Why Does Infidelity Happen? ... 227
- The Connection Between Sex and the Emergency Response ... 227
- Compromising When It Comes to Sex ... 227

First Responder Emotional Dangers 228
The Importance of Self-Care for First Responders 228
How the Spouse Can Help 229
Advice for Spouses Who Want to Help 229
Increasing Positive Relationships 229
Positive Words Start with You 230
Positive Actions Are Exciting 231
Can "Happily Ever After" Really Exist? 231
Using Intuition Versus Communication at Home 233
Senses and Equipment 234
Communication in Fires Versus Communication at Home 234
The Secret to Better Relationships Is Not What You Think 235
Pulling Weeds and Planting Seeds 235
The Mundanity of Excellence 235
The Mundanity of Relational Excellence 236
Basics of Relational Excellence 236
Reflection Questions 237

16 Cancer 239
Emotions of Cancer 239
Keys to Emotional Wellness in the Face of a Cancer Diagnosis 240
Breakdown of Relationships During Cancer Treatment 241
Cancer and the Family 242
When Children Are Asking About Cancer 243
Suicidal Ideation in Cancer Patients 243
Helping Those with a Cancer Diagnosis 244
Finishing Business 245
Reflection Questions 245

17 Live Like You Were Dying 247
The Death of a Situation 247
Escaping the Idea of Death 248
Living Like You Could Lose What You Have 249
A Premortem for Life 250
Reflection Questions 251

18 Retirement and Late-Onset Stress Symptomatology (LOSS) 253
Why Retirees Need to Focus on Emotional Wellness 254
Retirees Facing Depression 255

    Retirees Facing Substance Abuse and Addictive Behaviors 257
    Retirees Facing Suicide. 258
    Emotional Trauma. 259
    Late-Onset PTSD 259
    Late Onset Stress Symptomatology (LOSS) 262
    Why Do These Memories Show Up During Retirement? 262
    Erikson's Cognitive Stages 263
    Ron's Story 264
    What Can I Do to Overcome LOSS? 265
    Reflection Questions 266

19 Planning for Your Ideal Retirement 269
    Making the Most of the Retired Years. 269
    A New Definition of Retirement 270
    Retirement Planning. 270
    Changing Roles. 271
    The Five Cs of Successful Aging 271
    Financing a Healthy Retirement. 272
    Bridging the Gap from Full-Time Work to Retirement. 273
    Crossing the Bridge. 273
    Working for Your Well-Being. 274
    Motivated by Autonomy. 274
    Successful Retirement 275
    Reflection Questions 275

20 Epilogue: What Now? 277

Bibliography 279

Index 297

Foreword

Our world is full of unseen things. Everyone walking around us has thoughts, emotions, and memories hidden beneath the surface. Those who pass them by only catch glimpses of what is really going on inside of them.

First responders often hide their memories and emotions, keeping their experiences safe from exposure and safe from hurting anyone else. In my 47 years as a firefighter, I saw a lot of really bad things, and I saw a lot of firefighters handle their pain differently. Often firefighters do not talk about the death and destruction they have seen. Sometimes that is because they are processing it in their own way. Often it is because they do not really know how to deal with it.

My story has many layers of pain and loss, like most first responders. Trauma and loss of life are always hard to grapple with, especially when there is something about it that makes it affect you personally. Some calls hurt more than others, but tragedy always hurts a person emotionally and physically. Every first responder has a choice to walk through grief instead of burying it deep inside.

You see, grief is like a bridge. Behind you is the life you once had before your painful memory or tragic loss. Ahead of you is a long road that will hold a lot of tears. But on the other side of the bridge is a life where you are back on solid ground, resilient, and able to carry with you all the good things you have learned through the loss. It is a new life. It is not the same life, but it is a meaningful and hopeful life.

Building this kind of resiliency takes time, and first responders are beginning to realize that mental fitness is just as important as physical fitness. Today we are heading that direction. When I was the fire commissioner in Chicago, we did not begin to use the critical incident stress debriefing (CISD) until the early 1990s. We have always had counseling for drugs and alcohol, but treatment for mental fitness was not exactly considered a big thing. Now counseling and peer support are becoming more common, and I am so glad for it.

If we are going to be men and women who are powerfully effective at our jobs, then we need to be trained physically and mentally to do our jobs well. Our leaders need to know their people. Every individual needs to be informed, educated, and trained both physically and mentally, or they will fail. Part of the first responder leader's job is knowing his people and being almost like a parent to them because the world is full of hidden things.

Are your firefighters okay? As a leader, if you go to work, and if no one is talking to your firefighters about how they are really doing, that is not good. Unless you have been involved in a death, serious injury, or suicide, you do not really think about it. Is your firefighter reacting to a specific call or another on-the-job incident, or is his home life influencing his actions?

We used to prize "mental toughness" and being prepared, but that was much different than today's view of dealing with pain, talking about emotional stress, and becoming resilient. Today's perspective is so much more balanced and holistic, and I am glad we are focusing on getting our people well on every front.

Jada Hudson's book *First Responder Emotional Wellness* is valuable and well researched, and it will benefit you and your firefighter family. This book is a noble gift. Use it as reference to look for signs of challenges faced by a first responder.

I recommend this book to any first responder, fire officer, firefighter, and "probie" who has ever experienced loss or tragedy and may not know how to deal with it effectively. The tools in this book will help you at any and every stage of your first responder career.

Respectfully,

Robert S. Hoff
Fire Commissioner, Chicago Fire Department, retired
Fire Chief, Carol Stream Fire District, retired
Field Staff Instructor, Illinois Fire Service Institute, University of Illinois

Introduction

I love first responders. Growing up, I was surrounded by first responders. My father, who served in the US Army during World War II, worked as a correctional officer for a maximum-security penitentiary. My brother was a firefighter. My uncle was a New York City police officer. And more of my cousins than I could count were either in the military or were police officers. It is in our blood. I love first responders.

Being raised in a first responder family was a beautiful thing. I have never met people prouder of their jobs than the first responders in my family and their coworkers and friends. In fact, did you know that firefighters are the happiest workers in America, according to a Bloomberg study?[1] The study found that firefighters scored 9 out of 10 on the job satisfaction scale, ranking just higher than pediatricians, guidance counselors, and communications professors. This truly is the best job in the world!

Jobs with the highest job satisfaction rates often have a clear mission, social support, room for growth, respect from society, and a sense of self-actualization, and firefighting captures all of these traits. Firefighters have a clear mission. They know their job priority is to save lives and keep the community safe. There is no ambiguity with where to focus every day, and there is absolutely nothing like saving a life. On top of that, firefighters have an unbeatable community, offering unbeatable social support. The brotherhood is unlike what most people ever experience with their coworkers. You have their backs, and they have yours. Even in times of stress, firefighters bond together to weather hardship as a team, as brothers. Added to that, firefighters have countless opportunities for personal growth, ranging from trainings to advancing rank. There is no glass ceiling in the fire service. And beyond rank or income, firefighting is one of the most selfless acts a person could undertake. So many people do not live life to the fullest, but firefighters get the pleasure of helping humanity, which makes them feel empowered, connected, and alive. Firefighters know this even better than I do. The upsides of working for the fire service are unbeatable.

Growing up, I was proud of my first responder dad, and later my brother and uncle. I would not change a thing about my upbringing, but I must admit that, in addition to all the upsides of having a first responder dad, my childhood had some dark and confusing times. My dad was my hero, and our society held him and his colleagues in the highest regard. Yet there were some responders who held secret pain that they tried not to reveal. My dad was one of them. Without an awareness of first responder issues, particularly post-traumatic stress disorder (PTSD), my family was often left confused and isolated as we tried to support my dad. My oldest brother later committed suicide, and it had a profound impact on me and my family.

My love for my family and my deep concern for the pain first responders experience launched me on a quest to understand what first responders experience in their jobs, in their minds, in their hearts, and in their families. I began a search to determine how I could help make those experiences a little more beautiful and balanced.

As an adult, I have spent countless hours around first responders in the Chicago area. My counseling education has deeply examined the first responder experience, and I have attended national trainings related to emotional wellness for first responders. I have left no stone unturned because I want to help first responders as thoroughly as possible.

Firefighters are passionate. They chose this career because they want to help people. They are hardworking, courageous, and without a doubt, committed. I have watched them knock down walls and break through metal doors, all while still maintaining a clear awareness of their team around them. They are committed to the brotherhood. They can have fun and not take themselves too seriously. And I love that the environment around firefighters is changing in such a way that they can be honest with one another about their struggles and the things that they have seen.

I have spent the better part of my counseling career helping first responders sort through the issues that this unique field raises. Truly it is okay to be okay, and many first responders go throughout their careers without facing emotional wellness issues. But I have found that encountering emotional wellness roadblocks in the midst of a firefighter career can be surprising and disconcerting. It is my passion to help first responders find wellness again instead of remaining stuck in their pain. As a firsthand witness to my father's own lack of emotional wellness and mental health support, I am beyond excited to be able to put down on paper so many of the tools, stories, and methods of care that can help first responders around the world be able to live more connected lives. I am thrilled that first responders are now able to talk about their pain in a much more charitable atmosphere, where they can learn, heal, and grow instead of turning to secret vices such as alcohol, affairs, pornography, and other substance or behavioral addictions.

How to Use This Book

In the pages of this book, you will find truth, sometimes hard truth, about what it looks like to be an emotionally well firefighter. I have selected the issues that I think are most integral in the first responder experience, and I have delved into each of those to provide you with insight and tools so that you can overcome them if you should ever face them in your own life.

In these pages I explain what I have learned from psychologists, scientists, and first responder experts. I share stories of other first responders I know personally and the ways that they encountered each of these unique issues, so that you can see how they healed and reconnected with themselves to find emotional wellness. Each story shared in this book was shared with permission, and the names of these first responders have been changed to protect their privacy. I hope their stories will encourage you.

Feel free to skip around and dive into the chapters that apply most readily to your life. You might, however, find it helpful to read other chapters as well, because when a firefighter gets out of balance, it usually affects more than one area of his or her life. Emotional wellness issues are often interconnected. If a person is struggling with anger, he is likely struggling with depression also. If a first responder is dealing with substance abuse, he is more likely to begin thinking about suicide. If a firefighter is struggling under the weight of traumatic memories, he is likely to want to escape them via substances. Emotional wellness is more interconnected than you might think.

Even if you are not struggling now, you may encounter an emotional wellness roadblock at some point in the future. This book is full of tools for your emotional wellness toolbelt, which will help you if you encounter a dark time. You may find these ideas helpful for your brothers as well. Feel free to invite your first responder brothers to read the chapters that apply to them, and feel free to show chapters to your spouse or significant other to help them understand what you are going through. I encourage you to open up and discuss what you are learning from this book with a trusted friend, family member, coworker, or counselor. I highly recommend writing your thoughts down on paper, especially if you are more introverted, as a way to sort through experiences, identify your senses in each experience, and find truth amid the pain.

In my marriage section, I encourage you to spend some alone time with your fiancée or spouse to open up about the hardships that come with a role as a firefighter. I have seen many first responder marriages suffer because the first responder is unsure of what is appropriate and how to share with his or her spouse. This chapter includes activities that invite you to step out of your training, which is usually intuitive, and focus on connecting, which is usually conversive. Doing so can help you develop a deep, connected relationship with your significant other.

Think of this book as a training exercise for your emotional wellness. Pursuing emotional wellness and finding yourself in this career is a critical part of your success as a firefighter and human being, but it does not mean that you will have perfect understanding right away, if ever. What it does mean is that you will begin to find tools to help you grapple with what you have seen. It means that you will try to look into areas of your life that need some light to shine on them. It means that you will find hope and inspiration to resolve trauma, be honest and real, and thrive in your future as a first responder.

If you are reading this book, it is likely that you are a firefighter or are somehow connected to firefighting. Thank you for what you do. Thank you for the sacrifices that you have made to keep our communities safe. Thank you for every single person that you have touched and every single trauma that you have seen. I know it is not easy. This book is for you.

A Note about Gender Pronouns

I use male pronouns quite often throughout this book. This is not to make a normative statement that all responders are male. In fact, I celebrate the increasing gender diversity in the fire service. Currently, however, 93% of firefighters nationwide are male, while only about 7% are female, and around 65% to 70% of EMS personnel are male, while 30% to 35% of EMS personnel are female. Also, many of my counseling clients are male. Emotional wellness issues relate to women as well, especially women with instrumental personalities (more about this later), which most female firefighters have. I invite all male, female, and nonbinary individuals to engage with this material. Please recognize that all of my pronouns come from a place of profound respect and a desire to serve all responders.

Note

1. D. Rovella and A. McIntire, "Firefighters Are the Happiest Workers in America," Bloomberg, July 17, 2019, https://www.bloomberg.com/news/articles/2019-07-17/which-jobs-make-people-the-happiest-in-america.

Acknowledgments

To my daughter Kendall, my everything. My relationship with my daughter is the deepest connection I have ever experienced. You are the greatest gift I have ever received. You possess such intelligence and humor, and the way you care for others in everything you do shines bright. Thank you for supporting me on this journey in so many ways. You spent your school years with me when I began this journey. You have always been so curious and patient about my work. We have had some great memories traveling to firehouses all over the country and having all kinds of adventures. There is no one I would rather have had by my side. The best is yet to come, Baby Girl!

To my good friends Colleen and Cathleen, who have been through so much with me, always true, trusted friends. In the hard times in my life, they have been there to talk to. My hope is that all my first responders have these kinds of friendships, where you know that you are not judged and can be who you are and be supported. We all need people that we can just be human with, and that is what Colleen and Cathleen are for me. They accept me, flaws and all, and have been along this journey the whole time.

To my friend Paula, who helped me get into private practice. Thank you for providing an opportunity for me to develop my private practice and mentoring me through all of the beginning ins and outs of running my business. Thank you for being my trusted confidant and for being a safe place to be vulnerable. I cherish you.

To Fire Chief Bobby Halton (retired). You have believed in me and in the idea behind this book wholeheartedly, and I am overwhelmed with gratefulness for you. You are so genuine, intelligent, sincere, and FUNNY. I am so thankful for your friendship. Your support opened doors for me to be able to present at FDIC, and your belief in me truly motivated me in the process of writing this book. You are a treasure.

To Christopher Barton and Tony Quinn. Thank you so much for your patience. You have been so supportive, and your feedback gave me the confidence that this project is indeed worthwhile and valuable.

To Amy Bayer, my assistant editor, who is the wind beneath my wings. She is such a pure, sweet soul. She has given so much to make my business successful, and I could not have done any of this without her. Thank you for not giving up on me and for your immense patience. I believe that everything you do is done with your belief in God, and it is evident in your excellence. I am blessed to have you in my life.

To Nicole. I appreciate your meticulous editing for my bibliography. Your organization skills were exactly what I needed to wrap this project up with confidence. However, more importantly, you have become such an important part of Kendall's life. You are her teacher, mentor, role model, and big sister. "Thank you" is truly not enough to express what you have done to help the most important person in my life.

To Javier, who helps me run my business efficiently and accurately. It is because of you that I have even more time to work face-to-face with my first responder clients. You take so much off of my plate, which allows me to do the work that I love.

To Therese, my trusted friend and assistant. You are so very resilient, strong, and caring. Therese brings such respect, and such a loving heart, to the practical side of my business. My business would not be as organized or as productive without her care. My clients sense her warmth and sincerity in every phone call. I am so lucky to have developed a friendship with you over the years.

To Dan Shiradelly, the person who first suggested I think about working with first responders. It changed my entire professional life, which ultimately made beautiful changes in every aspect of my life. I have never been so passionate or purposeful about my career, and I owe it to him to have sparked that thought. People come into your life for different reasons, and his entire family opened their arms to me with love and care that will touch my heart forever.

To Matt Olson, who allowed me to grow to a place I never thought possible. He is my brother in this whole fire service environment. I have always been the listener, and I am a listener as a counselor, and so I never really had a voice until I met Matt Olson. I helped him with a presentation one day when he needed a favor, and from there, I have not stopped presenting. He gave me the confidence to relay all the information I am so passionate about. He also helped me to understand the fire service in so many ways and with so much patience—the politics, structure, legalities, emotional wellness issues, trauma, and really everything I could ever want to know.

To Dr. Robert Langman. Everyone who meets you instantly feels important and cared for. You are simply the best doctor in Illinois. I am so blessed to have a friendship with such an intelligent, interesting, and genuine person. I appreciate the time you spent validating the science in this book. I feel so much more confident in my approach to firefighter emotional wellness with your stamp of approval.

To Chief Hugh Stott, who along with retired Chicago Fire Commissioner Bob Hoff and Firefighter/Paramedic Matt Acuff took me under their wings and taught me all about firefighter trainings. They allowed me to sit in on any training I would ever want, and they helped equip me with an understanding of firefighter protocols during a variety of incidents. There is so much I have learned because they allowed me to take part in these trainings that not many people are allowed access to. It has been a total immersion, and I am so honored by their patience and mentoring.

To Dr. Thomas Joiner. I have immense respect for the research he has conducted and for his theories. Dr. Joiner has changed how I work with my first responder clients on a daily basis. I appreciate how he makes himself available for any question I may have regarding his work and research. He has done important work with the fire service and suicide research, and his willingness to communicate with me has been incredibly helpful.

To Dr. Stan McCracken. Dr. McCracken allows me to converse with him and extract some of the experiences of his long career with the military to explore how they might relate to firefighters. Thank you for giving me so much of your precious time to contemplate and discuss interesting connections between military and the fire service.

To Dr. Joel Fay, Dr. Marla Friedman, and Dr. Anne Bisek. They are all friends, psychologists, and absolute experts. They have taught me so much about what I know, just by allowing me to hang out with them for weeks at a time and observe them and ask any questions. They have given me, on a silver platter, so much of the knowledge that they have worked to obtain for more than 30 years. I am ever so grateful that they have taken me in.

To Jeff Dill. My respect for him is immense. He is the first person to collect research on firefighter suicides, and he has done a meticulous job. He makes his research available by sharing his numbers willingly and immediately, which is so generous.

To my friends Rob and Dawn Fox. They have not only helped me in this first responder world but have also become delightful friends that I have made through my work in this field.

To Dr. Diego Hernandez. Dr. Hernandez has given me such great tools by training me in therapeutic trauma treatments. Thanks to him, I have solid protocols that have helped me with so many first responders. He is the best instructor and has also become quite a good friend.

To Dr. Mayanil. I appreciate your help with my yoga chapter. I have a love for yoga, and I intuitively know that it will bring health and happiness to so many firefighters. You helped me to articulate the why. I am now more confident with the science of what I already knew but found difficult to form into words. I know you dropped everything in your very busy work to help, and I appreciate it greatly.

To Jerry Marzullo, attorney and lieutenant with the City of Berwyn Fire Department. You have been a very helpful resource for me in my quest to help firefighters. You are a very busy man, but for me you have always been so available, friendly, and powerfully insightful. I am grateful for your advice and for our time on the board of directors at Illinois Firefighter Peer Support. Your support has given me confidence in so many projects! Thank you.

To Rich Dory, Battalion Chief (retired), Chicago Fire Department. I value your friendship more than you know. Our conversations are always interesting, authentic, and memorable, and I am blessed by your perspective and refreshing insight into life.

***

I feel so blessed to have these people who are experts in their field at my fingertips. If I have a question, or if I want to run a case by someone, I have a whole Rolodex filled with many of the top experts in the country. I credit my success in this field to the people who are willing to teach me. Because of their generosity and lack of ego, I feel I have been given a head start in this work. It is a beautiful field to be working in, made up of a bunch of beautiful people, and I feel so privileged to be a part of it.

CHAPTER 1

RECONNECTING: WHAT IT MEANS TO BE EMOTIONALLY WELL

A few years ago, I received a call from a fire chief who was concerned about one of his firefighters, Pete. The chief had always considered Pete to be one of his best firefighters, but there had recently been some worrisome changes in Pete's behavior. Pete's productivity had dropped, and he had become angry and snappy. The chief did not call me in an effort to punish Pete; Pete was not in trouble. The chief was concerned, however, that Pete was heading for trouble if he continued down the same path. The chief wanted to help Pete, so he sent him to talk to me.

Disconnected first responders can be found in every county and in every department. Because of the shift-oriented lifestyle of a first responder, stress and trauma on the job, emergency calls that never seem to end, and the lack of time for self, many first responders walk around angry, disappointed, bored, frustrated, and simply burned-out. Because first responders are typically hard-working, uncomplaining people, the average friend or civilian might not notice the incongruity in behavior. But every once in a while, a lack of balance will show up as an emotional or physical health crisis, the collapse of a relationship, a snafu on the job, or another life debacle. That is when they often seek professional help.

For Pete, there was a dramatic drop in productivity paired with a surprising rise in cynicism. Pete was an excellent firefighter who was highly active in a number of extra assignments. He had always been a high achiever, but for the first time, he started feeling resentful of how hard he was working compared to those around him. On top of being a firefighter, he held a part-time job, and he spent his free time flipping houses, doing all the work himself.

So Pete decided to do a complete one-eighty. In one of our first counseling sessions, he explained to me that he had decided to drop to the level at which he perceived everyone else was working. As Pete explained, "I'm just going to work like everyone else does. I have been giving 200%, but now I will only give 100%." Since everyone around him was accustomed to seeing his 200% effort, however, his reduced performance was quite noticeable.

For a long time, Pete had been spreading himself too thin. He pushed his needs aside to focus on achievement, as many first responders do. He became resentful and isolated under the endless list of demands weighing on him. He felt disconnected from the person he wanted to be because of the person he currently was.

Pete needed to reconnect. He needed to find himself again and get back to what was most important to him.

In our work together, Pete began to realize that he needed to slow down. He began taking things off his plate. He resigned from his part-time job and started doing more things that he loved to do. He started working out more, spending more time with his wife, and traveling. He became increasingly comfortable with "doing nothing."

In his quest to reconnect, he began asking himself, "What, really, is my motivation for high achievement?"

For many first responders, achievement is driven by childhood abuse or neglect, but for Pete, bullying from his childhood peers motivated his actions as an adult. In more than 20 years of listening to people, I have discovered that many first responders have been bullied by peers in the past, and it can whisper a subtle lie into the heart of a vulnerable child: "You're not good enough. You need to prove yourself." These bullied kids often grow up to be highly busy adults who have more than proven that they are sufficient.

Many first responders believe that the more they achieve, the more loveable they are. This seemed to be true for Pete, who was working hard to achieve so that he felt worthy. Although his parents had been wonderful and supportive, his peers had been vicious. Pete was charismatic, funny, good-looking, and smart. Never in your wildest dreams would you think he had a history of being bullied, but bullying was the driver. As a child, Pete had wanted desperately to belong and to feel good enough, and so he filled his adult schedule with opportunities to prove he was capable and worthy of respect.

Achievement is like a drug, and when you slow down, you can feel odd or antsy. So next, I told him to expect discomfort with doing nothing. His job was just to sit with that uncomfortable feeling for a while. I am happy to say that after six months of therapy and slowing down, Pete is doing well.

Maybe you are reading this because you feel disconnected. In your first responder career, you feel a bit lost, bored, frustrated, unhappy, or burned-out. Where do you even begin? A useful place to start is to understand what it looks like to be "connected," so you can begin to envision what your life may look like if you reconnected.

What Is *Connected*?

A *connected* person is an emotionally well person. *Emotional wellness* is dealing with difficulties in a healthy, intentional way, so you can think about things clearly; live a life free from secrets, burdens, and shame; and be ready to continue living, taking on new dreams, and bouncing back in the face of adversity. Connected does not mean stuffing down the painful memories you have experienced. It does not mean you keep your struggles hidden from your spouse, friends, family, or coworkers.

Connected is the opposite of "Suck it up, Buttercup."

Connected involves balanced thinking, authentic conversations, being present and in the moment, and being flexible and adaptable throughout life's changes.

Connected is a confident ability to face hard days on the job with resiliency in the face of a tough call.

Studies from Benedek and others and the National Institutes of Health (NIH) have noted that connected people are able to handle life's stresses in healthy ways and are also able to adapt to change.[1] When a first responder is connected and self-aware, it pours out into everyday life, job performance, relationships, thoughts, physical health, and sense of well-being. According to experts at NIH, there are three signs of connected people:

1. They have more positive emotions and fewer negative emotions.
2. They are resilient amidst stress and change.
3. They can keep perspective and appreciate the good things in life.[2]

Emotional wellness can be defined as "the ability to be aware of and accept our feelings, rather than deny them, have an optimistic approach to life and enjoy life despite its occasional disappointments and frustrations."[3]

Does this sound like where you are at right now? If not, remember Pete. Pete knew he was out of balance. When he turned his attention toward becoming well again, he was surprised by how good life could be. When Pete reconnected with himself and slowed down his schedule, he found life to be so much more abundant, peaceful, and full of real connection with others than he had experienced in the past. He learned that achievement does not earn love. Reconnected people can see such truths emerge in their lives.

It is not that people who are connected with themselves do not have stress, loss, or problems in their lives; it is that they have learned how to manage stress, to process and grieve losses, and to resolve problems and conflicts in their lives in healthy ways. They *choose* to heal, resolve pain, and thrive. They *choose* to pay attention to their needs, so they can take care of them and become well for the long run.

If you are reading this book, it is likely that you are ready to do that, too. So get out your pen, and let us pinpoint some areas where you would like to grow. As you read "Actions of Balanced People," place a check mark next to the action that you want to grow in your life.

Actions of Balanced People

- Get enough sleep, ideally seven to nine hours a night.
- Reduce stress by being with friends and family.
- Exercise in a manner that is balanced, not excessive.
- Spend time on self-care, including healthy eating, taking breaks, and spending time outdoors.
- Deal with losses in a healthy way, including talking to trusted people and getting support.
- Surround themselves with community, including friends, church, sports, and other social networks.

- Make new friends and keep long-term friends.
- Think through decisions based on personal values and are comfortable with taking wise advice.
- Seek help when needed and work independently when needed.
- Enjoy reciprocal relationships. Sometimes you will be on the giving end of the friendship, and sometimes you will be on the receiving end.
- Try new things confidently.
- Are comfortable with being challenged by personal growth.
- Know that they are responsible for their actions.

Emotional wellness is about bringing yourself into balance. When one area is out of balance emotionally, it can affect other seemingly unrelated areas of life. That is why this book is set up to discuss the wide variety of issues first responders face. Many first responders discover that they are wrestling with something that seems disconnected from their most obvious issue—such as trauma, for example—when, in fact, seemingly unrelated struggles are often entwined. Feeling disconnected or out of balance is the direct result of a mind that is trying to come back to homeostasis.

I suggest viewing your emotional wellness as a web of support. The most "connected" first responders have a broad web of individuals and habits that sustain and help them recover in the midst of a busy, stressful career. Those who do not have a widespread web of support may find themselves barely coping when they encounter trauma, physical exhaustion, and emotional turmoil.

The Firefighter Support Web

If I could design a web of support around you to help you become balanced and healthy, here is what it would look like (fig. 1–1).

As you begin to work toward becoming a more emotionally well, connected first responder, one of your primary tasks will be to build a robust social support web for yourself, focusing your energy on the actions of balanced people which you circled above.

Looking at Ourselves from Seven Dimensions of Wellness

In 1976, Dr. Bill Hettler developed a list, "Six Dimensions of Wellness," which I like to use as a mirror for my clients to analyze their wellness.[4] A seventh dimension, environmental wellness, is often added with this list.

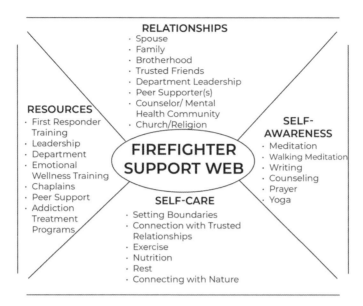

Figure 1–1. Circle the items on this firefighter support web that you know are lacking in your own social support web.

People are much more than physical beings, which means that their wellness is much broader than just physical health. A person's quality of life is largely affected by how well he or she is in seven specific areas of life: physical, emotional, intellectual, social, spiritual, environmental, and occupational. These areas of life are often called the "Seven Dimensions of Wellness."[5] Let me explain each of them.

1. Physical Wellness

Physical wellness means getting adequate sleep, physical activity, hydration, and nutrition. It means caring for your body and eliminating harmful habits like smoking, drug use, and excessive alcohol consumption. Physical wellness can be judged by how well you feel, how many injuries you have, and the results from your physical exam at the doctor's office.

You can build physical wellness through daily exercise, good sleep habits, wearing protective equipment, recognizing illness and seeking help when needed, eating a variety of nutritious foods, consuming appropriate portions for your body, stopping smoking, and drinking alcohol in moderation or abstaining from alcohol use altogether.

Based on this description, what grade would you give yourself in the dimension of "physical wellness"? _____

2. Emotional Wellness

Emotional wellness is heavily influenced by the other six dimensions of wellness. Emotional wellness means processing experiences around you, growing from experiences, and allowing yourself to feel the feelings associated with them. This starts with identifying your feeling. Then by addressing it properly, you can help yourself become emotionally well and balanced. Choosing a positive outlook and choosing resiliency both contribute strongly to emotional wellness.

You can build emotional wellness by identifying your thoughts and feelings, communicating these thoughts and feelings with others, choosing to look at things optimistically, practicing reconnecting with yourself, seeking support when needed, talking about pain with safe people, and accepting and forgiving yourself.

Based on this description, what grade would you give yourself in the dimension of "emotional wellness"? _____

3. Intellectual Wellness

Intellectual wellness means doing things that are creative and stimulating. Your mind is meant to be challenged, inspired, and illuminated. If you are intellectually well, you are open to possibilities and use your mind to expand your knowledge and continue to improve your skills. This involves staying up-to-date on the industry's latest best practices, choosing activities that stimulate the mind, and reading a variety of literature.

You can build intellectual wellness by attending courses and workshops, learning new languages, earning new certifications, reading, subscribing to interesting podcasts, choosing mentally stimulating hobbies, and learning to appreciate the arts.

Based on this description, what grade would you give yourself in the dimension of "intellectual wellness"? _____

4. Social Wellness

Social wellness means communicating in healthy ways, connecting intimately within safe relationships, creating a support network of friends and family members, and interacting with your global community positively. This means respecting others and yourself. It means building a sense of belonging in your family and community. Often a person will still consider a longtime friend "close," even though communication with that friend is infrequent. Yes, that individual is a part of your social support network. Building new friendships and continuing to connect with old friendships, however, will be critical for social wellness across your whole lifespan.[6]

You can build social wellness by investing time in healthy relationships, getting involved in community events and friendships, and communicating thoughts, emotions, and ideas with those around you. Choosing to reach out and spend time with old and new friends is like putting nails in the studs of a new house. The more nails you add, the stronger the house becomes. People who are socially well are hard to knock down in a storm.

Based on this description, what grade would you give yourself in the dimension of "social wellness"? _____

5. Spiritual Wellness

Spiritual wellness means assessing your values and priorities and choosing faith, hope, and commitment to a belief system. Those who are spiritually healthy are willing to seek meaning and purpose, to question and appreciate the things that cannot be understood, and to look outside the physical world for answers to their existential questions. Spiritual health is about harmony with others and connection with a higher power or forces outside life on this earth.

To build spiritual wellness, you can spend time reading spiritual books, meditate, pray, and ask questions that go beyond daily thinking. You can seek guidance from a chaplain, spiritual counselor, or pastor. You can choose to be fully present, to ask questions, to grow through life's challenges, and to explore your spiritual core.

Based on this description, what grade would you give yourself in the dimension of "spiritual wellness"? _____

6. Environmental Wellness

Environmental wellness means connecting with nature and the world around you. Nature benefits the other six dimensions of wellness. According to Colin Capaldi, Carleton University, and others, numerous studies have shown that *nature contact*—brief or intermittent exposure to plants, pictures of nature, paintings, or nature on television—and *nature connectedness*—longer-term interaction with the outdoors and natural surroundings—promote positive mental health. Capaldi and others also note that exposure to nature improves mental clarity and focus, physiologically reduces stress in the body, and creates a domino effect into other areas of life, influencing more creativity, generosity, connectedness, and resilience.[7] Nature touches almost every type of wellness struggle. It alleviates anxiety, depression, loneliness, fatigue, stress, and even physical illnesses.

A person can build environmental wellness through personal contact with nature. Visiting beautiful places releases dopamine in the brain that creates happiness, joy, and a desire to pursue goals. Going on walks outside, having a picnic, hiking local trails, camping at a nearby campsite, exploring national parks, going fishing or hunting, lying in the grass with your children, watching videos about outdoor locations, and even putting potted plants in your home or workspace will help balance all your dimensions of wellness. The natural oils released from trees send signals to the brain, reestablishing wellness. Perhaps one reason so many people struggle with emotional wellness these days is because we are surrounded by concrete. We spend less than 10% of each day outside.[8]

Being near the water is also rejuvenating for the brain.[9] Another way to build your environmental wellness is to spend time near a body of water—an ocean, bay, river, or lake. In a society that is overstimulated and constantly involved in social networking, it is easy for our minds and emotions to become overloaded. Nature is a welcomed break, beckoning you to move and breathe freely. Connecting with nature is a key piece to your wellness puzzle.

Based on this description, what grade would you give yourself in the dimension of "environmental wellness"? _____

7. Occupational Wellness

Occupational wellness means bringing the best of yourself, your skills, your passions, and your gifts to the table in the way you spend your time. Those who use their unique skills at work, at home, and in personal hobbies will feel more accomplished and fulfilled than those who are working outside of their skill sets. They will feel motivated and will be able to maintain a positive attitude about work. People who are occupationally well are satisfied with, though challenged by, their jobs.

You can become occupationally well by creating a vision for your future in the fire service, identifying personal skills, being intentional about sharing those with the brotherhood, dreaming about a retirement occupation, and being open to change and using other pieces of your unique skill set.

Based on this description, what grade would you give yourself in the dimension of "occupational wellness"? _____

**

If you are a first responder, it is likely that at least one of these areas feels disconnected from the rest of your life. When one area of our lives is not well, it puts strain on the rest of our lives. When we are out of balance, we are less resilient to the trials, challenges, or hardships of life. Instead of staying well, we fall into a cycle of physical, emotional, and social problems, compensating for those with unhealthy behaviors and eventually experiencing unwellness and functional decline (fig. 1–2).

Remember, people are holistic. You have so many layers and parts, and each of those plays into the bigger picture of how you are doing.

So, really . . . *how are you doing?*

Are you ready to learn about how to face some of the disconnected areas in your life?

I hope so. There is so much life and joy for those who reconnect.

Figure 1–2. Emotional wellness trajectories

Reflection Questions

1. In the "Actions of Balanced People," which actions on this list are easy for you? Which ones are difficult? Why?
2. Take a look at the Firefighter Support Web. Circle the quadrant in which you are strongest. What are some of the strongest supports in your life? Underline them.
3. In the Firefighter Support Web, draw arrows pointing to areas in which you are weakest. What are some supports you need to add to your life?
4. The "Seven Dimensions of Wellness" include physical, emotional, intellectual, social, spiritual, environmental, and occupational aspects of wellness. Which dimension of wellness do you think is strongest in your life?
5. Which dimension of wellness do you think is weakest in your life? Why is that? What can you do to build your wellness in this area?

Notes

1. US Department of Health and Human Services, National Institutes of Health, "Emotional Wellness Toolkit" (as updated December 10, 2018), http://www.nih.gov/health-information/emotional-wellness-toolkit; and D. M. Benedek, C. Fullerton, and R. J. Ursano, "First Responders: Mental Health Consequences of Natural and Human-Made Disasters for Public Health and Public Safety Workers," *Annual Review of Public Health* 28, no. 1 (2007): 55–68, https://doi.org/10.1146/annurev.publhealth.28.021406.144037.
2. US Department of Health and Human Services, NIH, "Emotional Wellness Toolkit."
3. Center for Health and Wellbeing, "The Seven Dimensions of Wellbeing," accessed August 10, 2021, https://yourhealthandwellbeing.org/about/sevendimensions/.
4. B. Hettler, "Six Dimensions of Wellness," National Wellness Institute (1976), https://nationalwellness.org/resources/six-dimensions-of-wellness/.
5. Center for Health and Wellbeing, "The Seven Dimensions of Wellbeing."
6. T. E. Joiner, *Lonely at the Top: The High Cost of Men's Success* (New York: Palgrave Macmillan, 2011).
7. C. A. Capaldi et al., "Flourishing in Nature: A Review of the Benefits of Connecting with Nature and Its Application as a Wellbeing Intervention," *International Journal of Wellbeing* 5, no. 4 (2015): 1–16, https://doi.org/10.5502/ijw.v5i4.1.
8. Capaldi et al., "Flourishing in Nature."
9. Capaldi et al., "Flourishing in Nature."

Chapter 2

RESCUER'S DEPRESSION

My father was a correctional officer for more than 20 years at Stateville Correctional Center, one of the toughest maximum-security prisons in Illinois. In the 1970s and 1980s, there were numerous gang riots at Stateville, and they had frequent lockdowns. My dad was an excellent shot, so they assigned him to the towers as a sharpshooter. To this day, I can only imagine the things he saw, heard, and did at Stateville. Like most first responders, he sheltered me from his pain by avoiding mention of it. I grew up entirely unaware of any sadness in him, only aware of his temper (see fig. 2–1).

Until my dad died in 2017, I did not understand the depth of trauma and depression that shrouded his career and life. He hid it well. In fact, he stayed so busy that no one could get a clear glimpse of any emotion in him—that is, except anger.

Like my dad, many first responders struggle with depression. They conceal it from the world around them and often conceal it from themselves. Masters of compartmentalizing, they keep work and emotions separate. Emotions get pushed to the back burner, while tasks, projects, and work remain in the fore. This is why a first responder can outwardly seem perfectly okay while actually walking around each day carrying a millstone of depression that threatens to grind him to a halt if he loses momentum long enough to catch his breath. He is busy because he is sad, and he does not want to feel it. He would rather keep moving.

Figure 2–1. My dad as a young man

Because first responders are excellent at compartmentalizing, symptoms of depression may never show up externally; nevertheless, they are present. In a quiet moment, the first responder may feel sad, anxious, empty, hopeless, guilty, worthless, helpless, irritable, restless, or uninterested in formerly pleasurable activities, according to the American Addiction Centers. They also note that the first responder may experience fatigue, difficulty concentrating, trouble remembering details, early-morning wakefulness, trouble sleeping or excessive sleeping, overeating or appetite loss, thoughts of suicide, suicide attempts, persistent physical pain all over, headaches, or other physical distress.[1]

Looking back, I think my dad had more of these symptoms lurking beneath the surface of his life than he would have liked to admit. Like most first responders, he had moments of strong anger that signaled something more was happening beneath the surface. He could be the most charismatic, friendly, big-personality person on the planet. He was often away from home, but when he was around, his communication was supportive. His eye contact and body language were focused on me like I was all that mattered. I was a daddy's girl growing up, and he could be funny, encouraging, and kind.

But Dad also had another side. Below the surface, he was a highly anxious, truly angry person. I was never scared of him physically because I knew he would not lay a hand on me or anyone in our family, but I was scared of his loud voice, his anger, and his unpredictability. What I understand now that I did not at the time is that he was living with depression and post-traumatic stress disorder (PTSD). My dad hid his pain well, but I realize now that when he got angry, it was because he was in crisis. Anger can feel like a way to fight instead of being vulnerable and victimized.

My dad always had three or four jobs going on at the same time. He owned a laundromat and a pool hall, and he rehabbed houses. When I was a young girl, the pool hall was bringing in a great deal of money. The laundromat was also successful and was responsible for laundering all of the towels for the local public schools. He also worked a great deal of overtime at Stateville. He was an incredibly successful man by all financial standards, but he was an incredibly impoverished man at heart.

My dad grew up in poverty. When he was a child, his family was regularly evicted from apartments when they could not pay their rent. He was born in 1923, during the Great Depression. His dad could not find work to support his family of five, which meant they literally went to bed hungry at times. This childhood poverty was accompanied by the vulnerability that my dad felt and the abuse that he experienced growing up.

Dad carried his poverty into adulthood. He oriented his real estate business around his ability to endure quite uncomfortable living situations. He bought houses that needed a great deal of work. Even though these houses were nearly uninhabitable to most people, my dad and mom and I would actually move in while he fixed them up. When the repairs were finally completed, he would sell the house, and we would move into another home, just to make that home livable for someone else. We kept doing this, and I kept trailing along on his wild ride. He was like a miser who has plenty of money but lives in a house that is falling apart. My dad's idiosyncrasies were like ashes from his traumatic childhood that were daily stirred up, just to leave a film of dust over his freshly built life.

The Unique Kind of Depression First Responders Develop

In my experience as a first responder counselor, I have discovered that there are two factors that can tip a person over from emotionally balanced to depressed. When these factors are combined, they can overwhelm the psyche. The first factor is negative childhood experiences, and the second is trauma.

First responders have a boatload of traumatic or potentially traumatic memories to draw from on a daily basis. Many first responders have childhood pain as well. Combine them, however, and the brain may link contemporary traumatic situations to similar feelings of helplessness experienced in the past from adverse childhood situations. Together these present and past experiences surround the first responder with a subconscious sense that the world is a dark place to live. Add to this our society's expectations for first responders, and the result may be what I term *rescuer's depression*.

Across the board—EMS, fire service, law enforcement, dispatch, and so on—depression rates are high among first responders. A study of Japanese emergency workers in the wake of the 2011 East Japan earthquake found that 21.4% had clinical depression.[2] According to another study, 6.8% of EMS professionals reported struggling with depression.[3] The list goes on, and I suspect these numbers are mere shadows of the true number of first responders who struggle with rescuer's depression. In my experience, rescuer's depression is more widespread than clinical depression. It is just harder to spot because sufferers try to hide the symptoms.

In my work as a first responder counselor, I encounter rescuer's depression frequently. It can be a hidden depression. Most of my clients are so busy that it takes a number of meetings together before we begin to pinpoint that there is more going on than meets the eye. Many of my clients come to me for more obvious reasons—marital problems, alcohol problems, job problems—only to discover the lurking depression that motivates many of their actions.

My dad was actually in therapy for quite a while, and he found it to be helpful. He initially went after getting a divorce from his first wife because he had so much guilt about leaving his two sons from that marriage. My brothers were much older; there is nearly a 15-year age gap between the younger brother and me. My dad was always working and was not around for them. Like many first responders, his thought was, "If I'm a good provider, that's what I need to do, and that's how I show my love."

He learned the hard way that making a lot of money is not the same as connecting with family, and I think it hurt him deeply to recognize that he was absent emotionally for his sons. I am proud of him for starting to address his sadness over his failed marriage and emotional absence by going to counseling. Pain needs to be talked about to be healed. It is like unwrapping a bandage on a flesh wound. The wound needs to be unwrapped, looked at, cleaned, and stitched back together correctly in order to heal. Emotional pain is the same, which is why keeping it a secret or avoiding it altogether by staying too busy to feel anything will prevent it from healing.

Secret pain, whether from childhood or adulthood, festers until it becomes darker and darker, and your heart becomes sicker and sicker. Running from it will not heal it. You need light to shine on it.

Avoiding Emotion in First Responder Jobs

My dad was the king of avoiding emotion. I understand now that this is a survival tactic. Avoiding emotion to focus on action is powerful for helping first responders be effective in high-intensity situations. It is essential for survival at work. When first responders continue to ignore emotions off the job, however, it can create a pattern of shutting down painful thoughts and memories, making it hard to connect with oneself and with others.

Emotion was not a part of how my dad raised his children. Both of my half brothers used drugs and alcohol from an early age. They did not live in our home, but they did live nearby, and my dad did whatever he could to help them, at least financially. He gave and gave and gave money to his boys, but tragically discovered that money does not solve everything.

Providing for his sons was my dad's way of showing love to them, but what they really needed was emotional availability. People need vulnerability to establish trust. Running away from emotion and vulnerability not only creates shallow relationships but also allows no outlet for pain. So first responders may develop a hidden sadness that they may find confusing. They may think there is something wrong with them, or they may feel numb to their pain, making it difficult to identify and express their emotions at all.

A literature review conducted by E. C. Nielson and others found that many first responders and military personnel feel they must adhere to society's picture of masculinity, which means self-reliance, strength, and avoidance of emotion. This same study found that restriction of emotion in both male and female firefighters has been linked to higher rates of PTSD. Nielson and others found that, in reality, the idea that first responders need to control their emotions to become the best they possibly can be at their jobs, relying only on themselves, sets them up for a very conflicted experience when they witness traumatic calls, various forms of abuse, or even moral injuries. They may feel powerless and helpless in the face of difficult experiences.[4]

Instead of getting help, many firefighters suffer from rescuer's depression in silence. With this flawed perception, military and first responder men (and women with instrumental personalities) often self-medicate to feel better or to avoid pain altogether, often with "exaggerated stereotypical male behaviors, such as aggression and increased sexual behavior, to compensate for the injury the trauma had on their identity."[5] But self-medication will never heal the pain. Running from pain only buries it deeper. It will eventually erupt as anger instead of surfacing in a way that allows the person to heal.

Firefighters: Heroes and Humans

Firefighters often carry around what I like to call a "hero mentality." They love helping people. They are wired for it. It invigorates them. Though they do not often like to be called "heroes," they love this job specifically because they get to help and rescue people. They truly are heroes! In a one-sided career like firefighting, individuals are accustomed to serving without getting anything in return. But this can fuel the idea that first responders do not need anything from others, which simply is not true. The truth of the matter was stated well by neuroscientist R. Douglas Fields, who noted, "People commonly refer to those who engage in dangerous or heroic actions as 'fearless,' but this is rarely the case. If you ask them, they will tell you that they *do* feel fear, but they persist by determination in spite of it."[6]

Yes, firefighters are heroes, but they are also humans. Humans have feelings. Humans can be strong, brave, and powerful, but they also can be hurting, afraid, and unable. Humans are both. Both types of emotions are perfectly acceptable. But because firefighters believe they have to be "tough," they often "stuff it" when it comes to emotions. They hide their emotions and run from feeling sad.

I have seen this in both men and women first responders. Sometimes the women have an even greater burden to hide their emotions on the job because of the fear of being seen as weak.

Instrumental Personality Type

Many firefighters—male and female—have a personality type called *instrumental personality*. This personality type is characterized by a fascination with things and how things work. This concept of instrumental personalities is discussed throughout the rest of this book. Instrumental personalities are less emotionally expressive and more independent, ambitious, and self-sufficient.[7] Instrumental individuals are focused, competitive, and great at making objective decisions. They gravitate away from emotion and toward competition and achievement.

Many individuals with instrumental personalities were athletes in the past. I have observed that this is particularly true of women with instrumental personalities. Often they have learned to train their bodies to deny pain and emotions and leverage adrenaline for increased performance and to block out unnecessary information. Frequently they have fathers, grandfathers, or other family members who were in the military or who had instrumental personalities themselves.

In their lives and careers, those with instrumental personalities avoid expressing feelings and see emotion as "weak." They "suck it up" and keep pressing forward independently toward their goals, impervious to pain and never showing vulnerability.[8]

The instrumental personality type, combined with cultural perceptions of what it means to be a first responder, a hero, or a "real man," make many first responders feel that they have no space to experience pain, sadness, hurt, fear, or any other sort of vulnerability.

When our children dress up for Halloween, they choose both superheroes and first responder costumes, as if the two are interchangeable. We teach our children what a hero looks like, but we forget that real heroes are not superhuman. Superheroes are not real. Real heroes are human. They are a powerful connection of strength and caring, and they do not have to be bulletproof.

Self-Medicating the Feelings Away

Instead of showing vulnerability within trusted relationships to keep themselves emotionally balanced, many firefighters choose independence and achievement.[9] In fact, researchers V. D. Ojeda and S. M. Bergstresser note that men have far greater trouble than women reaching out for help from medical and mental health professionals.[10] I believe the instrumental personality type is the real factor here. Men and women with instrumental personalities have trouble reaching out for help from others because their personality type is prone to self-reliance. When they are in pain, these independent firefighters turn to isolated ways to self-medicate.

One self-medicating behavior I see in first responders is adrenaline-seeking. Stan McCracken, Crown Family School of Social Work, Policy, and Practice, University of Chicago, calls it the "veteran dilemma,"[11] and I would add that it should also be called the "first responder dilemma." In combat, as on high-intensity calls, people's bodies ramp up for action. Adrenaline is one of the hormones released in response to emergencies, and it raises the threshold for first responders concerning what "alive" feels like. In those moments, first responders feel powerful and capable. They have enhanced performance resulting from adrenaline, and they feel on top of the world. But after the call is over and their bodies calm down, they often feel bored, purposeless, and uninterested. The rest of their lives may seem extremely dull compared to the adrenaline highs they experience at work.

In their bored moments, they often feel depressed, sad, and lonely. To fight that, firefighters often turn to workaholism, busyness, alcohol, gambling, affairs, illicit drugs, pornography, and other risky behaviors. These behaviors are effective in that they can help temporarily to ease the pain, calm them down, make them feel better about themselves, or fill emptiness. But numbing behaviors are an endless loop in which the individuals merely mask over their feelings without appropriately recognizing them and dealing with them.

Many of these self-medicating behaviors make the first responders seem even more like heroes. Workaholism usually looks impressive to others—an indication that workaholic is committed to the job. Excessive working out can seem like health consciousness because the individual is outwardly "taking care" of himself or herself. Pornography use is sometimes excused as, "He's just being a guy." Alcohol use may be viewed as, "She's just chilling out." Such assumptions can make it difficult to identify when a first responder is struggling with rescuer's depression. Under these masks, it can be difficult to recognize when the firefighter is suffering from hidden rescuer's depression.

Unfortunately, much of this behavior may have started long before the first responder's career began. According to psychologists Sissy Goff, David Thomas, and Melissa Trevathan, many children fail to develop the robust emotional vocabulary that they will need throughout their lives to communicate their feelings.[12] When children, especially boys, have not developed an emotional vocabulary, they often misidentify their needs and fail to communicate their feelings with others, which only increases their sense of isolation. First responders who did not develop an emotional vocabulary as children have to reprogram their ability to conduct emotion-based conversations as adults. This task is not impossible, but it requires intentional action.

I think this is what happened to my half brothers. Jerry, the younger one, got married and found his place as a fireman. He straightened out his act at that point and stopped drinking and doing drugs. He was proud and excited to be a firefighter. In his department, Jerry was known as the quiet guy who knew his stuff. When things were going well, he medicated with work. When things fell apart in his life, he was unable to express his emotions in healthy ways. He then medicated with drugs and alcohol.

The older of my half brothers, Jimmy, never really found his place. Jimmy used drugs and alcohol, and he never developed a connection with my dad like Jerry had. Jerry and my dad had bonded over first responder experiences. Lacking a commonality with my dad, Jimmy self-medicated with risky behaviors, alcohol, and drugs.

People who self-medicate with alcohol or drug abuse, addiction to pornography, affairs, gambling, and so on often do so because they just want to feel something—*anything*. Their hidden depression leaves them paralyzed by a feeling of numbness. Rather than feeling numb, they seek a "drug" or stimulant of some kind, whether it is a substance, sex, or gambling, to continue to mask their feelings of depression.

Both the negative and positive activities of people with rescuer's depression help them perform the ultimate defense and cover-up desired by society: to make the feelings go away.

Psychologist and suicide expert Thomas Joiner believes that male depression is often a symptom of man's pursuit of independence (again, I would include instrumental women in his assessment). In the process of accumulating wealth and accolades, he often neglects to develop deep relationships in which he is free to be honest, human, and self-reflective. Instead, he steels himself and comes across as a tough guy, depriving himself of the depth of connection that he could have if he were authentic with those around him. This "I've got this under control" mentality and pursuit of independent success often leaves men lonely and isolated when the going gets tough.[13]

Rather than allowing success to pull men away from deep, meaningful relationships, Joiner advocates a balance between the two:

> This balance, I'm arguing, is more of a male problem than a female problem because, in part, the allure of fierce independence woos men more successfully

than it does women. The balance is made all the more precarious by a companion process, represented well by the cliched but nevertheless fairly accurate image of men's reluctance—or is it failure?—to ask for directions. To 'keep your head when all about you are losing theirs.'[14]

The good news about people who walk themselves into depression as a result of a voracious pursuit of independence is that they have control over their choices as they move forward. They can invest in deep relationships and begin to heal, or they can shut down further and continue to wrestle with depression or worse emotional wellness struggles.

The Difference Between Clinical Depression and Hidden Depression

When we think of depression, we may think of acute and dramatic episodes that make it impossible for the individual to function. This is called *clinical depression*. In contrast, *hidden depression* is a state of depression that we find more often in men, but also in women with an instrumental personality. These individuals are motivated to medicate through action, even if they are not aware of it.

In clinical depression, individuals may avoid doing the things they formerly loved. Symptoms of clinical depression often include the following:

- Having low energy
- Overeating or not eating
- Having sad thoughts
- Losing interest in career or hobbies
- Sleeping more than usual
- Having a decreased sex drive
- Having increased thoughts of drinking
- Having insomnia
- Engaging in addictive behaviors
- Stopping an exercise program
- Avoiding social activities
- Having major feelings of boredom, irritability, or anger
- Experiencing crying spells
- Stopping normal activities like cleaning or work

But hidden depression is not the same. In hidden depression, a person often answers the problem with action. If the person keeps busy, he can disconnect from his feelings and will not have to deal with them. This is one way of basically "faking good."

In general, men have a societal pressure to be stoic, keeping their feelings hidden and remaining unshaken by circumstances around them. As a man, displaying feelings and sadness is frowned on. Add to that a paramilitary career, and firefighters feel like they must be invincible. The pressure to reject emotion is hammered into them their whole lives and is reinforced by their career choice. All that is left to feel is "faking good." To keep "faking good," they turn to working, compulsive gambling, substance abuse, and other behaviors.

Depression and Anger

The "real man" persona drives many men to feel angry instead of letting themselves feel sad. Constant anger—hidden or obvious—can be a sign of depression. It is easier to feel angry than to face other emotions. According to research conducted by William Riley and Frank Treiber, Medical College of Georgia, and Gail Woods, Medical College of Virginia, rather than feeling sad or lonely, many people choose to feel anger, while trying to keep it suppressed as much as possible. Riley, Treiber, and Woods note that this anger can erupt easily, however, because the emotion is right beneath the surface.[15] Anger shows that there is something more going on.

Men, and women with instrumental personalities, typically show anger instead of sadness when they are depressed. Riley, Treiber, and Woods note that the frequency, duration, and intensity of the anger are more indicative of their depression than the person to whom their anger is directed.[16] Although anger is effective at pushing people away, it has less to do with the people and circumstances and more to do with depression. But this is where the danger grows for firefighters. People who have difficulty coping with their anger may turn it inward, and it can become anger toward oneself. If left unchecked, this can stir up suicidal ideation.

Warning: Men are four times more likely than women to die by suicide.[17] If you are feeling particularly angry, talk to someone—a counselor, a friend, or even the National Suicide Prevention Hotline: (800) 273–8255. (More information about suicide is presented in chapter 5.)

Depression and Sleep Disturbances

One possible explanation for the prevalence of depression among firefighters can be found in first responder sleep patterns. First responders are often called on to wake up abruptly in the middle of the night for an emergency response. And they often have to work shifts that include being awake all night. Getting sufficient sleep is often a low priority for firefighters. After all, civilians need help at all hours of the day. But what does this irregular sleep do to the brain?

Researchers have asked the same question. In a report published in the *Journal of Clinical Sleep Medicine*, researchers studied 880 active and retired American firefighters, asking them to report their experiences with depression, insomnia, PTSD symptoms, and any difficulties in regulating their emotions. Of the 880 firefighters:

- 39% reported experiencing depression symptoms,
- 52% reported experiencing insomnia in some form or another, and
- 19% reported having nightmares.[18]

What they found was that insomnia and nightmares both strongly impacted the ability of firefighters to regulate their emotions. They also discovered a strong association between sleep disturbances and increased likelihood of experiencing depression. Further studies found that the greater the severity and frequency of sleep disturbances, the higher the likelihood of developing depression or suicidal ideation.[19]

Previous research linked insomnia to depression, but it was believed that depression caused insomnia. In fact, the opposite may be true for firefighters. The lack of sleep common to firefighters may be at the root of the depression epidemic. Add to that many years of emergency-level cortisol surges in the middle of the night, and retired first responders commonly have difficulty sleeping even when they are no longer being jolted awake by emergencies. Sleep disturbance potentially may become a lifelong problem.

Rest, as with all self-care actions in the first responder support web, is critical for emotional wellness (see fig. 1–1 and chapter 1 for additional discussion). This involves both rest in the form of good sleep and rest in the form of taking time off from work, projects, and the need to be busy.

What Happens When First Responders Stay Busy

Work keeps first responders extremely busy. They are constantly on the move. Even when they are sleeping, they are ready to hop up and go whenever a call comes in. When first responders are not working, even their days off are busy. Days off may represent an opportunity to go on an adventure with family or with the guys. They may offer a chance to finish a project around the house. First responders are always moving, and they are adept at switching from one task to the next. A busy life makes a first responder feel connected, empowered, and needed.

In our society, people respect those who are busy by saying things like, "You have so many talents!" and "I know you are busy, so thanks for taking the time to help me with that." Being busy is a badge of honor. And whenever first responders find a minute to squeeze in one more thing, they get a little boost of satisfaction.

There are so many ways to spend your time. How do you decide? Fun has a strong appeal, and so does productivity. But every time first responders say yes to something, they are essentially saying no to something else. Sometimes they feel great about it, and sometimes they are frustrated by all the demands people put on them. First responders often finish their days just exhausted. Being busy takes a lot of energy.

So consider this picture. At the end of a busy day, the first responder may sit down on the couch and watch a show so he can "let off steam" or "have some cave time." But what he never gets—the thing he is surrendering that he may not even realize is lost—is space.

The Downside of Being Busy

Being busy can be good. It can be productive and satisfying, but it lacks space for first responders just to be, to breathe, and to think.

Citing an assessment from researchers M. A. Bentley and others,[20] the Substance Abuse and Mental Health Services Administration (SAMHSA) explains the danger behind the quick-paced first responder life, noting,

> One of the core risk factors for first responders is the pace of their work. First responders are always on the front line facing highly stressful and risky calls. This tempo can lead to an inability to [mentally] integrate work experiences. For instance, according to a study, 69 percent of EMS professionals have never had enough time to recover between traumatic events. . . . As a result, depression, stress and posttraumatic stress symptoms, suicidal ideation, and a host of other functional and relational conditions have been reported.[21]

Without this space, first responders forfeit the chance to find emotional wellness and balance. Space is where your best thoughts arise. Space is where you gain clarity on your past memories. Space, if used wisely, allows people the opportunity to reconnect and find themselves again.

Powerful things happen when people open up their schedules and allow themselves to go to bed at a reasonable hour, ideally with time to lie in bed and think or to enjoy pillow talk with a spouse. When they slow down their pace and connect with their family, even allowing themselves to be a little bored, or when they talk to a friend over coffee or meet with a coworker to talk, they are giving themselves the opportunity to figure out their thoughts.

Space is where self-awareness grows. Self-awareness contributes to your resilience in the face of a highly stressful career. When you take time now to step back from the day-to-day stresses of your job and family life, you will find perspective and a new sense of who you are. You may find pain, but you will also find strengths and situations you are proud of handling. As Ian Robertson from the Center for Brain Health explains,

> Exercising self-awareness can be a painful process, and many people find subconscious avenues to avoid it, including throwing themselves into work,

drinking excessively or compulsively exercising. [To become more self-aware], make time to clear-mindedly assess your strengths and limitations. Taking time to self-reflect and gain perspective activates the brain's right frontal lobe, releasing noradrenaline—a powerful chemical that can build the brain's gray matter and give us the mental strength to solve new problems better.[22]

What If I Do Not Want to Sort Through My Thoughts?

A busy schedule can be an effective source of self-esteem and identity. After all, a busy body produces endorphins and adrenaline to complete all the things that must be done. This physiological "high" created by endorphins is a way that some first responders medicate pain and attempt to dodge rescuer's depression.

Eventually you will have time on your hands, however, and self-awareness will rise to the surface. It may occur when you retire or are sick or injured, but eventually you will have space, and your thoughts and emotions will rise to the surface. In the meantime, if you continue to "suck it up," you may find that you are irritable or downright angry, hard to be around, anxious, excessively focused on performance measures, or possibly not even able to contribute to your relationships in healthy ways when you finally do slow down.

Busy is not always helpful.

What would happen if you chose to become "un-busy" for even one week? Does it scare you to think about it? Studies show that children who have less programmed activity and are allowed the time and space to occasionally become "bored" are often more creative, better at solving problems, and stronger at self-motivation later in life.[23]

Staying Busy and Neglecting Yourself

People with instrumental personalities, the personality type of most first responders, enjoy projecting the image of competence and self-reliance. This can lead to an unhealthy denial of personal needs and problems. By nature, the career of a first responder requires that you "give-give-give." When you get home, you are still expected to love your family and give to them as well. Your whole life seems to become a measure of how much and how well you can give. It is nice to be needed, but what if you have nothing left to give? What if you have emptied the tank, and there is nothing left in it?

My dad's schedule was booked to the minute. His work shift as a correctional officer extended from 3 p.m. to 11 p.m. He liked that shift because it gave him almost an entire workday to take care of his businesses. His real estate ventures, laundromat, and pool hall all demanded daily ownership from him. Sometimes he let me follow him around when he was doing business. We went to the bank, carried out miscellaneous errands, and sat in many meetings where he cooked up new business deals. He was so busy dealing with three jobs, two sons who were struggling, and one housing remodel after another, that for the life of me, I cannot remember a time when he took a minute to sit on the couch.

I have no idea when my dad took care of himself. My guess is that he did not.

When you do not take care of yourself, it is like running on empty in both your job and your family life. When you are always busy without any rest or space, you end up wearing yourself out emotionally, spiritually, and mentally. You can hide from depression for a long while by staying busy.

I wanted my dad to know how much I loved him, so I made myself his busy little helper. I saw my role in the family as being well-behaved and being as helpful as possible. Many children of first responders feel that way. My dad taught me about courage, resilience, hard work, adapting to difficult situations, and finding pure joy in nature. But I saw him constantly neglect his own needs.

How to Stay Intentionally "Un-Busy"

Emotional wellness starts with protecting your time. Your time equates with your healing. Start by planning ahead. Before you jump into a project, allow yourself space to think about it from different angles, even allowing yourself to say no when someone wants something from you.

Did you know that one of the primary struggles of people with addictions is their inability to relax in healthy ways? Stress piles up in first responder careers. Your priority is finding ways to destress, relax, and give yourself space to be healthy.

Maybe you like to run. Maybe you like to take walks in nature. Maybe you like to hunt or fish. Maybe you like to play golf or basketball. Maybe you like to work on an old car. Maybe you like to read, or you just need to sleep. Maybe you need some time to meditate. Maybe you need to stretch or do yoga.

The main thing about giving yourself space is not what you *are* doing; it is about what you are *not* doing. What are you saying no to that will allow you to have the space you need to think, care for yourself, and heal? Your time is precious. Do not fill it all up.

The Overwhelming Experience of Facing Trauma On and Off the Job

A first responder is subject to many things that can lead to depression: bad calls, pressures from family and spouses, interdepartmental strife, potentially traumatic events, pediatric death, the death of a coworker, and city politics. Combine the events first responders see on a daily basis with the childhood trauma or neglect that many first responders grew up with (see chapter 3 for more information), and you have a whole group of people who are wrestling with psychological injuries that they need to talk about. They are unsure, however, about where to go or if they are safe to talk.

Joiner explains that male depression is a combination of vulnerability *and* a psychological injury, such as trauma or a childhood event.[24] First responders carry a heavy burden of responsibility to "save the world." Even when they are off-duty, they feel the responsibility to protect and save people. But the off-duty first responder who witnesses an accident or horrific event right before his eyes might be unable to do anything about it simply because he does not have his equipment and gear on at the time.

Deputy Chief Art Zern, Sycamore (IL) Fire Department, explains, "While the average civilian that witnesses an event is surely traumatized, the first responder also feels guilt, responsibility, and shame for being unable to 'save the day' as is expected of him from the public and himself. We can deal with losing a fair fight, but it's very hard to deal with when the deck is stacked against you."[25]

Healing means starting to talk about that psychological injury. In order to start talking about the psychological injury, however, a firefighter must move from the safe, socially accepted confines of his or her hidden rescuer's depression out into the vulnerable, somewhat uncomfortable, and unpredictable state of feeling the pain and loss. The injury may be from psychological trauma or neglect, either from the deep past or something more recent. Whatever it is, it must be acknowledged and discussed for healing to begin.

How Can I Begin Healing from Depression?

I believe the first step for first responders as a whole to break the bonds of rescuer's depression is for all of them to change their language toward emotional wellness. Joking, making fun of emotions, talking about people who cannot cut it as first responders, and so on only breed a culture that makes it unsafe for people to talk about their emotions. Instead of creating a toxic environment, we need to begin creating a safe environment, and that starts with the leaders. Leaders who share pieces of their story with others create a culture of vulnerability in which people can bond over their humanity, genuinely caring for one another and becoming stronger as a unit together. People who do not bond over vulnerability often bond over scapegoating, and one person becomes the outsider. Choosing vulnerability at the top is critical for the wellness of the whole department and the fire service at large.

In addition to changing first responder rhetoric about emotional wellness, you can begin to find personal healing by slowing down and allowing yourself to recognize your feelings; opening up with safe family members, your spouse, or friends; and starting to live an emotionally aware, connected life. This means you allow yourself to rest, restore balance in your life, get healthy sleep, address traumatic memories, and overcome the need to be needed. Connection is where healing lives.

I wish my dad had taken the time to slow down and face his pain. I think he would have been so much happier if he had given himself the grace to need others and the courage to be vulnerable. Right before he passed away, I told him that I was traveling all over the country presenting about emotional wellness and peer support to first responders, and he smiled a deep, relieved smile at me (fig. 2–2). I think he died happy, knowing that I was doing what I loved. I smiled back, hopeful and passionate that first responders can become genuinely happy, reconnected people as they begin to open up about their pain.

Reflection Questions

1. Have you ever felt the pressure to be unshakable or emotionless?
2. Based on this chapter, do you think you struggle with rescuer's depression?
3. When you are alone and quiet, what is your primary emotion?
4. Have you observed yourself acting out of anger this week?
5. Do you think your anger is an active emotion to mask a more passive emotion? If so, what emotion is hiding behind your anger?
6. What has your personal time looked like this week? Have you had any time to be alone?
7. When is a time this week that you could go for a walk?

Figure 2–2. My dad shortly before his death, 2017

Notes

1. American Addiction Centers, "Depression Among First Responders," *American Addiction Centers Blog*, August 8, 2014, https://americanaddictioncenters.org/blog/depression-among-first-responders.
2. S. C. Garbern, L. G. Ebbeling, and S. A. Bartels, "A Systematic Review of Health Outcomes Among Disaster and Humanitarian Responders," *Prehospital and Disaster Medicine* 31, no. 6 (September 19, 2016): 635–642, https://doi.org/10.1017/s1049023x16000832.
3. M. A. Bentley et al., "An Assessment of Depression, Anxiety, and Stress Among Nationally Certified EMS Professionals," *Prehospital Emergency Care* 17, no. 3 (February 15, 2013): 330–338, https://doi.org/10.3109/10903127.2012.761307.
4. E. C. Nielson et al., "Traditional Masculinity Ideology, Posttraumatic Stress Disorder (PTSD) Symptom Severity, and Treatment in Service Members and Veterans: A Systematic Review," *Psychology of Men & Masculinities* (January 27, 2020), https://doi.org/10.1037/men0000257.
5. Nielson et al., "Traditional Masculinity Ideology, Posttraumatic Stress Disorder (PTSD) Symptom Severity, and Treatment in Service Members and Veterans."
6. R. D. Fields, *Why We Snap: Understanding the Rage Circuit in Your Brain* (New York: Dutton, 2016).
7. N. Bozionelos and G. Bozionelos, "Instrumental and Expressive Traits: Their Relationship and Their Association with Biological Sex," *Social Behavior and Personality: An International Journal* 31, no. 4 (January 1, 2003): 423–429, https://doi.org/10.2224/sbp.2003.31.4.423.
8. T. E. Joiner, *Lonely at the Top: The High Cost of Men's Success* (New York: Palgrave Macmillan, 2011).
9. Nielson et al., "Traditional Masculinity Ideology, Posttraumatic Stress Disorder (PTSD) Symptom Severity, and Treatment in Service Members and Veterans."
10. V. D. Ojeda and S. M. Bergstresser, "Gender, Race-Ethnicity, and Psychosocial Barriers to Mental Health Care: An Examination of Perceptions and Attitudes Among Adults Reporting Unmet Need," *Journal of Health and Social Behavior* 49, no. 3 (September 1, 2008): 317–334, https://doi.org/10.1177/002214650804900306.
11. S. McCracken (Crown Family School of Social Work, Policy, and Practice, University of Chicago), in discussion with the author, October 22, 2019.
12. S. Goff, D. Thomas, and M. Trevathan, *Are My Kids on Track? The 12 Emotional, Social, and Spiritual Milestones Your Child Needs to Reach* (Grand Rapids, MI: Baker Publishing Group, 2017).
13. Joiner, *Lonely at the Top*.
14. Joiner, *Lonely at the Top*.
15. W. T. Riley, F. A. Treiber, and M. G. Woods, "Anger and Hostility in Depression," *Journal of Nervous and Mental Disease*, 177, no 11 (November 1989): 668–674, https://doi.org/10.1097/00005053-198911000-00002.
16. Riley, Treiber, and Woods, "Anger and Hostility in Depression."
17. "Recognizing Depression in Men," *Harvard Mental Health Letter*, June 2011, https://www.health.harvard.edu/newsletter_article/recognizing-depression-in-men.

18. M. A. Hom et al., "The Association Between Sleep Disturbances and Depression Among Firefighters: Emotion Dysregulation as an Explanatory Factor," *Journal of Clinical Sleep Medicine* 12, no. 2 (2016): 235–245, https://doi.org/10.5664/jcsm.5492.
19. R. A. Bernert et al., "Sleep Disturbances as an Evidence-Based Suicide Risk Factor," *Current Psychiatry Reports* 17, no. 3 (2015): 15.
20. M. A. Bentley et al., "An Assessment of Depression, Anxiety, and Stress Among Nationally Certified EMS Professionals," *Prehospital Emergency Care* 17, no. 3 (February 15, 2013): 330–338, https://doi.org/10.3109/10903127.2012.761307.
21. Substance Abuse and Mental Health Services Administration (SAMHSA), "First Responders: Behavioral Health Concerns, Emergency Response, and Trauma,"Disaster Technical Assistance Center *Supplemental Research Bulletin* (May 2018): 4, https://www.samhsa.gov/sites/default/files/dtac/supplementalresearchbulletin-firstresponders-may2018.pdf.
22. I. Robertson, "Stress Can Make Your Brain Stronger If You Know This," *Center for Brain Health*, University of Texas at Dallas, January 23, 2018, https://brainhealth.utdallas.edu/stress-can-make-your-brain-stronger-if-you-know-this/.
23. M. Ungar, "Let Kids Be Bored (Occasionally)," *Psychology Today*, June 24, 2012, https://www.psychologytoday.com/blog/nurturing-resilience/201206/let-kids-be-bored-occasionally.
24. Joiner, *Lonely at the Top*.
25. A. Zern (deputy chief, Sycamore [IL] Fire Department), in discussion with the author.

CHAPTER 3

NEEDING TO BE NEEDED AND LONELINESS

Mark's Story

Mark came to me in his midlife because he was not feeling like himself. He had recently retired from his fire department and taken on a different job. As he sank his muscular frame into the cushions of my counseling office sofa, he explained that he was "just feeling down and disconnected from his spouse." As I got to know Mark, I recognized a pattern that I have seen in many adults who were raised by first responders.

Mark was a second-generation first responder. He and his father had great difficulty connecting on anything deeper than a surface level. Growing up, Mark's family focused on success and work ethic, and Mark calibrated himself to please them by becoming goal-oriented and helping them whenever and wherever he could.

For many years, Mark blamed himself for his father's emotional distance. Mark felt like he was not good enough. He had a lot of pain about his relationship with his father, which is common in the fire service. Deep in his heart resided the lie that he was only lovable if he achieved.

In my years of listening to first responders, I have noticed that families with multiple generations of first responder, military, or paramilitary careers tend to raise children with limited emotional attention. I have seen many well-meaning first responder parents—whether as a result of their personality type or a result of their first responder training—avoid emotion in childrearing, resulting in children who may feel emotionally neglected by their parents. These children usually grow up and choose careers in the helping professions, like nursing, firefighting, law enforcement, or counseling because they find that they feel valuable when they feel needed.

As author Judith Viorst notes in her book *Necessary Losses*, "Another defense against loss may be a compulsive need to take care of other people. Instead of aching, we help those who ache. And through our kind ministrations, we both alleviate our old, old sense of helplessness and identify with those we care for so well."[1]

As previously discussed, I am the daughter of a first responder, and I have seen this phenomenon both in my personal life and in the lives of my clients at work. I have found that there is a loneliness that weighs on people whose parents were emotionally unavailable to them. There is also a deep-seated impetus to solve everyone else's problems in order to feel genuinely loved.

In a beautiful way, the children who were emotionally neglected and grow up to become first responders are redeeming their broken childhoods by making the world a better place for those around them. But behind this desire to help also exists a great desire to be needed—a sense of satisfaction in being a part of the team and a strong sense of identity derived from having others rely on you.

In first responder careers, interactions with others are often one-way by design. You are in your job to rescue, protect, and serve the community around you. But what about your personal life? Do you understand the value of reciprocal behavior? When you are conversing with your friend and ask a question, does your friend ask you a question in return? When you do something for them, do you allow them to do something for you?

In his adult life, Mark had rarely allowed others to help him. The way he saw it, his job was to protect everyone. In his mind was the thought, *I'm here to help you. I'm the guardian at home. I'm the guardian at work. I'm always there to help everyone.* Mark shunned the idea of being vulnerable. It is important to understand, however, that you do not actually have a relationship unless two people are mutually vulnerable. Always being the hero can be very isolating. It creates a false sense of connection but allows no real relationships.

One of the issues that Mark had with his wife, which is common in first responder couples, is that he was always trying to fix whatever problem she had instead of just listening to her talk. I explained to him, as I try to explain to all first responders, "You do not have to fix everything. You can just listen."

Mark was not alone in his quest to fix everything to feel needed and loved. In explaining what author Norah Vincent detailed in her book *Self-Made Man*[2], psychologist Thomas Joiner notes:

> He was really feeling the burden of being the safety net, the bread-winner and the "Mr. Fix-It of his household." The other of them said, "I guess I think that if I hold it all together, if I take care of everything and everyone that eventually I'll be loved. But the price is my life. I'm trying to do the impossible." . . . Men and women alike can feel this way, but men in particular struggle with the "eventually I'll be loved" part. Like everyone, men want to be foundational for their families; but more than women, men struggle with opening themselves up to reciprocated love, care, or help.[3]

Truth: You do not have to be everyone's hero.

Letting go of the need to fix everything can seem daunting, but I promise you, on the other side of this battle to be needed and to be loved awaits the freedom to let others own their own problems. There is relief from the stress of everyone else's emergencies and the freedom to be yourself and pursue what actually makes you uniquely you.

The Rescuer Identity

The foundation of the first responder's identity is often built on how the individual feels he is doing at meeting the needs of others, often at the expense of his own needs. He sees himself as a defender and rescuer, and he loves this job because he wants to save people and be their hero. But he is so wrapped up in the needs of others that he often forgets to acknowledge his own needs.

First responders are excellent at coming across as competent and self-reliant. They appear to have it all together—the job, the house, the marriage, the family—and everything looks great on the outside. But it is common for first responders to be insecure about where they stand with others because their self-worth is wrapped up in how well they think they are doing at caring for others. This is a shaky foundation on which to build one's self-worth, and it often results in even more striving to please those around them.

The effort is exhausting. In all their striving to help others, I often see first responders deny their personal needs. They become uncomfortable with receiving attention and have trouble accepting help from others.

Those who need to be needed often feel guilty and responsible for the suffering of others, though they themselves have little or nothing to do with it. They may feel like they have no personal time or are overburdened or stressed out.[4] Those who need to be needed feel like they are the healthier partner. It makes them feel important, in control, hardworking, and virtuous.[5]

In my time working with Mark, we dove into some childhood abuse that he had suffered, not from his parents, but from others who were bullying him. These were things he had kept private for a long time because he held a very masculine ideal of being strong and not showing emotion. Any show of emotion brought a great deal of shame in his family.

Most firefighters would not call their negative childhood experiences child abuse. They may say, "Yeah, I was just whipped when I did things wrong." But children and adults have feelings that they must be allowed to express.

Mark was not allowed to cry. He was not allowed to cry during his spankings, so from a young age he learned that key skill that firefighters learn, which is to dissociate himself from the pain. Firefighters must dissociate from their own fear to go into a burning building, which is almost like overriding your survival skills. Mark learned this emotional separation as a child in a first responder family. "I have an extremely high pain tolerance and an iron will," he stated in one of our counseling sessions. This is because he learned to flip that switch and override both physical and emotional pain. It may sound manly, but it is actually broken.

When people get in the habit of turning off their own pain receptors, they can easily apply that to their needs, ignoring what they need in order to focus on others. So those who need to be needed most likely have difficulty setting clear boundaries. According to author Rebecca Lee, they may sacrifice their personal needs in order to try to meet the needs of others. This can lead to passivity and feelings of shame, low self-worth, or insecurity. Lee notes that they may need approval from others, feel empty without others around, or unintentionally enable people with addictions because they want so badly to be the rescuer.[6] A first responder may even find himself in an unhealthy marriage because he has stepped in to be the hero for a woman who was not emotionally healthy. At times, needing to be needed can place firefighters in unhealthy situations. Needing to be needed often sets the stage for depression (for more information, see chapter 2).

Where Is the Line Between Wanting to Help and Needing to Be Needed?

If you wonder about the need to be needed, here is a quick assessment from Lee:

1. Do you feel solely responsible for someone even though that person has other avenues of support?
2. Do you often find yourself in the "savior" role?
3. Is it better to be with someone than alone?
4. If your gut tells you the opposite of what someone else is saying, do you first trust the other person?
5. Do you feel mean saying "no"?
6. Do you find yourself consistently resentful when others do not put in as much effort as you?
7. Will you settle for less so that you do not have to argue?
8. Do you alter what you say or look for approval from friends or significant others?
9. Have you lived with someone who has experienced a substance abuse/alcohol problem?
10. Have you lived with a physically or emotionally abusive person?
11. If nobody is around, do you feel inadequate?
12. Do you feel that the burden of others often falls on you?
13. Do you have trouble asking for help?[7]

If you answered yes to a majority of these questions, you exhibit "needing to be needed" behavior.

The problem with needing to be needed is that you miss out on living your own life. You are so entirely focused on the well-being of others that you miss out on discovering the attributes, interests, and passions that make you unique. Others around you are responsible for themselves. If they walked themselves into a problem, say an addiction or mental health issue, for example, they are responsible for getting themselves out of it. Yes, you can be supportive. No, you cannot make them better or healthy, and you do not have to be their savior.

Mark admitted to me that he had a natural talent and love for music and the arts, but these were not tolerated in the very masculine household in which he was raised. Being physically fit was an important value in his family, and he caused himself a lot of pain and actual damage to his body in pursuit of physical fitness. It also caused him to miss out on developing his own unique talents.

The priority that Mark was taught in his family was to never quit and never admit when you are hurt. That caused him to fall into depression over the years as he basically shut down his emotions and creativity. In Mark's family, there was a line between the feminine and masculine, and as his father would tell him, "You can't have both." He equated emotions and creativity with being feminine, and strength and productivity with being masculine. In his paradigm, there was no crossover. The message Mark received was to turn his back on his own talents, needs, and vulnerabilities in order to become worthy and special in his father's eyes.

Barbara Johnson, American professor of English and literary critic, noted that needing to be needed "means that when you die, someone else's life passes before your eyes."[8] When you need to be needed, you place your identity in someone with problems, someone who is needy. But helping them does not actually change them. In fact, it may enable them to continue their unhealthy behavior. Meanwhile, you continue feeling a sense of self-worth because you are their hero. It is a loop that is subtle to identify and tricky to extract oneself from. One of the biggest problems with allowing yourself to stay in unhealthy relationships is what it models for your children. Needing to be needed often transfers from one generation to another, so your children may learn unhealthy social boundaries as they watch you.

What Causes First Responders to Need to Be Needed?

Childhood experiences have a way of shaping us in unseen and often unacknowledged ways. As we grow, our personalities, passions, and paths take shape one step at a time. We find our unique sense of humor. We discover what angers us and what excites us. We encounter occupations and relationships that seem to fit us well. It is all part of becoming who we are. And as we live out our adult lives, we have innumerable past memories to draw from. These memories can cause us to feel the need to be needed.

If a child grows up in a tough situation where he is not well attached to a parental figure, he may feel that he is only as good as what he does for others. This can lead to a deep loneliness, whether conscious or subconscious, that drives him to live his life in light of making that loneliness go away. Some turn to unhealthy relationships, substance addiction, sex addiction, or even work to validate themselves and give themselves a sense of value. (Addiction is discussed in greater detail in chapter 4.) Instead of connecting with their parents, emotionally neglected children have learned to self-soothe their pain in isolation. As a result, they often turn to risky behaviors, substances, or other avenues to earn validation in times of stress.

Mark admitted to me that he had been emotionally numbing himself through alcohol and risky behaviors. After that, he had discussions with his whole family about what he was going through, for which I respect him profoundly. Once he started the therapeutic process, he began talking very openly. I met with almost everyone in the family, which I rarely do. I met separately with his wife for counseling and then also with his son. They were an intelligent and delightful family. Mark obviously wanted something different for his son, which is why he allowed himself to be vulnerable. He could see how this emotional neglect had been going on for generations, and he wanted to end the cycle. He had a great discussion with his son about being vulnerable, and about how you do not have to earn love and disconnect from yourself. He even mentioned to me that he had cried more in one month of working with me than he ever had in his whole life. This crying was a positive sign of powerful strength and healing.

The Unshakable Loneliness of an Emotionally Neglected Child

Kids who do not attach emotionally to their parents carry around a pathological and overwhelming loneliness. They consequently may work to find attachment in others because they suffer from *attachment trauma*. Attachment trauma is an emotional scar that children carry around deep in their limbic system. It can cause them to experience frozen emotions or sometimes frozen traumatic memories that are too painful to recollect. This attachment trauma can be expressed in unconscious ways, akin to the reflexes of a broken soul.

As we saw with Mark's story, kids who grow up with less emotional connection can struggle to feel a deep sense of belonging and to form trusted alliances as adults. But with help from a trusted friend or counselor, they can overcome loneliness and find connection and a strong sense of self as adults. It may take time to dig out painful childhood memories, but I truly believe that kids who have weathered hardship can become some of the most compassionate and connected adults.

What Is the Upside?

There is an assumption that if a family was marked by alcoholism, divorce, domestic violence, drug abuse, mental illness, or another traumatic experience, this adverse upbringing will produce a child who grows into a broken adult. It is true that in some cases, children with adverse childhood experiences (ACEs) may weather adult hardship with less force of will. In many cases, however, Friedrich Nietzche was right when he noted, "What doesn't kill you makes you stronger."[9]

In his book *Lonely at the Top*, Joiner explains:

Researchers reported on people's reactions to recent negative life events. . . . Unsurprisingly, those with very difficult pasts were further worn down by recent challenge. Intriguingly, those with little past adversity also reacted to recent difficulties by becoming more demoralized by them. Only those in the middle, with some prior experiences with adversity, but not too many, weathered recent challenge well. Just as muscles can be strained to a breaking point and can also atrophy with disuse but do well with moderate exercise, the psychological capacity to handle life stress can be overwhelmed by too much previous adversity, can be underdeveloped by too little, or can be optimized with moderate past challenge.[10]

Researchers Willem Frankenhuis and Carolina de Weerth of Radboud University, Netherlands, were early pioneers in studying the upside of adverse childhood experiences or "early-life exposures to stress," as they termed them. What they found was remarkable. People who grew up in "stressed environments" tended to be more adaptable than their peers who grew up in structured, well-provided-for environments.[11]

> **Truth:** Early adversity makes you adaptable.

A team including Chiraag Mittal and fellow researchers at the Carlson School of Management, University of Minnesota, found that these adverse scenarios trained children to adapt to their environments. With parents who were not guaranteed to follow through on their promises, these kids learned to react immediately when an opportunity presented itself. So as adults, they can tend to be categorized as impulsive or shortsighted, when perhaps they have been trained to respond to opportunities when they arise.[12]

Mittal and fellow researchers created two categories for assessing these individuals compared to their stable-upbringing peers: 1) inhibitory control, and 2) task-switching. When compared to peers, those who had "stressful childhoods" scored lower in "inhibitory control" and higher in "task-switching." These individuals thus become extremely adept at moving from one task to another, especially when it is somewhat related to their childhood experiences. They can move through uncertain situations without loss of accuracy. Mittal and others note that they seem more willing to leave things undone and ditch perfectionism for the sake of pursuing the most important task at hand.[13]

Mittal and others also note that those who endured stressful childhood experiences are better at noticing when something is "new, unexpected, or frightening."[14] According to Carlos Osório, King's College London, and fellow researchers, this is because they have greater presence of the chemical norepinephrine in their brains, an adaptive response to stress. Osório and fellow researchers found that when these individuals encounter a challenging event, they are better at making connections and learning to solve a problem.[15] They are also better at assessing threats. Interestingly, those with higher IQs tend to have greater presence of norepinephrine in their brains as well.[16]

These individuals become willing to take risks without hesitation. They can weather tough seasons at work without giving in. They tolerate ambiguity well. They are better at *working memory updating*, or forgetting information that is no longer useful and replacing it with updated information.[17] They tend to be stronger at novel solutions and thinking outside of conventional rules, which are traits shared by career first responders.

> **Truth:** Early adversity makes you empathetic.

According to clinical psychologist David J. Ley, the other beautiful outcome of a childhood riddled with pain is that it builds adults who genuinely care about others and their pain. Ley notes that the adult may think, "I know how they feel," or "After what my mom did to me, I would never do that to someone else."[18] Though this empathy has grown from a place of trauma, loss, and fear, it can yield beautiful tenderness toward the hurts of others.

David Greenberg and fellow researchers studied hundreds of people with childhood trauma and found that "childhood trauma increases a person's ability to take the perspective of another and to understand their mental and emotional states, and that this impact is long-standing." Greenberg notes that these individuals were better at *affective empathy*, or the ability to feel another's feelings.[19] They were also better at *cognitive empathy*, or the ability to think through what it must be like for the other person, according to Ley.[20] And those with childhood trauma are strong in measures of "social skills/sympathy," being motivated to take action to help the needs of others.[21]

The ability to care and reach out to help those who are hurting is a core piece of the first responder calling. Though childhood trauma and pain can leave scars, it can also bring empathy, compassion, and a willingness to help others that truly makes our world a better place to live.

Becoming Amazing Firefighters

Many firefighters find themselves strong in these areas, and whether they will say it out loud or not, some have closets full of hurtful childhood memories. For many, these memories came from their parents; for others, they came from peers who bullied them. But all this pain does not have the power to destroy the magnificent person they get to become.

Often first responders decide to pursue this line of work because of their rough pasts. Many have risen from the ashes of a broken childhood, and their experiences have prepared them to do this job well. One step at a time, they have learned to overcome their painful childhoods and embrace who it shaped them to be. Perhaps their pain is what has made them amazing at this job.

First responder careers require rapid task-switching, adaptability, tolerance of ambiguity, willingness to take risks, and outside-the-box thinking. These skills can be learned, but often they are developed in the crucible of a hard childhood. And they can even be what attracts first responders to this career. You could call it "making lemonade from lemons," or maybe it is just keeping others safe in response to a childhood that was not very safe.

The upside of adverse childhood experiences is that they can cultivate some of the most powerful first responder skills and prepare a person to become an amazing firefighter. The downside of adverse childhood experiences is that they can create people who are others-oriented to an extreme, with a need to be needed at the core of their identity. As with all areas of emotional wellness, we celebrate where first responders are strong, and we work toward balance in the areas of imbalance.

The Difference Between Needing to Be Needed and Compassion

Firefighters are compassionate people, and that is what makes them so remarkable. Needing to be needed becomes unhealthy when you ignore your own needs and wants in order to serve others or please them.[22] It is a form of addiction—*relationship addiction*—and it erodes your physical and mental health.[23] This is an automatic response of dropping your own agenda and needs to respond automatically to the real or imagined needs of the other person.[24]

Here are some key signs that you may be exhibiting unhealthy behavior:[25]

- You have trouble setting boundaries.
- The needs of others quickly trump your own.
- You are nervous that your partner, friends, or coworkers will desert you if you do not do what they want.
- You answer quickly instead of taking time to think about whether saying yes or no would be the healthiest for you in that moment.
- You feel resentful that people ask so much of you without considering your needs.
- You feel drained because of constantly responding to all the needs around you.
- You feel responsible for the needs of everyone around you, like the weight of the world is on you.
- You have an easy time giving to others, but you have trouble receiving from others.
- You feel guilty, worthless, burdened, lost, exhausted, and motivated to seek approval from others.

Instead of letting yourself stay in these behaviors, choose to shift toward compassion rather than needing to be needed. The good news is that you are never stuck permanently in unhealthy patterns.

Having compassion is to care about people, extending help to them as you are able while recognizing your own limitations. People who are compassionate are *empathetic*—they feel the other person's pain—and they want to alleviate the person's suffering.[26] As John Amodeo, licensed marriage and family therapist, notes, compassionate people do not bend over backward to be everyone's hero, and they do not feel insecure when they are not able to help. According to Amodeo, it can be very satisfying to help others with their needs, and we can do this while still meeting our own needs. He observes that compassion is demonstrated when "we extend ourselves without overextending; our caring lives in dynamic balance with caring about ourselves."[27]

When we choose compassion instead of needing to be needed, we get the best of both worlds. We get to contribute to the needs of those around us, but we stay filled and secure in ourselves and having our needs met as well. Balance is where health lives.

With this in mind, I love the concept of islands. On the surface of the water, a single island may look distinct, yet under the surface, it is interconnected in a chain of islands.[28] So are people. Individuals look separate at first glance, but when they help others and are helped by others, they thrive. That is because beneath the surface, people need one another. Even so, this interdependence must be carried out in balance so that your needs and the needs of others are equally respected.

Overcoming the Need to Be Needed

Overcoming the need to be needed is about reorienting oneself. Up until this point, the person who needs to be needed has focused all of his energy and emotion toward others. Being others-oriented is equally as out of balance as those who are completely self-oriented. Balance is the key to healing.

To shift from being completely others-oriented toward balance, begin by identifying your personal needs. Perhaps instead of going over to a friend's house to help finish a late-night house project, you could recognize that you need sleep. Allow yourself to say "no" to helping your buddy so that you can rest.

> **Truth:** Your buddy will not be mad at you because you said "no."

> **Truth:** Allowing yourself to focus on your own needs is not selfish. It is critical for balance.

In order to start to shift yourself from others-orientation toward a balanced self-orientation and healthy "interdependence" with others, I advocate self-care. Recognizing your needs is the starting point of overcoming the need to be needed. Taking care of your needs is key to this balance. (For ideas on how to do this, you can read more about self-care in chapter 11.) Then you will be able to find a "wise balance between caring about ourselves and being kind toward others."[29]

How Mark Overcame His Need to Be Needed

Mark recognized that he was an introvert, not an extrovert as he had always assumed. In one of our sessions, Mark and I started talking a little bit about the fact that many first responders are introverted, which means that they need time and space alone to recharge their batteries. If they do not get that time and space alone, they can sometimes become irritable or angry because they are pushing themselves past exhaustion. Mark had never really thought of himself as an introvert. But out of his desire to reconnect with himself, he read a book about introversion that I suggested. When we discussed it, Mark quickly recognized that he was an introvert.

His wife was an extrovert. When he got home, she would typically want to connect. By the time he got home, he was usually exhausted. He finally realized that he had been trying hard to be an extrovert for his whole life, and that it had been exhausting for him. He had a supportive, intuitive wife, and explaining this to her was all that was needed. She did not take it personally and just gave him some time to recharge, as well as time and space to think. She still wanted to talk, but they learned to walk and talk because these walks helped this type-A firefighter burn off some of the stress of the day, while simultaneously finding a way to connect with his wife. After he arrived home, they would go for a walk together to relax and talk in a natural way. I am a big believer in walking as therapy (more on this in chapter 15). Doing a little walking review of their day seemed to work for them as a couple.

For Mark, overcoming the need to be needed was about discovering who he really was, setting boundaries, and pursuing intentional self-care. I am happy to report that he and his wife are more connected than ever. He is much happier and has a greater sense of self and understanding of where he fits in this beautiful world we inhabit.

The Loneliness Epidemic

Cigna conducted a study of more than 20,000 American adults to find how many were lonely. Alarmingly, the study revealed that 46% of American adults reported feeling sometimes or always alone, 47% reported feeling "left out," and 27% reported feeling that they rarely or never had people who genuinely understood them. Two in five American adults identified feeling sometimes or always that their relationships are not meaningful and that they are isolated from others. Twenty percent reported that they rarely or never feel close to people, and 18% reported that they feel as if there are no people to whom they can talk. Living with others does seem to help the loneliness trend. Those who reported living with others had an average loneliness score of 43.5, compared to those who live alone, whose score was 46.4. Those with the highest degree of loneliness were the single parents, who scored 48.2. Only half of Americans (53%) say they have meaningful in-person social interactions with friends or family daily.[30]

People are not born lonely. Lonely people are built. They can come from broken parental attachments, as we discussed earlier, which create pathological loneliness. And they also can come from adult choices, like anger, as discussed in chapter 2, which will drive others away. The most heartbreaking thing about those who feel deeply alone is that they often feel worthless, unlovable, and existentially void. They feel lost.

Loneliness is not merely a feeling, either. It impacts the body, mind, and behavior of the lonely person.

The Physical, Mental, and Behavioral Impacts of Loneliness

Researchers at Brigham Young University discovered that loneliness increases the mortality rate of elderly adults by around 30%.[31] Stress and isolation trigger physiological responses in the brain, similar to physical pain. This affects the stress hormones, immune function, and cardiovascular function. The body literally responds to a sense of isolation by suppressing immune function and hormonal rhythms that determine an emotionally and physically healthy person.[32]

Loneliness affects people mentally, as well. People who are lonely have a harder time concentrating and are more easily distracted by unimportant events. As we discussed in earlier chapters, emotional wellness issues do not exist in isolation. Rather, they are all interrelated, and the body and mind work tirelessly to bring the individual back to homeostasis, which can tax a person's mental reserves and make him easily distracted.[33] Then, when the lonely individual makes small errors, he will often turn these errors into catastrophes, rather than taking them in stride.[34]

When people lack a sense of connection with others, it creates a mental environment that fosters unbalanced thoughts. People who lack connection with others are more likely to feel depression. Like horses, humans are pack animals, and we become anxious, depressed, and pathological when isolated.

Loneliness can be identified in the person's behaviors. People might come across as immature or inflexible when they have a hard time taking corrective action when things go wrong, but this is mostly a result of loneliness and the self-involved negative thought patterns that feed the reaction.[35] Accurately recognizing lonely behaviors has implications for first responder leaders. Having a consistent, developmental attitude toward a lonely person may help him move from inflexibility into corrective behaviors. (Paternalistic leadership is discussed in more detail in chapter 14.)

Joiner's Theory on Loneliness

In his book, *Lonely at the Top*, Thomas Joiner noted that men are uniquely lonelier than women, and loneliness can lead men down a road toward emotional disconnection, and at its worst, death by suicide. He believes that boys are biologically wired to be lonelier than girls. As boys grow up, the girls and women in their lives "rescue" them whenever they do not ask questions or dig deep into healthy conversation, so that boys become "spoiled" socially by women doing the work of connecting.[36] Joiner's foundational pillars about loneliness will be referenced throughout the rest of this book.

Joiner notes that in childhood, making friends comes easily to boys because of the many school and community programs in which they are involved. But as men age, they have fewer avenues for making new friends and much more pressure to achieve. According to Joiner, these men dig into careers that celebrate achievement and disregard social and emotional connection. Throughout their careers, their devotion to work causes them not to notice that they have less social and emotional connection with others, and they become unable to recognize when they are lonely.[37] As they near retirement, they have a weakened ability to connect with others from years of the women in their lives doing the work of connecting and years of ignoring connection in favor of achievement, and they find themselves "alone in a crowd." They may have family around them, but they are unable to connect deeply.

I do not believe this theory applies only to men. I believe that women with instrumental personality types—personalities that value strength, self-sufficiency, and limited emotional displays (see chapter 2)—also can experience the same loneliness that men often experience as they age. Because many firefighters have instrumental personalities, I have become passionate about first responders developing new friendships and investing in existing friendships throughout their lives.

The Link Between Broken Attachments in Childhood and Loneliness

Emotionally neglected children often grow up to be adults who are distrustful of others. They perceive that the entire world around them is connected and happy, and they perceive that they are on the outs with those around them. Not only do they feel like outsiders, but they make themselves outsiders by isolating themselves further from people. So the cycle continues, and they begin to lose hope that things will ever get better.

Lonely people may struggle with untruths such as:

"Maybe I'm just like this."

"Maybe I'm broken."

"All these people know how to get along with people and have these great friendships, but nobody wants to be my friend."

Scientifically, lonely people may have just as many social interactions as those who do not consider themselves lonely. The lonely are not present in the moment, however, contributing to their social isolation and negative self-thinking.[38]

It is not that they simply do not know anybody to be friends with. Maybe it is because they have weaker relationship-building skills, for whatever reason. It is perpetuated because they have allowed the consistent loop of negative thoughts, feelings, and behaviors to tell them that loneliness is the permanent state.[39] But it is not.

When Loneliness Shows Up at Work

In my experience as a first responder counselor, I have noticed that many firefighters experienced bullying in some form as a child. The pain from this often comes up again as they experience adversity in adulthood. So I feel the need to take a minute to pause here and recognize that it can be hard to connect with others when you are constantly afraid they will turn on you.

Did you know that high-stress situations tend to polarize people and can create a pack mentality? The problem with the pack mentality is that someone is always on the outside. The pack then bonds over shared animosity toward the outsider. That person may be able to survive alone for a short period of time, but loneliness actually weakens an individual's emotional and physical fortitude. In fact, loneliness causes the brain to release more of the chemical norepinephrine, which weakens the immune system similar to the way an infection would. Loneliness also increases white blood cell count, which triggers inflammation in the body.

As much as I would like to say that adults do not bully one another the way children do, I have encountered many firefighters who have been the targets of adult bullying. Dan, a client of mine, has been facing adult bullying from one of his officers. Dan truly is an excellent firefighter. He was in the military before he joined the fire service, and he really knows his stuff. He was severely bullied as a child. One way that Dan overcame childhood bullying is by becoming extremely capable and competent. He grew up seeking achievement in order to develop his own confidence. But then Dan began to be written up for multiple, insignificant things on the job, in what seems like an adult bullying situation.

The problems started when Dan questioned an officer about something that happened. In doing so, I believe he bruised the officer's ego. I am guessing that the officer probably is not confident and secure in himself, and he could not handle any questioning or discussion. As a result, Dan began to be targeted. When leaders are not confident in themselves, their "rock star" subordinates often can be targeted or bullied.

People who have been bullied as children often get bullied again as adults. I believe this is because those who have not worked through childhood bullying carry around a demeanor that can make them become targets for adult bullies. The way I have seen bullied children become confident and included adults is by working through their childhood bullying by talking to a counselor. In time, they build up their own self-confidence reserves, and their interaction with peers shifts, making them less likely to be targeted again.

Your View of Yourself

The way we view ourselves shapes the way we act, which in turn influences the way others treat us. The more you think of yourself as an outsider, as someone who has no deep relationships, the more you become trapped in a cycle of isolation.[40] It becomes a self-fulfilling prophecy that is often referred to as the *loneliness loop* (fig. 3–1).[41]

In the loneliness loop, individuals lose control and self-regulation and allow distorted thinking to take over in their minds. This paralyzes them from taking action to connect with others. So rather than doing something to make friends, like joining a local softball league, heading to the gym to take a group class, or heading out to breakfast with the guys, they choose to decline opportunities, and their activity level plummets. They become passive and negative, and they misread social cues from the people around them.[42] They think that although they are being invited to do things with people, they are really outsiders, and people would rather not have them there.

This could not be further from the truth. Believing all these untrue thoughts only isolates people more. It may be helpful for them to understand that it is not uncommon for people to feel lonely.[43]

> **Truth:** Most people feel lonely occasionally.

Breaking out of the loneliness cycle is easier than you might think. First, simply start by acknowledging that you are in the loneliness cycle. Then take action to bring people into your life. Identify five people in your life you would like to spend time with. Maybe you have a coworker or a buddy that you would like to go grab a pizza with. Maybe you have a local organization you would like to get involved in. Maybe you have a favorite

Figure 3–1. The loneliness loop can be a self-fulfilling prophecy.

hobby and know someone with a similar interest. Invite that person to join you. If they cannot come, it is likely not about you. Ask again at another time.

You have more control over your loneliness than you might think, so develop a plan to challenge the loneliness cycle in your life.

Reflection Questions

1. What is one truth from this chapter that you needed to hear?
2. What are some ways you have observed yourself making the needs of others a priority this week?
3. What is one personal need that you have neglected this week?
4. What can you do to meet this need today?
5. What boundaries do you need to draw in the future so that you do not neglect your own needs to meet the needs of others?
6. Who is one person you would like to spend time with? What is one thing you can invite that person to do with you to help you break out of the loneliness cycle?

Notes

1. J. Viorst, *Necessary Losses: The Loves, Illusions, Dependencies, and Impossible Expectations That All of us Have to Give up in Order to Grow* (New York: Simon and Schuster, 1986).
2. N. Vincent, *Self-Made Man: One Woman's Year Disguised as a Man* (New York: Penguin Group, 2007).
3. T. E. Joiner, *Lonely at the Top: The High Cost of Men's Success* (New York: Palgrave Macmillan, 2011).
4. I. Bacon et al., "The Lived Experience of Codependency: An Interpretative Phenomenological Analysis," *International Journal of Mental Health and Addiction* 18, no. 3 (August 21, 2018): 754–771, http://doi.org/10.1007/s11469-018-9983-8; and R. Lee, "Codependency: The Helping Problem," PsychCentral.com (February 14, 2018), https://psychcentral.com/lib/codependency-the-helping-problem/.
5. Lee, "Codependency."
6. Lee, "Codependency."
7. Lee, "Codependency."
8. Lee, "Codependency."
9. E. Weinstein, "Is It True: What Doesn't Kill You Makes You Stronger?" PsychCentral.com (October 28, 2017), https://psychcentral.com/lib/is-it-true-what-doesnt-kill-you-makes-you-stronger/.
10. Joiner, *Lonely at the Top.*
11. W. E. Frankenhuis and C. de Weerth, "Does Early-Life Exposure to Stress Shape or Impair Cognition?" *Current Directions in Psychological Science* 22, no. 5 (September 25, 2013): 407–412, https://doi.org/10.1177/0963721413484324.

12. C. Mittal et al., "Cognitive Adaptations to Stressful Environments: When Childhood Adversity Enhances Adult Executive Function,"*Journal of Personality and Social Psychology* 109, no. 4 (2015), 604–621, https://doi.org/10.1037/pspi0000028.
13. Mittal et al., "Cognitive Adaptations to Stressful Environments."
14. Mittal et al., "Cognitive Adaptations to Stressful Environments."
15. C. Osório et al., "Adapting to Stress: Understanding the Neurobiology of Resilience," *Behavioral Medicine* 43, no. 4 (2017): 307–322, https://doi-org.du.idm.oclc.org/10.1080/08964289.2016.1170661.
16. M. Hustad, "Surprising Benefits for Those Who Had Tough Childhoods," *Psychology Today*, March 7, 2017, https://www.psychologytoday.com/intl/articles/201703/surprising-benefits-those-who-had-tough-childhoods?amp.
17. Mittal et al., "Cognitive Adaptations to Stressful Environments."
18. D. J. Ley, "Surviving Childhood Adversity Builds Empathy in Adults," *Psychology Today*, September 18, 2020, https://www.psychologytoday.com/us/blog/women-who-stray/202009/surviving-childhood-adversity-builds-empathy-in-adults.
19. D. M. Greenberg et al., "Elevated Empathy in Adults Following Childhood Trauma," *PLoS ONE* 13, no. 10 (October 3, 2018), https://doi.org/10.1371/journal.pone.0203886.
20. Ley, "Surviving Childhood Adversity Builds Empathy in Adults."
21. Greenberg et al., "Elevated Empathy in Adults Following Childhood Trauma."
22. J. Amodeo, "Are You Codependent or Just a Caring Person?" *Psychology Today*, November 4, 2017, https://www.psychologytoday.com/us/blog/intimacy-path-toward-spirituality/201711/are-you-codependent-or-just-caring-person.
23. M. Beattie, *Codependent No More: How to Stop Controlling Others and Start Caring for Yourself* (Center City, MN: Hazelden Publishing, 1992).
24. Amodeo, "Are You Codependent or Just a Caring Person?"
25. Amodeo, "Are You Codependent or Just a Caring Person?"; and Beattie, *Codependent No More*.
26. Beattie, *Codependent No More*.
27. Amodeo, "Are You Codependent or Just a Caring Person?"
28. H. M. Jones, "The Philosophical Basis for Caring, Compassion, and Interdependence," in *The Pursuit of Happiness* (Cambridge, MA: Harvard University Press, 1953), 131–166.
29. Amodeo, "Are You Codependent or Just a Caring Person?"
30. E. Polack, "New Cigna Study Reveals Loneliness at Epidemic Levels in America," *Cigna.com*, May 1, 2018, https://www.multivu.com/players/English/8294451-cigna-us-loneliness-survey/.
31. J. Holt-Lunstad et al., "Loneliness and Social Isolation as Risk Factors for Mortality: A Meta-Analytic Review," *Perspectives on Psychological Science* 10, no. 2 (March 11, 2015): 227–237, https://doi.org/10.1177/1745691614568352.
32. Osório et al., "Adapting to Stress."
33. L. C. Hawkley and J. T. Cacioppo, "Loneliness Matters: A Theoretical and Empirical Review of Consequences and Mechanisms," *Annals of Behavioral Medicine* 40, no. 2 (October 2010): 218–227, https://doi.org/10.1007/s12160-010-9210-8.
34. S. J. Wilson et al., "Loneliness and Telomere Length: Immune and Parasympathetic Function in Associations with Accelerated Aging," *Annals of Behavioral Medicine* 53, no. 6 (2018): 541–550, https://doi.org/10.1093/abm/kay064.

35. Hawkley and Cacioppo, "Loneliness Matters."
36. Joiner, *Lonely at the Top*.
37. Joiner, *Lonely at the Top*.
38. S. N. Arpin and C. D. Mohr, "Transient Loneliness and the Perceived Provision and Receipt of Capitalization Support within Event-Disclosure Interactions," *Personality and Social Psychology Bulletin* 45, no. 2 (July 19, 2018): 240–253, https://doi.org/10.1177/0146167218783193.
39. Arpin and Mohr, "Transient Loneliness and the Perceived Provision and Receipt of Capitalization Support within Event-Disclosure Interactions."
40. Arpin and Mohr, "Transient Loneliness and the Perceived Provision and Receipt of Capitalization Support within Event-Disclosure Interactions."
41. Hawkley and Cacioppo, "Loneliness Matters."
42. Hawkley and Cacioppo, "Loneliness Matters."
43. C. Killeen, "Loneliness: An Epidemic in Modern Society," *Journal of Advanced Nursing* 28, no. 4 (October 1998): 762–770, https://doi.org/10.1046/j.1365-2648.1998.00703.x.

CHAPTER 4

SUBSTANCE ABUSE AND ADDICTION

Brett's Story

Brett grew up with little emotional connection with his parents. His father was a first responder, and like many children raised by first responders, Brett dealt with pain in his life by achieving and "sucking it up." Brett taught himself to self-soothe in isolation because nobody seemed to care. As he got older, Brett's favorite self-soothing measure became alcohol. At first it was just a tool he used to calm down or when he wanted to take a break or reward himself for making it through a hard day at work. But when he retired from the fire service, his schedule changed, his time with others decreased, and he had more time on his hands. He found himself increasingly turning to alcohol.

He had not intended to develop an addiction. It started as a "harmless" habit, but Brett's drinking quickly got out of hand, distancing him from his wife and kids, and ultimately leading him into an affair. When Brett came to me, he was ready to find freedom, so we began by recognizing what his struggle really was, what purpose it served in his life, and what he needed more than his relationship with alcohol.

What Is Addiction?

Addiction is a psychological as well as a physiological dependence on something. It is often the manifestation of a deeper problem, such as depression or childhood pain. Rather than confronting the pain, many firefighters escape to wherever they find comfort: substances, behaviors, or activities that will generate approval from others.

Seemingly productive activities such as overachievement at work, workaholism, excessive exercising, and obsession with diet become like drugs because they provide social approval. People think of these as "good" addictions, but they are most definitely harmful in that they distract from the pain instead of healing it. The addictions often take the place of the genuinely intimate relationships where healing actually resides.

For others, alcohol or drug abuse, pornography, gambling, extramarital affairs/sexual promiscuity, overspending, domestic or verbal abuse, violence, or multiple of these are the means they use to distract from their pain or to destress. Canadian physician Gabor Maté also includes shopping, eating, Internet usage, relationships, work, and extreme sports in his list of addictive behaviors.[1] These vices promise comfort but never truly deliver because there is never enough to make the pain go away.

Addiction is more than physical. It is a psychological response to life's stressors. As Maté stated in an interview with the late Richard Simon, clinical psychologist and editor of *Psychotherapy Networker*, and coauthor Lauren Dockett:

> I don't medicalize addiction. In fact, I'm saying the opposite of what the American Society of Addiction Medicine asserts in defining addiction as a primary brain disorder. In my view, an addiction is an attempt to solve a life problem, usually one involving emotional pain or stress. It arises out of an unresolved life problem that the individual has no positive solution for. Only secondarily does it begin to act like a disease.[2]

I agree with Maté, who identifies addiction as a solution to a life problem. For most people with an addiction, the substance or behavior is a way to escape temporarily from a problem in life. But addiction is a cycle of craving, experiencing the "high," suffering consequences from use, and having trouble giving up the substance/behavior.[3] This cycle leaves the individual exhausted, trapped, and still as unhappy as he was before.

Addiction is less about getting hooked on a substance or behavior and more about solving an emotional pain problem. Addiction is less about being genetically predisposed to addictive behaviors and more about the ways that people learned to self-soothe in response to their upbringing. Addiction is less about *what* the addiction is and more about *why* the person is addicted. Consequently, this forces us to admit that just about anyone is capable of developing an addiction.[4]

If it is possible for anyone to develop an addiction, I advocate we should be careful not to call those with addictions "addicts." These are people. They are humans who need support as they learn that they do not need their addiction anymore. They are people who need to fight to relearn how to meet their needs in appropriate ways, and they deserve respect, connection, and care as they find freedom.

What Does Addiction Do for the Individual?

In my experience working with first responders, I have found that firefighters like to keep themselves busy. When they have a lull in their schedule, when they have to take a break from work due to an injury or trauma, or when they retire, they are abruptly confronted with thoughts, memories, and feelings that they have never had to deal with before. This can make anyone uncomfortable, and many pursue an outlet for this discomfort. If alcohol, other drugs, or risky behaviors are within reach, they become tempting in times like this because they seem to promise comfort and distraction.

People do not just engage in addictions for the fun of it. Rather, as Maté advocates, addictions serve a highly specific purpose in the individual's life. People run to their addictive substance or behavior in an attempt to find comfort, pleasure, stress-relief, connection, and distraction from pain.[5] And if the individual does not learn to calm down or reduce stress in healthy ways, relapse becomes easy and frequent because his only outlet is the addiction.

In an addiction, an individual will seek to meet his intimacy and relational needs with his addictive substance or behavior until it becomes his most important relationship, according to researchers Chad Cross and Larry Ashley, University of Nevada–Las Vegas. Because the experience of using the addictive substance or behavior can change a person's mood from stressed to calm or from angry to happy, for example, the individual may perceive that his needs have been met. Cross and Ashley note that the individual will then begin to shut out people and turn to the addictive substance or behavior instead of friends and family. They observe that although the individual actually needs a connection with other individuals,[6] the addiction remains the individual's primary relationship.

Licensed clinical social worker and certified sex addiction therapist Robert Weiss notes that instead of trusting others who might potentially let them down, those with addictions begin to trust more and more in their addictive behavior or substance. According to Weiss, instead of turning toward people for connection and healing, they begin to mistrust people and become less accustomed and tolerant of working through the ups and downs of real relationships, preferring the addiction over people.[7]

Firefighters who struggle with addiction need to build a robust support web around themselves so that they can select a healthy means of destressing instead of turning back to their addictive substance or behavior. (See fig. 1–1 and chapter 1 for further information about the firefighter support web.)

When a firefighter quits drinking, it usually results in an entire lifestyle change. Not drinking puts new strain on the social life. Connecting with others is highly important in overcoming addiction, but first responders often hang out in alcohol-inclusive situations, such as at a bar, a barbecue, and so on. Most first responder social gatherings have alcohol connected with them, so often sobriety results in unintended isolation. Thus, a firefighter should be intentional to build up the relational part of his support web by spending time with others in healthy environments.

The Development of an Addiction

What starts out as a healthy relationship with alcohol—or another substance or behavior—can escalate over time. Firefighters may get in the habit of reaching for alcohol after a tough call. The intense situations faced daily by first responders leave residual *cortisol* (the stress hormone) and adrenaline in the body. This can cause lingering stress and anxiety, and many first responders tell me they were physically shaking after an intense call. This shaking is the body's way of working the excess cortisol out of the body and is actually a good thing. But it can feel like weakness, which first responders do not like, so it is common for them to reach for a drink as a way of "chilling out." When this happens repeatedly, the brain rewires itself for an alcohol-assisted chill out, and it can become increasingly more difficult to calm down without the use of alcohol in the future.

Additionally, I see many active first responders have a drink after work before their spouses get home. They believe that with the help of alcohol, they will be more relaxed and less likely to get in a fight with their spouses. I also hear firefighters tell me they drink to be able to sleep. Many first responders admit to drinking every night before bed. The problem with drinking before bed is that it interferes with *rapid eye movement (REM)* sleep, a period during which memory is consolidated and the body truly rests. When REM sleep is not achieved, it affects an individual's ability to process stress and trauma, avoid anxiety and depression, and feel rejuvenated. Patterns like these form dependencies in which the neurotransmitters in the brain literally adapt to the depressing and calming effects of alcohol and become addicted. They actually become unable to produce the regular balance of neurotransmitters that they had in their healthy brains. When they feel their neurotransmitters out of balance, they soothe with more alcohol.

But it is not only active firefighters who struggle with alcohol. I often see retired first responders come through my doors or check themselves into inpatient facilities to find freedom from addictive behaviors. Retired first responders often have flashbacks and dreams of things that happened decades ago, and they often experience sleep disturbances starting about a year into retirement. (More about retirement is discussed in chapter 18.) As the firefighter continues to turn to alcohol or other substances or behaviors for an escape from the anxiety, flashbacks, or memories, he or she becomes dependent on that substance or behavior. The delicate balance of firing neurotransmitters in the brain adjusts to find a new normal based on the inhibiting substance of alcohol. This provides only a temporary escape, however. People who did not intend to become addicted find themselves unable to stop.

The Biology of Addiction

All addictions have seven factors in common, according to Maté:

1. Compulsive behavior
2. Craving
3. Temporary pleasure or relief
4. Negative consequences
5. Denial
6. Shame
7. Brain circuits (dopamine released during the "hunt")[8]

An addiction is thus more than a short-lived act. Addiction is the process of thinking about the substance or behavior, desiring to act on these thoughts, feeling excitement or a rush during the "hunt," or the time on the way/just before engaging in the addiction. Then the person feels a temporary sense of pleasure or relief before an onslaught of consequences, denial, and shame set in. The person may especially feel shame over his or her inability to resist the addictive substance or behavior. In this process, the person continues to reinforce that the addiction is his or her primary relationship. The person's neurotransmitters orient themselves around the addiction, so the individual becomes dependent on the experience, unable to escape because of what it does for him or her.

Drugs and alcohol trigger a release of dopamine in the brain that causes a "high" feeling. The stronger and quicker the release, the more the reward centers in the brain remember that as a pleasurable event. Thus the more dopamine, the more likely to lead to physiological addiction. More exposure to this substance or behavior means the individual begins to build tolerance to it. The brain learns to release less dopamine, which means the brain needs more of the addictive substance or behavior for the same result. This is why people with addictions are always "chasing the high." They have developed tolerance to their substance or behavior, and the brain has adjusted to having that substance or behavior as the new normal.

This is why people with addictions feel a lack of control and compulsion when around their addictive substance or behavior. Their brains literally need the substance or behavior to feel a sense of normal, and these people begin to believe that they truly could not live without their chosen vice.

Risk Factors for Addiction

There are a series of factors that often place a person at risk of addiction. If you have addictive tendencies, or if you know someone in your life who is struggling with an addiction, it may help to start pinpointing some of the factors contributing to the onset of an addiction.

Place a check mark next to the factors that you identify in your own life or in the life of someone you care about:

- Experiencing trauma/traumatic incidents
- Negative peer influences
- Feelings of insecurity, loneliness, or being different
- Difficulty using positive emotions such as love, joy, or intimacy in times of trouble
- An inability to sit with stressful feelings or emotions
- Adverse childhood experiences/lack of connection in childhood/childhood emotional neglect[9]

Do any of these seem familiar to you? Many first responders have adverse childhood experiences. These experiences can leave firefighters exceptionally vulnerable to addictive substances and behaviors.

At the beginning of this chapter, I introduced you to Brett. Brett admitted that he could check a number of these boxes. He had grown up with an emotionally absent father. He did not have any negative peer influences at the time because he found himself quite alone after retiring. His feelings of insecurity, loneliness, and unproductivity weighed on him, and he had difficulty expressing his emotions because he had grown up with an emotionally absent family and a weak ability to articulate what he was feeling and express that to others in a heathy way. Every addiction has multiple factors contributing to it, but I would say that for Brett, his lack of connection in childhood had significantly played into his long-term struggle with alcohol.

When Brett came to me after his retirement, he had reached rock bottom, gone through a 30-day treatment program, and come to me for continued support. He wanted a way to rebuild his marriage after his affair. He desperately wanted connection. He wanted to find freedom, but he had no idea where to start. Instead of starting with the alcohol, we started with his relationships.

Childhood Emotional Neglect

According to clinical psychologists Jonice Webb and Christine Musello, behind every addict is a child whose emotional needs were neither identified nor met. If you can get people with an addiction to open up, it is surprisingly common to hear that they grew up with no one asking how they were doing or helping them pinpoint their feelings. This neglect was usually not overt. In fact, Webb and Musello note that children who are emotionally neglected are often well cared for physically, which can make it difficult to identify childhood emotional neglect (CEN).[10]

When a child's parents never care to hear why he was sad and never help him practice using tools to identify emotions, soothe, de-stress, or regain control of his emotions, it can be a dangerous path for him to travel. What this child needed was someone to ask about and listen to his feelings, but what he got was a closed door and a room to himself. So he "manned up" and coped the only ways he could find to cope.[11] As an adult, he now has less self-awareness to recognize which specific emotion he is feeling and a limited menu of self-care tools to downshift from "overwhelmed" to "at peace." Addictions were his self-soothing mechanism to face a world that just did not care.

The opposite of addiction is not sobriety. The opposite of addiction is connection.
—Johann Hari

Author and TED Talk speaker Johann Hari notes that connection with others is vital in the process of overcoming addiction.[12] Connecting with others and allowing yourself to be surrounded by your support community is the way to move toward freedom.

What Brett needed most was emotional self-awareness and safe relationships. Talking to me was a start, but rebuilding his marriage was his most challenging and fulfilling pursuit. As he fought for connection in his marriage and to deepen his friendships and his relationships with broader society, he simultaneously fought against his addiction.

Addiction is a relationship with a substance or behavior that served as a coping mechanism for the individual to face a tough and uncaring world. But connection with real people in deep, meaningful ways is the way to move toward freedom.

What Scientists Discovered About Rat Parks and Addiction

Scientists have been perplexed by addiction for decades. Many have endeavored to study addiction by observing how rats have interacted with addictive substances. In study from 1969, researchers provided rats with two water feeders—one with cocaine-laced water and one with plain water.[13] Hands down, all of the rats chose the cocaine-laced water over the plain water and became addicted.

A more recent study looked at the rats and asked, "What would happen to the rats if they were placed in a cage with other rats?" Rather than putting the rats in isolated cages and offering them an addictive substance, they wanted to see what would happen if they were surrounded by community and then offered an addictive substance.[14] To their surprise, the rats who were in cages surrounded by other rats, and all the things a rat society needs to be interesting, almost always avoided the cocaine-laced water. Those who tried the cocaine-laced water almost never used the drug regularly or developed an addiction.[15] So what was their conclusion?

Rats are social animals. They need a community and relationships around them to remain emotionally well. Denying them their social connections drove them to self-soothe in isolation, and addiction set in. But those who were surrounded had no need of addiction because it offered them nothing. They did not need an addiction.

Obviously humans are much more developed than rats. But consider that humans are also social creatures. We were made for deep connection and meaningful relationships with others. When we are taught to isolate ourselves and ignore emotions, we still need somewhere to turn to find comfort, so we can turn to addictive behaviors or substances. But when we seek community, talk to someone, and develop supportive relationships, we can live in freedom from addictions.

Instead of looking at someone with an addiction and thinking that there is something wrong with that person, consider thinking instead that the person likely feels all alone and needs someone to trust on the path toward freedom. The thirst for an addictive substance rightly should be seen as a thirst for real connection.

How to Identify an Addiction

To help determine whether or not your behavior is an addiction, answer the following questions with either "true "or "false."

- ____Before engaging in the behavior/substance, I feel low, insecure, or anxious.
- ____The substance/behavior makes me feel better about myself.
- ____When I am craving the substance/behavior, I am usually bored, stressed out, or lonely.
- ____Without the substance/behavior, I am just angry all the time.
- ____It stresses me out to think about my life without the substance/behavior.
- ____I think about the substance/behavior every day.
- ____When I am alone, the substance/behavior comes to mind.
- ____I do not know what I would do without the substance/behavior.
- ____Whenever I am around the substance/behavior, I have to do it.
- ____The substance/behavior keeps my demons at bay.
- ____I just feel calmer and more relaxed when I engage in the substance/behavior.
- ____This substance/behavior is the one thing that gives me a chance to check out from the world around me.

If you answered the above statements with more "true" answers than "false," you may have an addiction. I always recommend talking to a trusted friend or counselor if you think you may have an addiction. Talking about it is the first step to being able to find freedom.

The second step is to begin taking active steps toward freedom from addiction. The two most powerful tools in the journey toward freedom from addiction are connection with others and self-care.

Connection: What People with Addictions Actually Crave

I am happy to say that Brett has been sober for more than a year now, and he is reestablishing his relationships and reconnecting with his family. He is back together with his wife and children, but he is having to rebuild his self-image at the same time. Brett struggles with shame for his drinking and for having an affair, so he is choosing to focus now on talking, being more present, and enjoying his family, rather than focusing on his past.

What we have found now is that Brett is afraid to have a voice in his family. He cannot put his foot down or be assertive when he needs to because he feels that if he makes one false move, he will be kicked out. He realizes how important family is to him, so now we are working on ways to handle some of the lingering resentment his family feels. Resentment is one of the main reasons that relapses can happen. His wife is very hurt and angry. They have been married for more than 30 years. He is sober now, which helps with the disconnect that initially led to the affair, but he is having a hard time finding his voice. I have respect for his self-awareness. He is aware that a relapse could happen as a result of resentment, and he is choosing connection and self-care instead of shame and escaping to alcohol.

Brett is also working on his developing friendships. Like many first responders, he struggles with social anxiety. He gave up a significant part of his life when he retired. He lost the brotherhood and the connection of the fire service, which left him lonely. His wife still works. Retirees need to connect with a group of other retired first responders or peer supporters. You still need friends, maybe even more now than you did while you were working.

Building trust with others can be a very healing part of overcoming an addiction. And trusting others in some form is essential for recovery. For those who have experienced neglectful or adverse childhoods that left them with deep emotional wounds or an inability to deal with emotions in a healthy way, trusting others can be new and scary territory. It can be easier to trust a substance or addictive experience than a person. But *connection with others is the key ingredient in recovery*. This could be a therapist, accountability partner, partner, peer supporter, or a higher power. As Simon explained, "Unless the [person with an addiction] learns to connect with themselves and others, long-term recovery will not occur. Understanding emotions is foundational to learning how to connect."[16]

Not talking about or dealing with emotions in childhood is a major contributor to addiction. But building your "emotional vocabulary," as authors Sissy Goff, David Thomas, and Melissa Trevathan call it,[17] in adulthood is a powerful way to connect with yourself and others.

Here are some categories of emotions that you may feel on a daily basis (fig. 4–1). Take a minute to look over the Feelings Wheel, which is an updated version based on one originally created by Gloria Willcox. Try to think of a time when you felt some of these emotions. Sometimes it is easiest to start at the core and work your way out. Start by identifying the core feeling and then work that out into the specific feeling you are experiencing. Sometimes it is easiest to start at the outer part of the Feelings Wheel and work your way in, identifying your core feeling later. Either way, the more you practice expanding your emotional vocabulary, the more you will be able to identify what you need in that moment and meet that need in a healthy way. This will also help you to communicate your feelings and needs with a spouse, trusted friend, or even fellow firefighter.

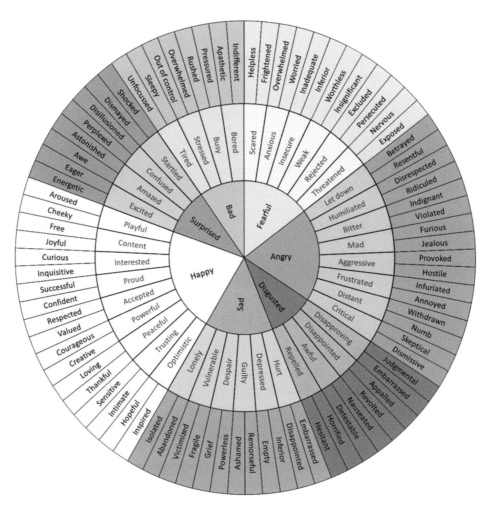

Figure 4–1. The Feelings Wheel can help you communicate with family, friends, and fellow first responders.

Identifying and naming feelings and learning to express them to others is essential for emotional wellness. If you have an instrumental personality, you may need to make a conscious decision to allow yourself to respect emotions. This may be a drastic shift from the past. Building your emotional vocabulary is the first step in learning how to connect with yourself and others and learning to regulate your emotions in a healthy way. Having a robust emotional vocabulary—being able to recognize when you are angry, sad, happy, insecure, ashamed, afraid, bored, anxious, stressed out, calm, content, proud, frustrated, rejected, burdened, surprised, disgusted, disappointed, resentful, or hurting, for example—is a foundational part of addiction recovery.

Here is an activity that might help you build your emotional vocabulary. Next time you are watching TV or reading a book, pause for a minute and ask yourself, "What is this character feeling?" This practice will help you begin to expand the language you use to talk about emotions.[18] Connecting with others starts with being able to recognize and talk about your feelings, and connection is key to healing.

With the help of a support team, the next step is to remove the addictive substance or behavior. I must warn you that when the substance or behavior is removed, an inevitable depression or feeling of loss will set in as the brain no longer receives the excitatory or inhibitory messages. But research shows that neurotransmitters can and do regulate over time. Eventually the individual can resume a normal, addiction-free life if given the space to process childhood issues or a difficult trauma and the support to walk away from an addiction.

Self-Care: What People with Addictions Need to Prioritize

The second greatest contributor to success in addiction recovery is self-care. Individuals struggling with addiction are "masters of self-neglect," according to Webb and Musello, and in order to recover, they must learn "consistent and effective self-care."[19] Research shows that exercise, sleep, healthy eating, and mindful breathing increase self-control (more about self-care is presented in chapter 11). Those with addictions "demonstrate an almost complete inability to relax and enjoy themselves."[20] Individuals need to give themselves permission to slow down.

Self-care requires dealing with uncomfortable feelings in healthy ways, healthy relaxation, and having fun apart from an addiction. It also means practicing self-compassion and ending self-criticism because a "self-critical mindset decreases one's ability to successfully change," according to psychologist Kelly McGonigal. Self-criticism increases chances of repeating behavior, and self-critical people have less control and motivation. McGonigal explains that "it's unlikely that true change will occur when self-criticism remains the default."[21]

Healthy brains are a large part of healthy living. When people take care of themselves and their minds, they can succeed at breaking free from an addiction. This includes healthy movement (exercise), healthy rest (sleep and mental breaks), healthy nutrition, healthy play (laughter), healthy learning (learning new things, growing), and healthy relationships with healthy people. (For more information on self-care for first responders, see chapter 11.)

Relapse and Triggers

When people struggle with an addiction, there are risks that can increase the likelihood of relapse. The first risk factor is boredom. Those who stop engaging in addictive behaviors may find that lives can seem uneventful and lacking in emotions. Add to this the fact that their brains are not being stimulated by the substances/behaviors used in the past, and their lives may begin to seem boring to them. So be aware when they start spending a lot of time alone, lacking structured activities, or saying things like, "I have nothing to do," or "Life just seems so boring." When individuals have these thoughts and possibly too much free time, they may be at risk of returning to their addiction.

The second risk factor for relapse is illness. Illness makes a person weaker and drains the body and mind of energy that is important for focusing on recovery. Simple illnesses like colds, the flu, or other infections can make it much more tempting for a person in recovery from an addiction to want to relapse. Even surgery or dental work can require recuperation that saps their energy and tempts them to use.

The third risk factor for relapse is a major life change. Major life changes are disruptive to the structured lifestyle that is critical to a successful recovery. These can include divorce, break up of a relationship, the death of a friend or family member, moving, a child moving away to college, or even the change of a job or retirement. When people experience change, the new patterns created can set them up for excellent recovery, or it can be a risky time for them to relapse.

The fourth risk factor for relapse is anger. Often the process of getting clean results in increased emotions and irritability, which can instigate rage and anger. When those who are struggling with addiction say things like, "It just gets me so angry," or "I feel so pissed off," it is a warning sign that their emotions are becoming more intense. When people are not able to control their feelings or to calm down, they are making it abundantly clear that they need support and are at risk of relapse. Remember, their addictive substance or behavior is most likely what they formerly used to calm themselves down. (For more information on anger and depression, see chapter 2.)

The fifth risk factor for relapse is *impulsive sex*, which is sex that is not related to intimacy but rather is a compulsive behavior. It can become an addictive behavior either on its own or as part of a relapse from another addiction. This sex is irrelevant to the person's emotions and has little or no relational involvement with the sex partner. Impulsive sex sends a red flag that a person is at risk for relapse.

The sixth risk factor for relapse is high stress. Stress can take a toll on many areas of life, increasing the opportunity for triggers. When an individual who is in recovery begins talking about sleep or stomach problems, chronic illness or headaches, fatigue, moodiness, or difficulty concentrating, it is most likely related to a high-stress situation or season of life. Those who face high-pressure life situations or life events can find it enticing to return to drugs, alcohol, or other behaviors in order to escape, calm down, or reward themselves.

Signs of Relapse

Emotional signs of relapse

According to Terence Gorski and Merlene Miller, authors of *Staying Sober: A Guide for Relapse Prevention*, when a person in recovery begins to have a shift in emotions, it can often lead toward relapse.[22] Identifying the signs of emotional relapse are crucial. Recognizing them makes it easier to avoid relapse. When you catch relapse in the early stages, it is easier to disengage. In the later stages, the pull of relapse strengthens, and the sequence of events accelerates. Emotional signs of the potential for relapse include the following:

- Anxiety
- Intolerance
- Anger
- Defensiveness
- Mood swings
- Isolation
- Not asking for help
- Not going to meetings
- Poor eating habits
- Poor sleep habits

Mental Signs of Relapse

Authors Gorski and Miller note that in mental relapse, there is a war going on in the minds of the individuals who are addicted.[23] Part of them wants to use, but part of them does not. In the early phase of mental relapse, they are just idly thinking about using. But in the later phase of relapse, it is harder to avoid using the addictive substance or behavior because the enticement grows stronger. Mental relapse could include the following:

- Thinking about people, places, and things you used with
- Glamorizing your past use
- Lying
- Hanging out with old using friends
- Fantasizing about using
- Thinking about relapsing
- Planning your relapse around other people's schedules

Triggers

Triggers are specific instances that can cause a person to use or engage in addictive behaviors.[24] Learning what the triggers are is important to a successful recovery. It will be necessary to identify the triggers beforehand and then avoid them in order to stay clean. Here are some questions to help identify potential triggers:

1. What are the times when you want to use most?
2. The last time you used, what was happening right before?
3. When do you find yourself thinking about engaging in addictive behaviors the most? What are the circumstances?

Another way to identify potential triggers is to identify any excessive behaviors. Behaving in certain ways can create increased opportunities for a person to feel the need to use or slip from recovery. Has the person in recovery excessively been doing any of the following things?

- Working all the time
- Using prescription medications
- Abusing other illicit drugs
- Drinking too much caffeine
- Smoking more
- Engaging in compulsive sexual activities
- Exercising to the extreme
- Compulsively masturbating
- Gambling
- Spending too much money
- Eating foods high in sugar

These and other unhealthy self-soothing practices may be signs of addictive behaviors that could lead the person to relapse by blurring the lines between sobriety and addiction.

Often, when a person relapses, it is because of a few key triggers in their lives: the influence of others, specific purposes, and risky situations. Perhaps one of the most compelling triggers is the negative influence of others around the individual. Gorski and Miller note that the person may explain how a relapse happened, saying things like: "It was offered to me. What could I do?" "An old friend called, and we decided to get together," "I was cleaning my house and found drugs I'd forgotten about," or "I was in a bar, and someone offered me a beer." According to Gorski and Miller, a person who is influenced by specific purposes may say things like: "I'm gaining weight and need stimulants to control what I eat," "I'm out of energy. I'll function better if I'm using," "I can't enjoy sex without using," or "I need drugs to help me with social situations."[25] Risky situations that serve as triggers can include situations such as seeing needles or drug paraphernalia, provocative ads, commercials with alcohol or parties, lottery ads, scratch-off games in the store, or even smelling cigarette smoke or perfume.

But people *can* avoid relapse. They can hold their ground and find complete and lasting freedom from addictions. Many people do the work, get the support, and find the healing they seek. A few basic strategies may help when a person is tempted to relapse. First, the individual should avoid triggers that can make him want to use. He should escape urge-provoking situations, thinking about foreseeable temptations in advance and steering clear of them for his own victory. He should distract himself with other activities and develop a vocabulary filled with coping statements to help him speak into motion the freedom he is pursuing. He can say to himself, "Even though it is difficult that I have to deal with this problem, drinking isn't going to help me."[26]

What Does True Recovery Look Like?

As noted by researchers for SAMHSA, "Recovery is the process of change through which individuals improve their health and wellness, live self-directed lives, and strive to reach their full potential."[27]

The three main reasons a person turns to a substance or behavior, according to physician Stephen Melemis, are to 1) escape, 2) relax, and 3) reward oneself.[28] Most often these are used to relieve tension. In order to change one's life and addiction, the individual must first change the way he or she relieves tension. Normal, healthy adults all need to escape, relax, and reward themselves. They just choose to do so in healthy ways, rather than turning to addictive substances or behaviors. Without healthy means of relaxing, addicts will let tension mount, and tension then becomes the primary driver that takes them right back to their addiction.

Perhaps people think they are too busy to relax. If so, they should consider how much time their addiction takes up. Often people make excuses for not relaxing, but this allows tension to build, and they relapse. It would be better to take the time they would have spent on their addiction and use it to relax in a healthy, enjoyable way.

Learning to relax is a critical part of an individual's ability to deal with stress in healthy ways rather than turning to an addiction to find comfort. People use substances to escape, relax, or reward themselves. Finding healthy ways to escape, relax, or reward oneself is a powerful means to freedom. If the individual cannot find healthy ways to relax, tension will build, and the individual will likely relapse. People should find something they can do every day to de-stress and take their minds off the mental busyness of life. Some examples of daily de-stressors include activities such as meditation, prayer, stretching, walking, exercise, massage, and pursuing a relaxing hobby or project.

In order for individuals to have a successful recovery, they will need the foundation of a home that is consistent and safe. Building on that, they need health to be a priority. They must learn to address discomfort in healthy ways and build a healthy lifestyle with regular self-care. Next, they need to be surrounded by a community. Healthy relationships and social support are a core part of connecting with others to overcome addiction. Finally, they need a purpose—a meaningful occupation to fill their time and use their creativity, the ability to choose their own motivation for doing things, and the satisfaction of being able to make personally significant contributions to society (fig. 4–2).

Supporting Someone with an Addiction

Perhaps you do not have an addiction, but you know someone—for example, a peer, family member, or friend—who does. There are ways you can support them in their recovery without enabling them to relapse back into their addiction.

Figure 4–2. Recovery needs include home, health, community, and purpose.

10 Ways Family, Friends, and Peers Can Help Someone in Recovery

1. Provide compassion, nonjudgment, support, acceptance, connections, and useful feedback.
2. Help them follow treatment recommendations.
3. Encourage total abstinence.
4. Help them build good coping skills.
5. Provide social support.
6. Encourage participation in peer support groups.
7. Help them create a sober peer network.
8. Know the signs of relapse.
9. Encourage their involvement in meaningful, structured activities.
10. Keep hope alive.[29]

First, offer help and guidance to peers/family members struggling with addiction based on mutual respect and understanding. Engage your loved one or friend in collaborative and caring conversation. Remember, one of their greatest needs is safe relationships because connection, rather than abstinence, is the best means to recovery. Provide support by validating the person's experiences and feelings, conveying hope to them about their recovery, celebrating their efforts and accomplishments, and providing concrete assistance to help them accomplish their tasks and goals.

If you have a story of a stronghold or addiction in your life, you can share your own experiences of recovery. Relate your own recovery story with theirs, and with permission, share the recovery stories of others to inspire hope. Talk about your ongoing personal efforts to enhance health, wellness, and recovery. Recognize that there are times to share your experience, and there are times to listen. If possible, describe your personal recovery practices and help them discover recovery practices that work for them. Finally, consider linking them with resources, services, and supports that can help them move forward and find freedom. This may mean doing some research on community resources and services and getting them up-to-date information about what is available to them. It can also mean helping them investigate, select, and use needed and desired resources and services. You can help loved ones or friends to find and use health services and supports. Finally, participate with them in community activities and accompany them to appointments when requested.

When you are supporting someone with an addiction, there are some things you should always do and there are some things you should avoid altogether. In the positive, do what you can to understand them. People with addictions are going through physical, physiological, and psychological changes. Learning about their addiction and recovery will help you provide meaningful support. This support, simply put, means offering to be there for them, especially when they are having a craving and need someone to talk to. Recovering from an addiction takes a team. People cannot find freedom on their own, so be available to them, but also be realistic that it will take time and the right support team before this person breaks free. Then as they get settled in their post-addiction life, they will need ongoing friendship and support. Relapses can happen and are more likely to happen in the first two years, so continuing to support your friend is critical for long-term success. Whatever you do, as you provide positive support to a recovering person, continue to encourage them, direct them toward the support they need, and believe in them as they walk toward freedom.

Relationships are the key to overcoming an addiction. Please be careful not to promise to be there for a person only to break that promise later. As you try to support someone with an addiction, be mindful of how you communicate with them. Rather than jumping into teacher-mode, try to empathize and listen to them. Feel free to share your story but be mindful that you are not out to make them feel guilty or ashamed. They already have enough of that weighing on them. Also be aware that you do not own any of the responsibility for their behavior. The individual will do best if they take full responsibility for their addiction and the consequences of that addiction. And if you ever encounter them while they are under the influence, choose not to argue with them. If you want to be helpful, be available, set boundaries, listen, try not to lecture, and do your best to communicate when they are sober.

Treatment for Addictions

As neurotransmitters in the brain regulate themselves, withdrawal can set in. This can be accompanied by other physical side effects or even depression. But the brain eventually does return to normal.

Recovering from an addiction is possible. There are some proactive steps to take as you begin to work toward recovery.[30] First, I recommend working with a therapist. If possible, join a support group or find an inpatient program for other first responders seeking freedom from an addiction. Admitting you have an addiction is the first step toward moving forward. Making a commitment not to engage in the addiction in the future is a key part of recovery. Then choose each day to identify emotions and deal with them in healthy ways, connect with others, explore the world and nature, and live out healthy practices to meet the needs formerly met by the addiction.

Reflection Questions

1. What is one thing you learned about addiction that you did not know before?
2. Do you think you might have an addiction?
3. Did you recognize any of the statements about adverse childhood experiences or childhood emotional neglect as relating to your childhood? Which ones?
4. Would you consider yourself to be good at identifying and talking about your emotions?
5. Who is one person you trust that you can start to talk to when you feel stressed or overwhelmed?
6. What are some triggers that you can recognize in your life that could make you crave an addictive substance or behavior?
7. What can you do to redirect yourself when you start to have a craving?
8. Who do you need to tell or what group or program do you need to join to get the support you need to find freedom?

Notes

1. L. Dockett and R. Simon, "The Addict in All of Us: Gabor Maté's Unflinching Vision," *Psychotherapy Networker,* July/August 2017, https://www.psychotherapynetworker.org/magazine/article/1102/the-addict-in-all-of-us.
2. Dockett and Simon, "The Addict in All of Us."
3. Dockett and Simon, "The Addict in All of Us."
4. J. Hari, "Everything You Think You Know About Addiction Is Wrong," TED Global London video, June 2015, 14:33, https://www.ted.com/talks/johann_hari_everything_you_think_you_know_about_addiction_is_wrong?language=en.
5. Dockett and Simon, "The Addict in All of Us."

6. C. L. Cross and L. Ashley, "Trauma and Addiction: Implications for Helping Professionals," *Journal of Psychosocial Nursing and Mental Health Services* 45, no. 1 (January 2007): 24–31, https://doi.org/10.3928/02793695-20070101-07.
7. R. Weiss, "Why Do People with Addictions Seek to Escape Rather Than Connect? A Look at the Approach to Addiction Treatment," *Consultant 360*, 56, no. 9 (September 2016), https://www.consultant360.com/articles/why-do-people-addictions-seek-escape-rather-connect-look-approach-addiction-treatment/.
8. Dockett and Simon, "The Addict in All of Us."
9. Dockett and Simon, "The Addict in All of Us."
10. J. Webb and C. Musello, *Running on Empty: Overcome Your Childhood Emotional Neglect* (New York: Morgan James Publishing, 2019).
11. Webb and Musello, *Running on Empty*.
12. Hari, "Everything You Think You Know About Addiction Is Wrong."
13. C. R. Schuster and T. Thompson, "Self-Administration and Behavioral Dependence on Drugs," *Annual Review of Pharmacology and Toxicology* (1969): 483–502, https://doi.org/10.1146/annurev.pa.09.040169.002411.
14. B. K. Alexander, "Addiction: The View from Rat Park," accessed August 9, 2021, http://www.brucekalexander.com/articlesspeeches/rat-park/148-addiction-the-view-from-rat-park.
15. Alexander, "Addiction: The View from Rat Park."
16. Dockett and Simon, "The Addict in All of Us."
17. S. Goff, D. Thomas, and M. Trevathan, *Are My Kids on Track?: The 12 Emotional, Social, and Spiritual Milestones Your Child Needs to Reach* (Grand Rapids, MI: Baker Publishing Group, 2017).
18. Goff, Thomas, and Trevathan, *Are My Kids on Track?*
19. Webb and Musello, *Running on Empty*.
20. P. J. Flores and L. Mahon, "The Treatment of Addiction in Group Psychotherapy," *International Journal of Group Psychotherapy*, 43, no. 2 (1993): 143–156, https://doi:10.1080/00207284.1994.11491213.
21. K. McGonigal, "The Willpower Instinct," YouTube video, February 1, 2012, 54:02, https://www.youtube.com/watch?v=V5BXuZL1HAg&t=1s.
22. T. T. Gorski and M. Miller, *Staying Sober: A Guide for Relapse Prevention* (Independence Press, 1986).
23. Gorski and Miller, *Staying Sober*.
24. Gorski and Miller, *Staying Sober*.
25. Gorski and Miller, *Staying Sober*.
26. Gorski and Miller, *Staying Sober*.
27. Substance Abuse and Mental Health Services Administration (SAMHSA), "First Responders: Behavioral Health Concerns, Emergency Response, and Trauma," Disaster Technical Assistance Center Supplemental Research Bulletin (May 2018), https://www.samhsa.gov/sites/default/files/dtac/supplementalresearchbulletin-firstresponders-may2018.pdf.
28. S. M. Melemis, "Relapse Prevention Plan and Early Warning Signs," *Addictions and Recovery* March 15. 2017, https://www.addictionsandrecovery.org/relapse-prevention.htm.

29. "Nine Strategies for Families Helping a Loved One in Recovery," Behavioral Health Evolution, Hazelden Foundation, 2016, http://www.bhevolution.org/public/family_support.page.
30. "Step by Step Guides to Finding Treatment for Drug Use Disorders: Treatment Information," US Department of Health and Human Services, National Institutes of Health, National Institute on Drug Abuse, June 4, 2020, https://www.drugabuse.gov/publications/step-by-step-guides-to-finding-treatment-drug-use-disorders/if-your-adult-friend-or-loved-one-has-problem-drugs/treatment-information.

CHAPTER 5

SUICIDAL IDEATION

There is nothing that breaks my heart more completely than when a first responder takes his own life. Suicide is often seen as an escape, but the truth is that it does not stop the pain, it only passes the pain on to others. A deep, horrible pain that does not go away becomes part of the story of that family from that point onward. I know from personal experience. Suicide is not a real solution.

In the famous words of Dr. Martin Luther King Jr., "Darkness cannot drive out darkness; only light can do that. Hate cannot drive out hate; only love can do that."[1] I would add that pain cannot drive out pain. Only healing can do that. Death cannot drive away death. Only life can do that. Talking to someone and bringing light into the dark places of a hurting heart is the only way to stop the pain. Things hidden deep inside you will hurt you from the inside out if you keep them hidden. But when you allow light to shine on them by talking to a trusted friend or counselor, you begin to see that this world can be a beautiful place to live in, and that light truly is what drives out darkness.

Ryan Elwood's Story

In 2016, I was honored and heartbroken when the family of firefighter Lieutenant Ryan Elwood asked me to sit down with them to hear his story. Ryan Elwood took his life in the fall of 2015, and his parents and aunts shared with me their experience with losing their son and nephew. Through their tears they told me about the wonderful person Ryan was. Ryan grew up as a smiling, funny, quick-witted kid. He was the youngest of three children, brother to Tommy and Meghan. He was tough. He played every sport possible and especially enjoyed basketball and volleyball. From a young age, Ryan loved helping people.

He was social and extroverted and never showed any signs of depression. His intuitive nature made him able to sense if someone was having a bad day. He cared about people and always looked for ways to help. His friend's father even remarked on a time when 12-year-old Ryan stood up for a homeless man. Ryan's peers made jokes about the homeless man for asking for money, but Ryan intervened. He told his peers, "You don't know what he's going through. Stop teasing him." This act left an impression on his friend's father. How many of us have this kind of compassion and courage?

As Ryan grew up, he became increasingly passionate about the idea of helping others, and he joined the fire service to be able to put his compassion and courage into action for others. He loved acquiring more knowledge and training as he worked for both the Hometown and the North Palos Fire Departments. He tirelessly earned every certification possible, and in 2012 he became a Medal of Valor recipient. Ryan was one of the youngest firefighters to be in line for lieutenant and was promoted to lieutenant posthumously in September.

Ryan's family is tight-knit. He was extremely close to his grandfather, who passed away in 2015. This loss hit Ryan hard, and he grieved for his grandfather. Soon thereafter, Ryan lost a close friend in a car accident. Still reeling from these losses, Ryan continued to serve and care for others in his work in the fire service.

Two weeks before Ryan took his life, he went out on a call to help a man who had attempted suicide. The man had unsuccessfully tried to end his own life, and Ryan was able to resuscitate him in the ambulance. But upon arriving at the hospital, the man did not make it. Ryan was quite disturbed and affected by this incident, and his family believes this incident had an even greater impact on him than anyone understood.

Sadly, in the fall of 2015, Ryan never traveled on the trips he had planned or participated in all the groups and events he had lined up. Ryan had seemed okay. Aside from mourning the loss of his grandfather and friend, in addition to this traumatic call, he had healthy relationships and a lot going for him in his future. Perhaps this decision to end his life was a somewhat quick decision. His family wonders.

Tragically, like Ryan Elwood, some first responders experience deep enough emotional pain to drive them to suicide. First responders have the highest suicide rate of any career group, even higher than military service members. Among EMS professionals, studies have shown that up to 28% feel that life is not worth living, 10.4% have serious suicidal ideation, and 3.1% have attempted suicide in the past, according to research conducted by Ian Stanley, Melanie Hom, and Thomas Joiner, faculty at Florida State University.[2] When an individual has both EMS and firefighting responsibilities, the risk goes up by six times! Another study discovered that 37% of those with firefighter and EMS responsibilities have contemplated suicide.[3] Then there are the completed suicides. My colleague, Jeff Dill, has meticulously tracked first responder suicides in his organization. The suicide rate for firefighters is 18/100,000 people, whereas for the general population, it is 13/100,000.[4]

Why is this? As you have seen in the chapters of this book, this career often causes firefighters to become disconnected. They develop places in their lives where their emotional wellness is out of balance. Some face depression, some face addiction, and some face trauma and PTSD. In response to these mental health issues, some consider ending their lives by suicide.

Firefighters are constantly exposed to death and injury. In combination with the typical first responder instrumental personality type (as detailed in chapter 2), which is self-sufficient, nonemotional, and nonexpressive emotionally, a situation is created in which pain stays hidden and false beliefs can take root.

Cultural expectations of masculinity, which have strongly asserted themselves into first responder and military cultures, require men to avoid vulnerable emotions, exhibit toughness and emotional control, value autonomy, and avoid seeking help.[5] The fear of shame and guilt drives many men to stay quiet about their pain.

Valuable, capable firefighters literally crumble under the weight of a career that shoulders the tragedy of so many people around them. Their minds become a battlefield, where there is a war between the pain they feel and the desire to make the pain go away. It takes a lot of energy to hide the pain. It is exhausting, and the consequences are detrimental.

Some days the pain is manageable, but other days it feels overwhelming, like death is the only way out. But I promise you, death is not the only way out. Here are four quick starting points that can help the pain go away:

1. *Talk to a counselor.* This will shed some light on dark situations and memories, helping you think more reasonably, so you do not do something drastic.
2. *Intentionally connect with someone.* Connection helps people find healing from depression, trauma, and other pain.
3. *Talk to a trusted friend or peer supporter.* Having support and someone to speak truth into your life will help you face your pain and move forward with your life.
4. *Call the National Suicide Prevention Hotline.* If you are to the point where you are considering ending your life, call someone! Give yourself one more day. Your life can turn around and become beautiful. The National Suicide Prevention Hotline is (800) 273–8255. They will be happy to talk with you.

I am glad to say that the culture of first responder emotional wellness is changing. The amount of emotional support available to firefighters is growing and becoming more thorough and effective, but not every department has gotten there yet. When I was serving as the clinical director of program development with the Illinois Firefighter Peer Support team, our slogan was "Make It Safe." We need it to be safe for firefighters to talk about their pain.

It is becoming safer by the day as we begin to see the importance of dealing with pain in healthy ways and supporting first responders as they reconnect with their emotional wellness. But there are still many first responders carrying the burden of pain that they think they cannot share. And thus there are many first responders walking around with suicidal thoughts and desires. Let us explore how these thoughts take root in a person's life.

Joiner's Theory on Suicide

Thomas Joiner proposed the theory that in order for an individual to commit suicide, the person must have both the desire to commit suicide and the ability to commit suicide (fig 5–1). This desire to commit suicide develops under the specific psychological situation in which an individual not only feels that he or she is burdensome ("perceived burdensomeness"), but also feels a low sense of belonging with others ("thwarted belongingness").[6] This social alienation and feeling burdensome then drive the individual to undertake high-risk behaviors that expose the individual to pain and numb his or her sense of self-preservation. The individual may engage in fights, self-injury, or other accidents that serve to desensitize him or her to pain and make the ability to commit suicide a reality.

One of the greatest factors in first responder suicides is the number of traumatic or horrific situations they see on a regular basis. First responders often have deep emotional scars. There are things that they will never forget and calls that they will never be able to un-see. The sense of emotional distress can eventually drive them to feeling like they need to seek help, but many fear that they will be seen as weak or considered unfit for the job if they speak up about their painful memories. They feel like they are a burden.

If individuals who are struggling do not seek support, this feeling of being a burden may become overwhelming. They may feel disconnected from people around them as they shut down and hold in their pain rather than talking about it. These two factors combined—the sense of being burdensome *and* a low sense of belonging—contribute to the individual's *desire* to commit suicide.

In his book *Lonely at the Top*, Joiner differentiates between men who are "lonely in a crowd" and men who are "lonely and unaware." Joiner, whose own father committed suicide, believes that men in pursuit of their careers become increasingly more focused on tasks, and as they age become increasingly lonelier. Joiner explains, "Much more so than women, men seem to be under the impression that friendships will always be provided for them, just as they were in grade school."[7] So instead of investing in friendships and close relationships, they dig into their work. To their way of thinking, providing for the family is a sufficient way to show love.

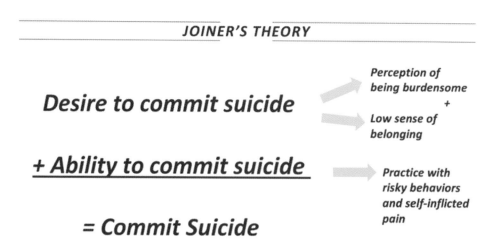

Figure 5–1. Joiner's theory

The problem is, and this is what my father learned the hard way, money is no substitute for connection. And men who do not practice connecting with others slowly lose their ability to connect. Over time, as they focus on achievement at the cost of relationships, their relational muscles weaken, and they find themselves alone in a crowd. Though they are sitting among loved ones, they feel alone, disconnected, and perhaps bored. This is what Joiner believes happened to his father. Joiner's father spent a lovely evening with his family the night before he took his life, but because he had lost his ability to develop deep attachments with others over the years, he sat there feeling desperately alone, although surrounded by loved ones.

Joiner's second type of loneliness, "alone and unaware," is a phenomenon typically for younger men who are so focused on tasks and achievements that it is as if their social gas gauge breaks, and they do not notice that they do not have many relationships around them.[8] They are so busy that they do not even notice they do not actually have many friends anymore.

Given that a low sense of belonging is one of the factors that enable a person to commit suicide, I believe that we should pay remarkably close attention to first responders' social lives, constantly encouraging them to continue to pour into existing friendships and build new friendships. As Joiner notes, "The cultivation and maintenance of friendships are literally investments in the safety net of the future."[9]

If we could train first responders to build strong relational networks and invest in them regularly, I truly believe we would see fewer first responder suicides. Joiner states,

> A large portion of suicide deaths are of this character: men who, on the inside, feel desperately alone, even though others are there with them who are expressing love and support. The men tell no one, and they stun everyone with their deaths, which come from out of the blue to everyone except the men themselves, now deceased. Of course, people can be 'truly alone'—both desperately lonely and actually alone, and a fair number of suicide deaths have this signature as well.[10]

When Suicide Hit My Own Family

I knew someone who felt desperately alone. My oldest half brother's name was Jimmy. He was 6'4" tall, but awkward and lanky, not like a football player. He was broad-shouldered but not coordinated or athletic. He was not a cool guy in the way my other half brother, Jerry, was. It was very obvious that Jerry had a stronger connection with my father. Jimmy was on the fringe. He just did not have a place where he fit. There is much I did not understand and did not know about him, but I found him incredibly sweet. He was a sensitive guy.

It happened when I was 12. I was walking home from school and noticed that Jerry was at our house. My dad was at home, too, when he should have been at work. They were talking when I walked in, and Dad said, "Honey, let's go for a walk." As we walked, he explained, "Jimmy's dead. He killed himself." And then he started crying. The first time I had ever seen my dad cry was when he told me that my brother had taken his own life.

With trauma, there are moments in time that change everything. There is life before that event, and then there is life after it. When my brother took his own life, I never grieved for myself; I grieved for my dad. I was young, and I was just so sad seeing that much emotion from him. He was Superman to me and always had been, but then I saw his vulnerability and was overcome by how hard it must be for him. I did not even think about my own loss until many years later, but I did notice a drastic shift in my family life from then onward.

After Jimmy took his life, our family started spiraling downward. My dad started drinking a lot, becoming an alcoholic almost immediately afterwards. My brother had taken his own life in a hotel in Aurora, IL, and the fire department that responded was the one my other brother worked at. This was in 1981. We may think there is a stigma about suicide now, but in 1981, it was even more intense. Since the department knew all of Jerry's business, I think it is possible he was shunned a little bit for that. My brother Jerry started acting out with a lot of anger and began using drugs and alcohol again.

My dad carried a great deal of guilt for the rest of his life for not being more present and available. As a consequence, my dad started spiraling downward. More anger and more drinking followed, and there were not as many good days as bad. We started butting heads. I called him on his choices, and I was probably the only person who had the courage to call him on anything. I genuinely cared about him, and everyone else was scared of him. He would respond very angrily. If it would have been anyone else, I am not sure how he would have responded. At the very least, he would have washed his hands of them. Even so, I was calling him out on many of his behaviors without really understanding them.

My brother started doing some of the same things as my dad. He ultimately got divorced. He fought to get custody of his daughter, but then could not take care of her with all of the shifts at the fire department and the difficulty of finding childcare. He had to give her up. That is when he snapped. He did not care about his life anymore. I remember very clearly that he destroyed his entire house with his bare hands—every appliance, every wall. My dad called the fire chief and asked him to talk my brother off the ledge. My brother told the fire chief to go to hell, and he quit his job right then and there. He never led a productive life after that, and he ended up dying of a drug-related heart issue. He was estranged from us for some time before he reconnected with my dad about 10 years before my father died. Looking back on it all, my brother Jerry's years at the fire department were the best years of his life. He was working and feeling productive and happy in the field (see fig. 5–2).

My dad quit drinking in his mid-80s, and he died at age 93. There was a deep sadness there. He thought that money would solve all problems, and in reality he had to bury both of his boys. But he was resilient at his core. He kept engaged in life, kept learning, kept growing, and kept doing his best until the day he died.

Figure 5–2. My brother Jerry Hudson

Situations That Contribute to the Desire to Commit Suicide

Injury/sickness

Stanley, Hom, and Joiner note that the most vulnerable times for a first responder to feel like a burden or to experience a low sense of belonging is during injury, sickness, or post-traumatic stress, and also during the transition to retirement. When a first responder is put on leave due to an injury, sickness, or post-traumatic stress injury, he is often left alone to deal with thoughts that he has not yet had to face. He is also bombarded by his sense of inability to complete the tasks that were formerly easy for him. So he begins to feel that he is inadequate and a burden to those around him, according to Stanley, Hom, and Joiner.[11]

Loss of Significant Relationships

Healthy attachments are very protective for the emotional wellness of first responders. These often include a first responder's spouse or partner, parents, or other close family members. So first responders who lose an attachment through a divorce, breakup, or death will often feel a loss of belonging. They may begin thinking about where they fit in and who they can trust.

One study by Thomas Buckley, University of Sydney, Australia, and others found that people who had recently lost a loved one had a significantly higher heart rate than the control group and also showed decreased *heart rate variability*.[12] Physician Marcelo Campos, Tufts University School of Medicine, defines heart rate variability as "the variation in time between each heartbeat." Campos notes that heart rate variability has been connected to cardiovascular health, better ability to handle stress, and decreased risk of heart disease.[13] Loss of significant relationships literally changed the functioning of the heart. In a way, broken attachments make you heartbroken. Remember, a low sense of belonging is one of the contributing factors to a desire to commit suicide. Those who have experienced the loss of a significant relationship should be closely watched and supported so they do not begin to consider suicide.

Anger
When a first responder is angry, it may be a warning sign for both depression and suicidality. This is because anger is an indication of something deeper going on, and it is often linked with perceived burdensomeness or an increased capability for self-harm, or both. Addressing anger can be critical for helping a first responder overcome suicidal ideation.[14]

Cancer
Research conducted by Nicholas Zaorsky, Penn State Cancer Institute, and others indicates that cancer patients commit suicide at about two times the rate of noncancer patients. Zaorsky and others note that the greatest risk period for cancer patients committing suicide is in the first year after diagnosis as they grapple with death and still have the energy to take their own lives.[15] Later stages of treatment can be difficult psychologically, but the greatest suicide risk is early in the cancer process. When individuals receive a cancer diagnosis, it is common for them to become depressed and overwhelmed by loss of functions in their body and by feeling like a burden to their caretakers.[16] If these patients are first responders, it is common for them to be accustomed to being on the helping side, and they may feel uncomfortable needing help. This may contribute to the perception that they are a burden, which is one of Joiner's factors that contribute to committing suicide.[17] (More information on cancer is presented in chapter 16.)

Financial Problems
First responders who experience the overwhelming feeling of excessive debt, bankruptcy, or other financial problems may feel like a burden to those around them. As mentioned previously, the instrumental personality type, which is common among firefighters, values self-sufficiency. Firefighters expect to be self-sufficient and see themselves as "the rescuers." Firefighters who find themselves needing help from others are often disoriented because it breaks their sense of self. Feeling like a burden is one of the contributing factors to the desire to commit suicide.

PTSD
A first responder who sustains a post-traumatic stress injury, as with an illness, may feel frustrated by his inability to function as before. With PTSD, a person often dissociates from himself. Dissociation makes it difficult to connect with oneself, let alone other people, so he may feel a noticeably low sense of belongingness. This is one factor that contributes to a desire to commit suicide.

The second factor in the desire to commit suicide is perception of being a burden. If the individual is not able to work or needs help from others, he may feel like a burden. This sense of being burdensome may be exaggerated by a first responder's past experience. Children who grow up in the face of adversity, such as occurs with the death of a parent, bullying, emotional neglect, *adultification* (having to grow up too soon), physical abuse, and sexual abuse, often feel compelled to help others. They need to be needed. When PTSD prevents them from being able to help others, they feel like they are no longer needed. This can make the responder feel absolutely worthless, lacking a will to live. This is what it looks like when a person has a perception of being burdensome. PTSD contributes to the low sense of belongingness and the perception of being burdensome, both of which are factors needed for the desire to commit suicide.

As mentioned previously, according to Joiner, individuals must have both the *desire* to commit suicide and the *ability* to take their life by suicide.

Practicing Self-Harm

Joiner explains that the second factor contributing to an individual taking his own life is his *ability* to commit suicide. Hurting oneself is not natural. People are mentally wired for self-preservation and survival. But as a person struggles with depression, traumatic memories, or other psychological burdens, he may begin to make riskier choices in a subconscious attempt to disregard the value of his life. He may begin to practice self-harming behaviors, such as driving extremely fast, to numb his sense of self-preservation, according to Joiner.[18]

Suicidal people develop the ability to hurt themselves over time. Day in and day out, first responders are exposed to danger, injury, and death, and it is common for them to encounter people who have attempted suicide. Reckless behavior is part of the "work up to suicide" noted by Stanley, Hom, and Joiner. They will gradually accumulate experiences of self-harm, making them fearless about death. This happens over time, and they become sufficiently desensitized from their own pain, injury, and self-injury, as well as witnessing other people in pain, to take their own lives. Going against their survival instincts, they no longer have any fear about death, and they complete their own suicides.[19]

First Responder Personality Type and Risk

The combination of repeated exposure to injury and the typical first responder personality type, which is predisposed to adventure and daredevil behaviors, sets first responders up for self-harm. Risk-taking behavior is, sadly, very natural for first responders. First responders typically fit the Myers-Briggs personality profile of Introvert/Sensing/Thinking/Perceiving (ISTP) or Extrovert/Sensing/Thinking/Perceiving (ESTP). (*Note:* If you would like to take the Myers-Briggs personality test, you can access the test at https://www.mbtionline.com/.)

These ISTP/ESTP personality types are self-sufficient, excel in high-pressure or emergency situations, enjoy fast-moving or risky hobbies, and are often "closet daredevils." These risk-taking individuals thrive in high-intensity situations such as those often faced by first responders on the job. They find them exciting, fun, and invigorating. But this risk-taking personality type may actually be the unfortunate characteristic that enables first responders to go through with taking their own lives.

Why Does It Feel Like I Hear About One Suicide After Another?

One surprising phenomenon that has become obvious over the past few decades is *contagion suicide* or *cluster suicide*. This occurs when multiple people, sometimes friends, coworkers, or even local first responders, commit suicide within a short time period of one another. Robert Olson, Centre for Suicide Prevention, explains that *mass-cluster suicides* occur when people in a large geographic area commit suicide within a relatively short time period. Olson further explains that *point cluster suicides* may occur when people within a small geographic distance of one another—such as the same hospital, prison, school, or even fire department—take their lives within a short period of one another.[20]

Joiner believes that whenever people in a social "cluster" witness a "suicidal stimulus"—like suicide in the media or the suicide of someone they know—it opens the door for all of the members of this cluster to become suicidal.[21] It is impossible to blame the media for publicizing suicide and thus causing cluster suicides. It is highly possible, however, that people who commit suicide following the suicide death of someone else may have already been in a vulnerable state and acted in response to the story of another's suicide.[22] This is because exposure to suicide breaks the taboo of suicide and makes it seem feasible for those who have witnessed it to take their own lives.[23] This is truly tragic.

Point cluster suicides occur regularly in indigenous communities. These individuals live in close proximity, experience the same social problems, and are personally aware of another member of their social circle's suicide. They are often influenced by the death of another individual.[24] It is true that living in a tight-knit community of local pride and mutual care is usually a protective factor against suicide, and there are many indigenous communities in which the suicide rate is almost zero. In groups where a suicide occurs, however, there often is a cluster of deaths by suicide located there in a short time period.

Events in Micronesia offer one example of this. There is a suicide phenomenon in Micronesia in which young boys take their lives at an excessively high rate compared to the global rate of suicide. This story is detailed in *The Tipping Point* by Malcolm Gladwell. Gladwell explains that Sima, a 17-year-old boy, took his own life when his father got angry with him for not being able to find a pole knife to harvest breadfruit. His father yelled, "Get out of here and go find somewhere else to live." Sima left, wrote a suicide note, and took his life later that day.[25]

Word of Sima's death spread rapidly in this small community, and many young boys began taking their lives for a wide range of seemingly trivial reasons. Suicide almost became a form of teenage self-expression. Some kids were even found to have attempted suicide because they "wanted to try it."[26] The closeness of this community, and the notoriety that Sima's death received, made suicide spread much faster in this point cluster than in other societies.

Point Cluster Suicide in Fire Departments: Chief Hojek's Story

Fire departments are in a similar situation. They have the protective factors of mutual care, ownership, and closeness with other members of the community. They also experience point cluster suicide more often than some communities, however. I think this happens because they identify so strongly with their fire brothers that when they witness one of their brothers dealing with pain in this way, it makes it feasible for others to follow. As you will see in the story of Chief John C. Hojek Jr., the way firefighters can break the power of cluster suicides in their communities and departments is by distancing themselves from the suicide choice of one of their brothers.

Chief Hojek serves the Hometown Fire Protection District (FPD) in Illinois. He and I copresented at a fire symposium, and I have a great deal of respect for him and the way he handled multiple suicides within his own department. Chief Hojek first experienced the suicide of one of his men in September 2015, when Ryan Elwood, whose story I shared earlier, took his life. Ryan had been a firefighter in the fire departments of both Hometown and North Palos, IL. Because these departments shared many personnel, the chief of the North Palos department was close to Chief Hojek.

That fateful morning in September 2015, Chief Hojek received a call from the chief of North Palos, informing him that Ryan was dead. Hojek had never experienced anything like this before. He sat, stunned, asking, "How do I handle this? What am I supposed to be doing right now?" As he gathered his thoughts, he decided he would reach out to other fire chiefs for advice. But first, he needed to tell his department. He summoned them for a meeting, where he told them that Ryan had taken his life.

Ryan's best friend, who also worked for the Hometown FPD, lashed out immediately, punching a wall and shouting, "I was just there last night!" The response around Ryan's death was sadness and regret, and Chief Hojek made sure that Ryan was honored with a full firefighter funeral.

Ryan truly was an extraordinary individual. He had received the Medal of Valor for a heroic rescue he made in which he crawled through a fire under terrible conditions to save an older lady. Ryan was amazing at saving people, but this made him feel as if he could save anyone. So when his friend was in a horrible car accident, he kept saying, "If I had just been on shift, I could have saved her." Ryan heaped a lot of blame on himself for his friend's death.

As you read earlier in this chapter, Ryan went on a suicide call in which he was able to stabilize a man who later died in the hospital. This left a profound impact on Ryan, and he never healed from that event. Ryan did not let people know he was in pain, however. The men of his department kept saying that Ryan was the happiest guy they ever knew. He completely hid all other emotions.

Chief Hojek received a great deal of support from other fire chiefs in the state as he processed the death of Ryan. He felt an overwhelming mixture of hurt, disappointment, and grief while remembering Ryan's truly amazing life. Meanwhile, the other guys in the Hometown FPD had to do their own processing.

Ryan's best friend was one of these firefighters. He had come from a dysfunctional family, and his mother had committed suicide when he was young. When Ryan committed suicide, it was the beginning of the end for his best friend. Ryan's friend had fallen in love with a woman and fathered a child with her. He held a good job in the Chicago Fire Department while still working part-time for the Hometown FPD. Everything looked good for this friend. He wrestled with depression and excessive drinking, however, and had not allowed himself any time or space to process the death of Ryan.

Beneath the surface, Ryan's best friend was consumed with anger, grief, and rescuer's depression. He had a few intense conflicts with his girlfriend and ended up moving out and living with another firefighter from the Hometown FPD. The firefighter who hosted him was hospitable, but Ryan's best friend was unstable, hurting, and disconnected, so they too had conflict.

One fateful night, just 15 months after Ryan Elwood took his life, Ryan's best friend was extremely drunk, and his host was gone for the evening. Ryan's friend set fire to his host's house and took his life inside that house. A year after that, this man's brother also committed suicide.

Each of these men lived within the jurisdiction of the Hometown FPD or the North Palos FD. Each of these men was drunk at the time he took his life.

Chief Hojek's response to Ryan's best friend's suicide was shock, horror, and huge disappointment. The firefighter brothers responded similarly. When Hojek informed them that another one of their brothers had committed suicide, they responded with shock and sadness, but they almost immediately shifted toward anger. They were mad at the firefighter for putting them through this pain again.

I believe that this anger was an outward reflection of their choice not to identify with his decision. In cluster suicides, there is often one individual who commits suicide, followed by a number of other vulnerable individuals who identify with that first individual and take their lives in response to the first. I believe Ryan's best friend never recovered from Ryan's death, and he identified so much with Ryan that he dealt with this pain the same way Ryan had—suicide.

However, the shift in response between Ryan's death and his best friend's death showed the remarkable boundary these firefighter brothers had set up between suicide and themselves. Instead of saying, "This is my brother, and this is how we deal with pain," they essentially said, "This is not an acceptable way to deal with pain. Doing this to us again was completely messed up, and I refuse to do this to my department again." In most cases, anger is unhealthy. In this case, anger was a healthy boundary between life and death.

Because Ryan's friend was working for the Chicago FD and Hometown FPD at the same time, there was controversy over who should go first in this man's funeral procession. The Chicago FD ended up going first, but the Hometown FPD resented Chief Hojek for not speaking up for Hometown. The drama was yet another factor Chief Hojek had to process in the wake of losing two of his men, along with the brother of one of them. On reflection, Chief Hojek will tell you that he did the best he could, but part of the reason he did not fight harder was because he, too, was grieving.

Leaders in first responder careers may become a type of father figure in the lives of their subordinates. (See chapter 14 for more discussion of paternalistic leadership.) I believe Chief Hojek was a very paternalistic leader, so his grief was more than just the loss of an employee. His grief was the loss of not one, but two, *sons*.

After these three deaths, the Hometown FPD added a new slogan to their onboarding process for new firefighters: "We're all out of f**** to give." By that they meant, "If you are going to take your life, we will not be able to do anything to honor your memory." They had spent their emotional energy grieving two major losses. Now they were choosing to heal and to set a boundary between themselves and suicide.

Together, as a department, their choice for mutual social support will be one of their most protective factors going forward. Social support often makes the difference between unhealthy choices and healthy ones, and between death and life.

Shneidman's Theory on Suicide

Suicidologist Edwin Shneidman believes suicide stems from an individual's deep, unbearable psychological pain, which he calls "psychache."[27] Pain overrides all other thoughts and feelings, making it hard to adapt. Judith Viorst, author of *Necessary Losses*, believes that early childhood trauma and losses cause triggers later in life, and many first responders see death and loss regularly on the job.[28] Individuals often begin a downward spiral toward suicide, which often starts with a major stress or trauma.

Most responders have an empathic temperament (emotional depth, empathy, creativity, deep thinking, and deep feeling) that causes them to feel deeply. These empaths are ready to jump in and rescue other people from their pain. But they also may feel that they need to hide their private pain from others, developing a different outward persona. It is common for first responders to portray strength on the outside while grappling with pain on the inside. Everyone in this job sees horrible things. Choosing to face them and talk about them, however, is the road to freedom.

Those who keep their feelings hidden because of fear or shame for their emotions can begin to form a habit of negative self-talk. Negative self-talk can become an internal "narrator" that makes the person feel shame for having negative emotions, which can cause the individual to disengage from family or friends.[29] Rejecting this negative self-talk can be a strong starting point for healing. It is okay to not be okay. It is okay to feel things, to remember horrific calls, and to be sad. You do not have to be strong all the time. The perception that you need to hold it all together often drives first responders to isolate and suffer alone, which paves the way for suicidal thinking.

According to Shneidman, suicide is seen as a solution or an escape from perpetual and excessive shame, guilt, humiliation, and so on, and it occurs among the rich and poor, and mentally ill and mentally healthy alike.[30] When people are prevented from seeking help by their own personal fear of shame, suicide may begin to look like the only escape route from pain. A military study on PTSD and masculinity explains how masculinity's expectations can lead men to suffer alone:

> Although these [masculine] values (self-reliance, emotional control, and concealment of perceived weakness) can promote self-confidence and skill-building, the emphasis on mental toughness and self-reliance, even in the face of physical or mental injury, can create an environment in which these injuries are seen as weakness.[31]

When physical or mental injuries are seen as weaknesses, it can be hard for men, and women with instrumental personalities, to come forward for help.

The pain of a trauma can feel suffocating. Individuals who have experienced a traumatic event should talk to someone immediately and begin processing what happened. There are many evidence-based treatments for trauma. Getting help right away can prevent the person from considering ending his life. (Read more about trauma, stress, PTSD, and post-traumatic growth in chapters 7 through 10.)

Shneidman recognized that most suicides come with warning signs. In fact, about 80% of those who commit suicide talk about it beforehand.[32] And as they consider suicide, they experience strong feelings of ambivalence. They are torn between wanting to die and not wanting to die. Often their desire to commit suicide is concentrated and most intense for a short window of time, during which they should be monitored closely. After this period, the likelihood of them committing suicide goes down significantly.

Acute Suicidal Affective Disturbance (ASAD)

Some individuals seem to turn quickly toward suicide. Out of nowhere, they seem to develop a distinct combination of symptoms, including "suicidal intent, thwarted belongingness, perceived burdensomeness, disgust with others and oneself, agitation, irritability, insomnia, and nightmares."[33] These symptoms are different from typical anxiety or depression symptoms, and they usually indicate a unique and highly time-sensitive risk for the individual committing suicide.

Joiner identified, characterized, and named this condition, which is called *acute suicidal affective disturbance (ASAD)*. As he explains, "ASAD is a time-limited arousal state. You can't sustain it for more than an hour or a few hours. It'll abate with the passage of time."[34]

ASAD can also be called "suicide mode." It is the state of mind during which people are highly suicidal and must be monitored very closely. During ASAD, people have extremely high perceived burdensomeness, exceptionally low sense of belonging, hopelessness, self-disgust, insomnia, nightmares, irritability, agitation, social withdrawal, and other emotions.[35]

The term acute suicidal affective disturbance or ASAD is new, and psychologists are working to introduce it into the upcoming edition of the *Diagnostic and Statistical Manual of Mental Disorders* (*DSM-6*). According to Bruce Jancin, contributor to MDedge, the following are the criteria for ASAD:[36]

- "A sudden surge in suicidal intent occurring over the course of minutes, hours, or days rather than weeks or months."
- "One or both of two alienation criteria: a) severe social withdrawal defined by extreme disgust with others or perceived liability to others; b) marked self-alienation manifested as self-disgust or the view that one's selfhood is an onerous burden."

Jancin also notes that a person who is in ASAD may express harsh self-criticism, comments about having no one around, mental illness, physical illness/injury/pain, and guilt (moral pain).[37]

Depending on recent life events, social community, physical health, any mental illnesses, and religion and values, people who have been suicidal in the past may recover and go on to live productive, happy, and healthy lives. Or they can quickly jump back into suicide mode and will need calm, focused care to support them as they seek out help.

The difficulty with ASAD is that it is hard to detect and is highly dangerous. This is why suicidologists are working diligently to get it included in the *DSM-6*. The more we can learn about "suicide mode," the more we can help people who find themselves in this state.

The bottom line about ASAD is this: If you or someone you know is suicidal, get help from a professional immediately!

Retirees' Risk of Suicide

Retired first responders can experience a sense of being burdensome to their families as they leave the fire service. Their occupational commitments have decreased along with their financial contributions to the family, which may lead to a sense that they are a burden. If this feeling grows, the individual may feel that his death would be more valuable than his life. This could in turn lead to a desire to commit suicide.

Add to this the loss of camaraderie he had while on duty, and the individual has both indicators necessary for suicidal tendencies. Without social support, he may indeed end his own life. By talking to someone and finding a group of peers to connect with, however, he can build a new sense of belonging, validating his memories and first responder career experiences, and reminding him that he is not a burden but a blessing. If you know someone who is recently retired from a career in the fire service, encourage that individual to get connected into a community. Recently retired first responders should consider talking to a counselor to help sort through some of the transitions and memories they have experienced throughout their careers. (See chapter 19 for more discussion about retirement.)

The second group at the highest risk for committing suicide consists of those with a recent cancer diagnosis. You can read more about suicide and cancer patients in chapter 16. If you know someone who has recently been diagnosed with cancer, encourage that individual to join a cancer support group and consider talking to a counselor to sort through the questions and transitions of a cancer diagnosis.

As I mentioned before, along with those who have been diagnosed with cancer, those who live with chronic pain are also high on the list of individuals likely to commit suicide. In fact, those with chronic pain are *twice* as likely to think about or commit suicide, according to research conducted by Mélanie Racine, Schulich School of Medicine and Dentistry, Western University, Ontario, Canada. It is important to recognize the risk factors that come with chronic pain. For instance, Racine notes that those who are unemployed or disabled are at a higher risk, as are those struggling with "mental defeat, pain catastrophizing, hopelessness, perceived burdensomeness and thwarted belongingness."[38] So patients with chronic pain can be at increased risk of committing suicide and should be monitored closely.

The third group of first responders at highest risk for committing suicide are those who fit Shneidman's criteria of having experienced a trauma, having an empathetic temperament, and having an accusing "internal narrator." Individuals who have experienced a traumatic event should talk to someone immediately and begin processing what happened. There are so many evidence-based treatments for trauma. Getting help right away can prevent the person from considering ending his life. More about trauma, stress, PTSD, and post-traumatic growth is presented in chapters 7 through 10.

The Truth for Those Who Consider Suicide

Everyone who considers suicide needs to hear the truth and receive support. Even the most devastated people can find new hope and new life with truth and support. When individuals receive support and can process feelings of pain, loneliness, isolation, and shame, they can often heal and find hope to continue living. If they give themselves just one more day, it can make a big difference toward moving them along a healing trajectory. There are many people living healthy, emotionally well lives who at some point in the past reached a point of despair but decided not to give up. The world is full of people who have chosen to live one more day and have discovered that the world truly is a beautiful place. Life and hope can be found.

In truth, those with suicidal thoughts do not actually want to die. They simply feel like the world would be better off without them, and their death would be more valuable than their life. But this is a lie.

I wish I could sit down with every first responder who is thinking about ending his or her life and say . . .

- You matter.
- Give yourself one more day.
- There is a unique place in this world for you, and nobody else can fill it.
- Your laugh, your smile, your heart to help others, your relationships, your talents, and everything you know and care about are all unique to you.
- No one wants you to die. They want you to live!
- If you were gone, the world would mourn you, miss you, want you back, and think about you all the time.
- Your life is so special.
- Your life is worth living.
- There is no one else like you in this whole world.
- Even if you never worked another day or saved another person, even if you just sat around and painted pictures that no one ever saw, the world is a better place because you live.
- You are loved exactly for who you are.

Let those truths sink in for a little bit. Pick a few and write them down on an index card and carry it around with you. Repeat it to yourself every time you are tempted to doubt, every time your internal narrator starts accusing you. Through truth and support, I believe you can come out of the dark and find new hope and a beautiful life.

Are Addiction, Depression, and Suicide Linked?

Is there a link between addiction, depression, and suicide? The short answer is yes. Depression and mood disorders are the number one risk factor for suicide.

People who have addictions are six times more likely to take their lives.[39] Addiction is the second-highest risk factor for suicide. The strongest predictor of suicide is, specifically, alcoholism. Alcoholism is connected to a whopping 50% of suicides. And one in three people who die by suicide are under the influence of either opiates or alcohol. So if you are struggling with alcoholism or other addictions, seek help and live! Freedom is life-giving.

Where to Turn for Help

If you or someone you know is considering suicide, please get help. There are so many people, including counselors like myself, who can help you. There are even people available 24/7 to talk to you and support you at the National Suicide Prevention Hotline: (800) 273–8255.

Reflection Questions

1. What is one thing you learned about suicide that you did not know before?
2. Is there anyone in your life who you think might be at risk of committing suicide? Tell someone!
3. What truths listed above stood out to you? Write them down and repeat them to yourself as often as possible.
4. Thanks for being you. Thanks for being alive!

Notes

1. M. L. King, Jr., *Strength to Love* (United States: Beacon Press: 2019), 57, https://www.google.com/books/edition/Strength_to_Love/FHKEDwAAQBAJ?hl=en&gbpv=0.
2. I. H. Stanley, M. A. Hom, and T. E. Joiner, "A Systematic Review of Suicidal Thoughts and Behaviors Among Police Officers, Firefighters, EMTs, and Paramedics," *Clinical Psychology Review* 44 (March 2016): 25–44, https://doi.org/10.1016/j.cpr.2015.12.002.
3. C. Abbot et al., *What's Killing Our Medics?* Ambulance Service Manager Program (Conifer, CO: Reviving Responders, 2015), http://www.revivingresponders.com/originalpaper.
4. M. Heyman, J. Dill, and R. Douglas, "Ruderman White Paper on Mental Health and Suicide of First Responders" (White paper, Ruderman Foundation, 2018), https://rudermanfoundation.org/white_papers/police-officers-and-firefighters-are-more-likely-to-die-by-suicide-than-in-line-of-duty/.
5. E. C. Nielson et al., "Traditional Masculinity Ideology, Posttraumatic Stress Disorder (PTSD) Symptom Severity, and Treatment in Service Members and Veterans: A Systematic Review," *Psychology of Men & Masculinities* 21, no. 4 (January 27, 2020): 578–592, https://doi.org/10.1037/men0000257.
6. T. E. Joiner, *Why People Die by Suicide* (Cambridge, MA: Harvard University Press, 2007).
7. T. E. Joiner, *Lonely at the Top: The High Cost of Men's Success* (New York: Palgrave Macmillan, 2011).
8. Joiner, *Lonely at the Top*.
9. Joiner, *Lonely at the Top*.
10. Joiner, *Lonely at the Top*.
11. Stanley, Hom, and Joiner, "A Systematic Review of Suicidal Thoughts and Behaviors Among Police Officers, Firefighters, EMTs, and Paramedics."
12. T. Buckley et al., "Effect of Early Bereavement on Heart Rate and Heart Rate Variability," *American Journal of Cardiology* 110, no. 9 (November 1, 2012): 1,378–1,383, https://doi.org/10.1016/j.amjcard.2012.06.045.
13. M. Campos, "Heart Rate Variability: A New Way to Track Well-Being," *Harvard Health Blog*, Harvard Health Publishing, October 22, 2019, https://www.health.harvard.edu/blog/heart-rate-variability-new-way-track-well-2017112212789.
14. K. A. Hawkins et al., "An Examination of the Relationship Between Anger and Suicide Risk Through the Lens of the Interpersonal Theory of Suicide," *Journal of Psychiatric Research*, 50 (March 2014): 59–65, https://doi.org/10.1016/j.jpsychires.2013.12.005.

15. N. G. Zaorsky et al., "Suicide Among Cancer Patients," *Nature Communications* 10, no. 1 (January 14, 2019), https://doi.org/10.1038/s41467-018-08170-1.
16. D. Daneker, *Counselors Working with the Terminally Ill*, VISTAS Online, 2006, https://www.counseling.org/Resources/Library/VISTAS/vistas06_online-only/Daneker.pdf.
17. Joiner, *Why People Die by Suicide*.
18. Joiner, *Why People Die by Suicide*.
19. Stanley, Hom, and Joiner, "A Systematic Review of Suicidal Thoughts and Behaviors Among Police Officers, Firefighters, EMTs, and Paramedics."
20. R. Olson, *Suicide Contagion and Suicide Clusters*, Centre for Suicide Prevention (2013), https://www.suicideinfo.ca/resource/suicidecontagion/.
21. T. E. Joiner, "The Clustering and Contagion of Suicide," *Current Directions in Psychological Science* 8, no. 3 (June 1, 1999): 89–92, https://doi.org/10.1111/1467-8721.00021.
22. Olson, *Suicide Contagion and Suicide Clusters*.
23. Joiner, "The Clustering and Contagion of Suicide."
24. Olson, *Suicide Contagion and Suicide Clusters*.
25. M. Gladwell, *The Tipping Point: How Little Things Can Make a Difference* (New York: Little, Brown, 2002).
26. Gladwell, *The Tipping Point*.
27. A. A. Leenaars, "Review: Edwin S. Shneidman on Suicide," *Suicidology Online* 1 (March 8, 2010): 5–18, http://www.suicidology-online.com/pdf/SOL-2010-1-5-18.pdf.
28. J. Viorst, *Necessary Losses: The Loves, Illusions, Dependencies, and Impossible Expectations That All of Us Have to Give Up in Order to Grow* (New York: Simon and Schuster, 1986).
29. Leenaars, "Review."
30. Leenaars, "Review."
31. Nielson et al., "Traditional Masculinity Ideology, Posttraumatic Stress Disorder (PTSD) Symptom Severity, and Treatment in Service Members and Veterans."
32. Leenaars, "Review."
33. M. L. Rogers, M. A. Hom, and T. E. Joiner, "Differentiating Acute Suicidal Affective Disturbance (ASAD) from Anxiety and Depression Symptoms: A Network Analysis,"*Journal of Affective Disorders* 250 (2019): 333–340, https://doi.org/10.1016/j.jad.2019.03.005.
34. B. Jancin,"AAS: Acute Suicidal Affective Disturbance Proposed as New Diagnosis," *Psychiatry*, MDedge.com (May 28, 2015), https://www.mdedge.com/psychiatry/article/100017/depression/aas-acute-suicidal-affective-disturbance-proposed-new-diagnosis/.
35. Rogers, Hom, and Joiner, "Differentiating Acute Suicidal Affective Disturbance (ASAD) from Anxiety and Depression Symptoms."
36. Jancin, "AAS: Acute Suicidal Affective Disturbance Proposed as New Diagnosis."
37. Jancin, "AAS: Acute Suicidal Affective Disturbance Proposed as New Diagnosis."
38. M. Racine, "Chronic Pain and Suicide Risk: A Comprehensive Review," *Progress in Neuro-Psychopharmacology and Biological Psychiatry* 87, part B (December 20, 2018): 269–280, https://doi.org/10.1016/j.pnpbp.2017.08.020.
39. N. Seay, "How Are Addiction, Depression and Suicide Linked?" *The Rehab Journal*, American Addiction Centers, November 4, 2019, https://www.rehabs.com/blog/how-are-addiction-depression-and-suicide-linked/.

Chapter 6

Pediatric Death

When I was in my early 20s, I worked as an art and play therapist with terminally ill children in a well-known children's hospital in downtown Chicago. Working with children was highly rewarding for me because I felt like I was helping parents in a traumatic season of their lives to have little glimpses of relief, rest, and hope. But my whole world was rocked one night by a little British girl.

I typically worked with children who were undergoing treatment, and most were bald and had tracheotomies and IV pulls. I paid no attention to their medical equipment, however. To me these were just kids who needed a chance to play. On my floor of the hospital, most of the children had cancer, but none of them appeared to be experiencing pain. One of my patients was a 6- or 7-year-old girl with dazzling blue eyes and long, curly blonde hair. Unlike the other kids, who were bald, this little girl looked perfectly healthy. She was radiant and full of life. She seemed out of place in a hospital full of significantly ill children.

One Friday afternoon when I walked into the little girl's room to begin our play therapy, I was met with the most gut-wrenching screams I had ever heard. The little girl was in agony. Her mother looked at me with an expression of desperation, silently begging me to do something to make it better for her daughter. The daughter's cancer had spread throughout her body, and it was severely painful. Her medications were not alleviating her suffering.

I wished deeply that I could do something to help, but all I could do was leave the room to find her some medical assistance. I felt overwhelmed by her helplessness and by my own helplessness. I would have given anything to be able to change the situation for that precious little girl, but there was nothing I could do. My friends and I were supposed to go out that night, but I cancelled my plans with them. All I wanted to do was go home, lie down, and be alone.

When children die, it is the most unnatural thing in the world. Children are not supposed to die before adults. It is not the natural order of things. So on a call involving a child, it is common for first responders to have unrealistic expectations for themselves. In my years of counseling first responders through traumatic memories, I have noticed that pediatric death calls often cause them to carry around unnecessary self-blame and guilt. Even if they understand logically that the incident was not their fault, they often blame themselves because they cannot find a reason why something this horrible would happen to a precious child.

Though I did not blame myself for this girl's pain, I wrestled with the feeling that I had let her mom down. Instead of being able to save the day by providing the mom some respite while I played with her daughter, I could only seek the assistance of a medical professional. I wanted to help, but I simply could not. Perhaps you have experienced something similar.

If you have, know this: It was not your fault. Some things are out of your hands.

Why Pediatric Death Is So Hard to Process

Processing the death of a child can be particularly difficult because the standard coping mechanisms that first responders use to process the death of an adult do not apply to children, according to clinical psychologist Anne Bisek. It is a normal and subconscious process for first responders to employ techniques to help get their minds around the tragedies they have witnessed. With an adult death, first responders may use rationalization to identify why a similar death will not happen to themselves or someone they know, according to Bisek.[1] But when a child dies, there is often no way to rationalize or make light of anything about the death. In this way, the death of a child becomes uniquely unsettling.

First responders also rely on their training to deal with death situations. They use whatever they know to be able to save the victim, but if there was nothing that could be done to prevent the child's death, they may feel powerless.[2]

Jack's Story

Jack came to me after experiencing a very traumatic call in which a little boy died. He struggled with the memories of what happened, and he was questioning something he did. Jack questioned if something he had done had caused this boy to die. He went over and over it and asked many people, all of whom told him that he did everything by the book. Everything was carried out correctly with that call. But because Jack could not understand why it had happened, he reasoned that it had to be his fault.

After this boy's death, Jack became extremely anxious, sad, and depressed, and he was having a hard time functioning. Jack's fire chief gave him some time off work, and he came to see me. As we explored his reaction to the trauma of this boy's death, he recognized how vulnerable it made him feel to not be able to save this child. This opened the door for us to discuss previous traumatic situations in his life. In my experience, when someone has a strong reaction to trauma, it can often be tied back to an earlier trauma. The same vulnerable feeling is present in both situations.

Jack grew up in a firefighter family. During his childhood, his father had a terrible accident on the job that left him unable to go back to work. After that, Jack's family struggled financially, and his dad was in a lot of pain. As Jack and I explored that time of his life and grieved his loss, he began to realize that from that time forward, he had experienced one crisis after another. He had never really spent time processing the first trauma that changed his family forever.

Not dealing with trauma often means strong reactions to new traumatic situations. This is true even if you cannot remember the details of an earlier event. It is the same feeling, which in this case might be a powerless, vulnerable feeling. This particular pediatric death call took Jack right back to the emotions he felt as a young boy.

Jack ended up healing very nicely by taking some time off at his favorite spot, a family cabin out of state. He spent time with his family in the woods, away from everything. Often the key to healing from trauma is just giving yourself time off to process what happened and what it means. The mind will naturally do a little life review of traumatic things in the past, and just going over them again can bring about new meaning, healing, and learning.

Factors That Make a Pediatric Death Harder to Process

The death of a child is always hard for firefighters to process, but there are specific situations that make it harder. The worst pediatric death situations for first responders are ones that have details to which the first responder can relate, according to Bisek.[3] A first responder who has a child of a similar age as the child who died may become anxious at the idea that it could have been his child. If a first responder is of the same background—for example, race, religion, demographic—as the victim, it may feel relatable. When a victim lives in the same neighborhood as the first responder, it may cause the first responder to feel that his "safe place" could become a crime scene. These similarities may haunt the first responder.

A firefighter who has something in common with the victim may perceive that the boundary of first responder and civilian has been broken. As we saw with Jack, if the responder experienced a traumatic event when he was a child, the way he felt back then will resurface to the forefront of his mind. Whenever a victim is in a similar life stage as the first responder, the first responder may worry, "Will I have to face this?" Worst of all, if a firefighter has lost a child of his own, he may think, "I know exactly how it feels to lose a child."

Mike's Story

Mike and his wife met through a mutual friend and got married within a year of meeting. Soon after, they decided to try to start a family. Although they did not know it at first, Mike's wife had endometriosis. After the condition was diagnosed, she eventually underwent surgery, and later in vitro fertilization (IVF), in order to conceive. During the IVF process, Mike noticed his anxiety started to surface. Feeling helpless, worried about his wife, and anxious about their ability to have children only compounded the anxiety he was already experiencing from the events he had seen at work.

Anxiety, anger, and helplessness rushed over him one day after his wife's procedure. He drove home from his shift, went directly to the backyard, and began smashing a stump. He kept envisioning his wife dead on the operating table, and it was invading his thoughts and ruining his ability to sleep. That is when his wife called me, and we met together for a session of accelerated resolution therapy (ART), which is a no-talking therapy to help resolve traumatic memories. It is amazing, and I truly recommend it for everyone who has invasive memories. (More discussion about accelerated resolution therapy is presented in chapter 12.)

After their first round of IVF, Mike and his wife received a positive pregnancy test. To his wife's joy, they were going to have a baby. But Mike did not feel that same happiness; he felt terrified. He started having nightmares, reliving traumatic calls, and thinking of harming himself. One call that kept coming back to him was a call in which he and his team responded to a house purportedly to save a 60-year-old person in full arrest. On arrival, they discovered the call was actually for a six-day-old baby who was not breathing. Mike and his partner rushed the baby to the ambulance and tried everything they could to save her, but they were unsuccessful. Later they learned that this child's mother was a known drug user, and that the baby had a hole in her heart and could not be saved. Though this call had happened years earlier, Mike experienced flashbacks to this event as he and his wife met with her OB-GYN. When the doctor explained the potential complications of IVF and possible heart defects in IVF babies, Mike immediately flashed back to this pediatric death call, and he cried uncontrollably.

A second call that resurfaced as Mike and his wife walked through pregnancy was a call in which Mike had helped deliver a 21-week premature baby in the ambulance. He had turned to reach for IV supplies, heard a blood-curdling scream, and saw the baby being delivered on the cot. The subsequent loss of this baby's life was difficult for Mike to process emotionally. Later when Mike's wife was 20 weeks pregnant, she called him about an hour before his shift change to tell him that she was cramping and leaking fluid. Mike left work to drive his wife to the hospital, with flashbacks flooding his mind of this previous traumatic call.

These two pediatric death situations caused Mike to have strong anxiety, and he was eventually prescribed Zoloft. He also began to realize that he needed to confront his trauma and PTSD. That is when he began meeting with me regularly for accelerated resolution therapy.

There is something about the death of a child that leaves a mark on the mind. Because firefighters care a lot about children, and many first responders are parents, the death of a child at work can feel like a threat to one's own family. The death of a child can be difficult to process.

Processing Pediatric Death

One coping mechanism first responders often use in non-child-related tragedies is humor, according to Bisek. Humor is a healthy way to keep perspective on the traumas and tragedies first responders face. Humor can reduce stress and make it easier to be okay in the light of sorrow. But it may become impossible to joke if the victim is a child, Bisek notes.[4]

Along with humor, first responders also use *blaming* to help process fatal calls. Blaming the adult victim for his or her death can be a simple way to frame the potentially traumatic event as "something that won't happen to me." But children cannot be blamed for their deaths. So the tragedy cuts deep into a first responder's heart.[5]

Processing pediatric death can also be exacerbated by the coping mechanism of *magical thinking*, in which first responders who cannot blame the victim blame themselves for the tragedy.[6] It is in these situations that first responders grapple with a weight that is too heavy to bear. Though first responders are there to help and rescue, sometimes there is nothing that can be done. Even so, it can be hard not to blame yourself when anyone dies, and especially when a child dies.

> **Truth:** Your attempt to help, whether or not you could save the child, laid the foundation for grieving for the relatives. Even if you were not able to save the child, knowing that you tried everything you could is the starting point for the parents and family members to grieve and process their loss.

> **Truth:** You do not ever need to blame yourself for a victim dying on your watch. Your effort did more than you know to support everyone involved. It helped them frame the experience in light of love, care, and fighting for the person as long as possible. Your effort paves the way for grieving in the context of love, which makes their future resiliency easier to reach.

When Pediatric Death Becomes More Than You Can Bear

Sometimes even when you know that you did everything you could and your effort provided the platform for healing for the parents, pediatric death can still be too heavy to bear. The most important thing to understand is that you should not try to process it alone. Talk to someone about what happened, get extra rest, and allow yourself to process what happened. Remember, if you keep things hidden, they will hurt you from the inside out. Bringing light to the situation is the only way to heal it.

Rick's Story

Rick is a great example of healing from the trauma of a pediatric death. Rick came to me as a young paramedic. An infant had died on a call, and Rick's wife happened to be pregnant with their second child when it happened. Rick is an excellent paramedic, but after that call, he began to cry on his way to every call. He was concerned and started thinking something was wrong with him.

His chief gave him some time off of work. Rick spent his time sleeping a lot and isolating, and his wife was terribly worried about him. She thought he might be suicidal. But when I asked him, he denied any suicidal thoughts. With isolating behavior, we must make a distinction between suicidal ideation and normal introversion. Introverts need rest in order to process and restore. Introverts take longer to process things, and they sometimes need to process alone, which can be confused with a major depression and suicidal behaviors. It can be quite confusing, so talking to a counselor can be beneficial.

Rick and I only met three or four times, and much of the time was spent in education. Our discussion included giving him permission to rest and to grieve the situation in which a mother lost her child, understanding the connection this had with his sickening realization that this could happen to his child.

Pediatric death is horrible, and first responders who face it will almost always need time to heal. So Rick took a little bit of time off of work. I educated his wife about what introversion really means and about giving him some time and space to rest without taking it personally. She had been feeling like she was not a good wife because he was not talking to her. I also explained how important it is for firefighters to rest. Sleep can resolve so much with trauma.

After taking the time he needed, Rick was ready to get back on shift, and he is doing great. Making sure he was emotionally well included asking questions such as, "Are you feeling bad?" "Are you feeling suicidal?" "Are you feeling like you are going to take your own life?" And his answer was, "Absolutely not. I'm just tired."

It is okay to feel sad. It is okay to need rest. It is okay to need time to process. You will feel better again as you give yourself time to process and heal. (You can read more on trauma and PTSD in chapters 7 and 8, and you can read more about resiliency in chapter 9.)

It is also okay to be okay. In some situations, you may be able to think about the loss of a child and not have it overly affect you. Whatever you experience after a pediatric death call is okay. Just be sure you take the time to listen to yourself and do a check-in with a counselor if you need one. You do not have to face trauma and tragedy alone.

Reflection Questions

1. Have you experienced a pediatric death call?
2. What was your initial response to the call? How did you cope with it?
3. Are there parts of that call that you still need to deal with? If so, what are some things you can do to heal yourself and develop a healthy understanding of that event?
4. Who is someone you can talk to about that event?
5. Exercise: Write a letter to an aspiring first responder to read on the very first day of his career.
6. In your letter, what would you tell him about pediatric death?
7. If you sent that letter to yourself today, what would you highlight?
8. Write that phrase down and repeat it to yourself.
9. If you experienced pediatric death and are dealing with PTSD or suicidal ideation, give yourself one more day. Take a risk and talk to someone! You may discover that life has much more hope than you are feeling at the moment.

Notes

1. A. Bisek, "When a Child Dies: Understanding Emergency Responder's Reactions," 2012, http://www.whenachilddies.com/.
2. Bisek, "When a Child Dies."
3. Bisek, "When a Child Dies."
4. Bisek, "When a Child Dies."
5. Bisek, "When a Child Dies."
6. Bisek, "When a Child Dies."

CHAPTER 7

STRESS AND TRAUMA

Trauma is that scarring that makes you less flexible, more rigid, less feeling, and more defended.

—Gabor Maté

Trauma is not what happens to us, but what we hold inside in the absence of an empathetic witness.

—Peter A. Levine

You may not consider stress and trauma to be related, but they are actually close brothers. Everywhere from the grocery store to the boardroom, people talk about how stressed they are. Stress is so common that it is often overlooked. But did you know that stress and trauma are quite similar physiological responses? Stress is a milder version of what happens physiologically to a person experiencing a traumatic event.

Stress occurs when the body responds to demands from the world around it, both good and bad. When people feel stressed by something going on around them, their bodies react by releasing chemicals in the blood.[1] These chemicals ramp up the body's ability to either fight or flee from a situation. The same stress hormones that are secreted during a hectic day in the office are also secreted in emergency situations. The only difference is the amount of stress hormones released.

Despite the close connection between stress and trauma, there is an important distinction between them. You can experience stress on a daily basis without being traumatized. Stress does upset balance, but it is not usually harmful on its own. Trauma, however, is almost always harmful.

What Is Stress?

According to resiliency expert Joel Fay, stress is the effect of anything in life to which people must adjust.[2] *Stressors* are things that have the effect of causing stress. A person reaches his stress capacity when he maxes out the amount of stress he can tolerate at a given time. Everyone's stress capacity is different, and the amount of stress an individual can tolerate is based on a person's wellness and current life circumstances. *Stress load* is the amount or quantity of a person's stress, according to psychologist Frances Figueroa-Fankhanel.[3]

In my years as a first responder counselor, I have seen people with every symptom of stress walk through my doors. First responders are stressed people. Here are some common signs of stress in first responders:

- *Emotional:* Numbness, anger, depression, confusion, derealization, mood/reaction incongruent with the situation, feeling isolated, decreased sense of self-identity.
- *Cognitive:* Changed perspectives, decreased focus, hyper focus on one particular issue, cynicism, tendency toward "replaying the tape."
- *Physical:* Digestive issues, weight changes, fatigue, aches/pains, constant illness (cortisol-related issues).
- *Spiritual:* Questioning or abandoning belief system, questioning one's purpose.
- *Occupational:* Increased errors, shift call-ins/tardiness, strain with coworkers/management, intoxicated at work.
- *Relationships:* Increased arguments, decreased communication, breakup/divorce, isolation (emotionally and physically) from partner.
- *Behavioral:* Outbursts, substance use, adrenaline initiating behaviors, isolation.

When an individual is stressed, everything is thrown out of whack.[4] The person may be able to tolerate stress for a while and may even be able to tolerate a certain level of stress. But eventually stress will show up in the way the person feels, thinks, acts, believes, works, relates to others, and behaves. At that point, the person has the choice to either let stress take over or to choose self-care practices instead. If stress rules, the results will be emotional wellness issues such as depression, anxiety, substance abuse, marital issues, or suicidal ideation. Physical wellness issues may also result, including injuries or possibly even cancer. A much better choice for first responders overwhelmed by stress is to dive into self-care practices to create a path out of the stress overload.

Chronic Stress

The kind of stress I typically see with first responders is *chronic stress*. Chronic stress produces all of the same symptoms as regular, acute stress, but not always at such a high level. When first responders are routinely and constantly stressed on the job, their bodies release stress chemicals, and they become used to having these chemicals in their bodies. Their bodies never fully return to their baseline levels of stress-related chemicals, leaving them susceptible to disease, angry outbursts, and breakdown of relationships.

First responders face trauma almost daily, and it is imperative that they pursue intentional self-care activities to allow themselves to return to their baselines, enabling them to remain emotionally well. (For more information about self-care, see chapter 12.)

What Is Trauma?

Although stress and trauma are related as previously discussed, trauma is a more specific event that produces a greater volume of stress hormones in the body. Trauma goes beyond stress to include mental, physiological, and physical components that make the memory harmful and harder to heal.

According to Catherine Ashton, a yoga instructor specializing in trauma recovery, trauma can be identified as an event that posed a significant threat to the individual, or a series of events that are intense, continuous, or recur and seem to have no means of escape. It can be an injury, wound, or hurt that caused the individual to have a sense of helplessness, loss, and fear afterward, making it impossible to feel safe, confident, balanced, and happy. When trauma occurs, the body releases a profound quantity of survival hormones and energy, but it lacks the space or circumstances to use that energy for effective action. Ashton notes that elements of trauma can include a perpetual sense of terror, despair, hopelessness, and disconnection after an event or injury.[5]

Erin Wall compiled the following list of the most traumatic events for first responders based on research conducted for her master's thesis at the University of St. Thomas, Minnesota:

- Catastrophic loss of life
- Incidents involving children
- Individuals with significant blood loss or horrible pain
- The death of a fellow first responder
- Presence of emotionally-evocative contrasting details (i.e., a "Just Married" sign on a car in which the newlywed occupants have been killed)
- Preventable tragedies involving human error
- Events involving unknown substances or causes
- Conditions of prolonged uncertainty, where the worst is yet to come (i.e., aftershocks of an earthquake)
- Prolonged contact with dead/injured
- Loss of life following intense rescue efforts
- Unusual or distressing sights or sounds (i.e., falling bodies at Ground Zero)
- Lack of opportunity for effective action (i.e., the search for bodies at Ground Zero)
- Knowing the victim
- Family members on the scene[6]

After experiencing a traumatic event, firefighters can show immediate signs of distress, but signs can also begin to show up hours, days, weeks, or even years after the event. Signs, symptoms, and effects of trauma could include hypervigilance/hyperstartle reflex and unprovoked rage or violence, which can indicate diminished capacity to self-regulate and self-soothe. Other signs include impulsive responses such as intense emotions that drive actions/reactions and ability to process information, feelings of numbness, feelings of helplessness and hopelessness, fear and terror, intrusive memories and flashbacks, anxiety and panic attacks, depression, insomnia, night terrors or sleepwalking, lack of openness, repetitive or compulsive destructive behaviors (including sexual behaviors), interpersonal and intrapersonal challenges, body image issues, feeling like the "living dead" or "walking wounded," difficulties focusing and concentrating, intrusive memories and flashbacks, and dissociation.[7]

According to the Northern Illinois Critical Stress Management Team, trauma is recognized by the "aftershocks" the individual experiences.[8] Sometimes these aftershocks occur right away, while at other times they may start to occur hours after the event. They can show up days or weeks after the traumatic event or even a longer time later. Some retired adults will begin to experience aftershocks after they retire, even though they never experienced them previously during their working years. This phenomenon is called *late onset stress symptomatology (LOSS)*. (More about LOSS is presented in chapter 17.)

Reactions to Trauma

There is a spectrum of reactions that people have when they experience a traumatic event. Some are normal reactions to trauma, which include fatigue, wanting to be alone to think about it, wanting to be surrounded by loved ones, or wanting to be in safe places. Many may even need to get some sleep in order to be able to recall with accuracy exactly what happened during the event. This is because going through a sleep cycle can help them consolidate the memories and can improve their ability to remember certain facets of their experience.[9]

It can be very normal for those who have experienced traumatic events to be able to report, with accuracy, the sensory elements of the event. The rest of the event, such as the timeline, can get hazy because of the way the brain has encoded the memory. During a traumatic event, the brain encodes memories in intense fragments, so the context and sequence part of the memories may be less accurate.[10]

Clinical psychologist and trauma expert David Lisak notes that when assisting an individual in reconstructing what happened during a traumatic event, it may be helpful to ask him or her about details such as the sights, sounds, smells, and feelings of the event. According to Lisak, the individual then can be asked additional questions to work backward to establish a timeline: "What happened before that?" "And what happened before *that*?" The inability to remember the sequence and context of a traumatic memory perfectly is a normal reaction to trauma. It also should be noted that people who are traumatized may come across as if they are lying because they are unable to construct the full timeline and every detail of the event. In reality, this is a natural neurological response to a traumatic event, and the individual should be given support and sensory questions to help him or her reconstruct the memory, according to Lisak.[11]

Remember, it is okay to be okay after a trauma. Sometimes first responders can process and move on. Sometimes they just need to take a while to regroup, and they are fine. Many first responders can witness a traumatic event and walk away unaffected personally by the event.

But there are other events that disturb the individual more deeply. Responding to a call with multiple factors has a way of pushing first responders beyond their ability to cope. These events can shatter the responders' beliefs about themselves, others, or the world around them and cause them to begin experiencing symptoms of trauma. Stan McCracken, Crown Family School of Social Work, Policy, and Practice, University of Chicago, notes that the helplessness and horror alone are not enough to predict the development of PTSD. Instead, it is *avoidance* that predicts PTSD. When individuals avoid thinking about and dealing with trauma, they are much more likely to develop PTSD.[12]

For individuals who begin having the above symptoms, the experience has taken them beyond "normal reactions to trauma" and caused them to begin experiencing *acute stress disorder (ASD)*. ASD is similar to post-traumatic stress disorder (PTSD), but it is generally less severe and shorter in duration. According to the criteria developed by the American Psychiatric Association, "An ASD diagnosis requires that a person experience or witness a traumatic event and have symptoms immediately after but persistent for three days to one month."[13]

PTSD goes beyond acute stress disorder. The most traumatic experiences often lead an individual into a season of overwhelming anxiety and a shattered perception of the world. Before we dive into the specifics of PSTD, let us explore what happens in the brain that causes such extreme reactions to trauma (fig. 7–1).

Trauma and the Brain

The human brain functions as a system of interconnected networks, chemicals, and parts. The three parts most involved in memory are the prefrontal cortex, the hippocampus, and the amygdala. Within a healthy human brain, the frontal lobes (which make up the prefrontal cortex) are responsible for thinking, reasoning, planning ahead, controlling impulses, and containing emotions. When memories are made, memory is encoded with the help of the frontal lobes and the hippocampus so that experiences are in context and in sequence.[14]

When an individual is in danger, however, the fight-or-flight response kicks in, and the body releases adrenaline to empower action. If the danger is surmountable, the individual will gear up for fight. If the danger can be outrun, the individual will gear up for flight. According to Lisak, however, if fear is present, the amygdala secretes norepinephrine and dopamine (neurotransmitters) in response. Norepinephrine and dopamine handicap the frontal lobes, making them unable to encode experiences in normal context and sequence. Thus during trauma, memory is encoded in intense fragments—sights, smells, sounds—that can come back as flashbacks or nightmares afterward, notes Lisak.[15]

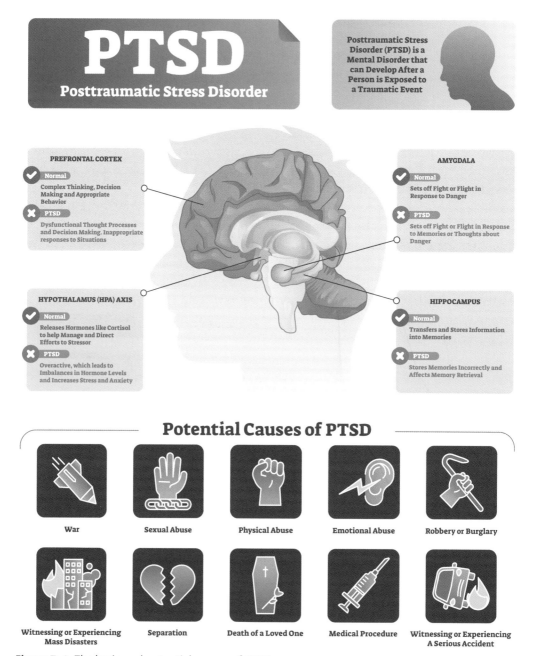

Figure 7–1. The brain and potential causes of PTSD
Courtesy: Shutterstock.

Even first responders may experience this type of traumatic memory encoding. It is not that they do not know what happened, but rather that their brains encoded the memories in sensory fragments rather than sequential events. Individuals who experience traumatic events may not be able to reconstruct an accurate timeline of what happened because the frontal lobes were not actively working at the time of the trauma. But they will be able to identify all five senses and what they were experiencing at the time of the trauma.

Mirror Neurons

The fight-or-flight response is the perfect example of how the human brain is wired for survival. Another way the human brain is wired for survival is through neurons called *mirror neurons*. Both human and animal brains have mirror neurons. These neurons fire when a human or animal acts and sees someone else doing the same action. The other person's action triggers mirror neurons in the brain that tell the body to mimic that other person's body language and emotional state. This can be powerful for sensing a danger through the emergency responses of others around us. It also can be trauma-inducing in some situations, such as when first responders arrive at a stressful call and observe the victim or the victim's family members experiencing trauma. Because mirror neurons cause us to adopt the emotional state of those around us, they can trigger secondary trauma for first responders. Empathy is valuable, but first responders who do not remain in control of their thoughts can find themselves developing secondary trauma rather easily.

The Freeze Response

Psychologists of the past used to sum up the body's emergency state as the fight-or-flight response, believing that this provided the complete picture of how an individual's brain ramped up for survival action. But they overlooked a third type of response to traumatic situations—the *freeze response*. Instead of responding by running or fighting, truly severe events can cause the individual to freeze, like a deer in the headlights.

Captain Andy Perry, a firefighter peer supporter from the Peoria (IL) Fire Department, shares his story of experiencing the freeze response:[16]

> In March of 1998, in addition to having the greatest job in the world, I was married, owned a home and two cars, had a savings account, was putting money aside for retirement, and was on the brink of planning a family. Just after two o'clock in the morning of a warm spring day and nineteen hours into my shift, my life changed forever. I became lost and trapped in the attic of a multi-family

dwelling on fire. I was alone without a radio, running low on air, trapped in place by rising impinging flames, and holding on to the nozzle of a fire hose that had burnt in half two stories beneath me. I panicked. I lost control of my mind. I froze in place, went to my knees, and embraced my certain demise. How could I have so many things going for me only to have it all end like this?

A month later I was sitting on a couch in a shrink's office for my third visit in just a week. This meeting followed my being pulled to safety by my Captain and four weeks of hell. I wasn't happy with my life. I couldn't make heads or tails of the fire. Why did everything go so wrong when we did everything right? Why did I escape completely unscathed physically and feel as if I would have been better off six feet under as a result of that awful night? I had a million questions and no answers.

Now I felt as if I had become crazy and found myself here. Loaded up with anti-depressants and anti-anxiety medications, [I was] encouraged . . . to return to work. The Union had done an amazing job of covering my shifts for me and fabricating a story of my absence, so I was on board with my return. And scared to death, I went back to work with only a handful of people aware of my situation.

I remember turning a corner as I rode backwards on the engine and seeing the plume of black smoke rise in the distance. This was about a month after coming back. My brain immediately went back to the attic. I felt my mind repeating its dismemberment from my body again. *I can't do this*, I thought. I reached inside my coat and clumsily unpinned my badge from my shirt. *When we get there, I'll just give this to the battalion chief.*

The final turn we made near the end of the cul-de-sac confirmed what little bit of rational thought that I had left, when I saw the flames erupting from the roof of a split level home. *If we are ordered into that building, I'm walking up to the chief and quitting. No ifs ands or buts about it.* As fate would have it, we were not ordered into the building but to protect an exposure. And that night in my bunk, I tripled my dose of medications and slept until shift change. *I won't be back*, I thought, as I drove home, *Ever.*

The shrink visits increased but the time off from work was far from a picnic. I was forced to use my accrued time as the City took a stance against my claim that I was off for a duty related injury—this would later be settled in court. I was diagnosed with PTSD, depression, and anxiety. I was losing weight, having panic attacks, and becoming extremely irritable.

At my worst, sometime in the summer of 1998, I reached a point where my depression was so bad that I could barely make it out of bed and my anxiety was so intense that it was nearly impossible to live in my own skin. The mere thought of returning to work made me nauseous. I was experiencing a high level of suicidal ideation, and my alcohol intake skyrocketed. I couldn't wrap my mind around how fifteen minutes of my life could leave me so desolate, fearful, ashamed, and bewildered. Why do I even want to get better? So I could return to work only to be faced with the exact same stressors all over again?

Little did I know, that question, to stay or not to stay, was exactly what I had to face. I asked my shrink, "Why would I want to go back? I'm just going to have to face this again." He smiled at me and this of course angered me. He told me that if I ever decided to go back to work that first of all, I would be ready. Second of all, if I returned to work, I would naturally be hypersensitive to my surroundings. Most importantly, though, if I returned to full duty, I would be afforded the opportunity to conquer what had beaten me, an opportunity many discharged war veterans never had. I went home that day and thought about that for a while. I could be afforded the opportunity to conquer what had beaten me. The opportunity. A chance to overcome. I experienced a glimmer of hope.

My return to full duty was 10 months from that original fire. I knew I had a lot to prove to myself. I still felt like I had to prove myself to others. The shame of being off for 10 months for a psychological reason terrified me as much as anything. After a week or so back on shift, I had not received any grief for my injury. This was a step in the right direction. To this day, over 18 years later, I have yet to hear anything negative from another firefighter regarding that absence. The firefighting aspect of things came slowly as anticipated, but after the first year back, I thought about the small opportunities that I had overcome.

I didn't realize until about 10 years later that I wasn't going to master or dominate or overcome or defeat my past at one fire. I wasn't going to be able to claim victory after just one incident. I realized that, although a bit hypersensitive to my surroundings, every incident that I went to, I was regaining composure. I had begun to rebuild one brick at a time.

Today, I look back at what happened in my career. It has been a whirlwind of a ride, and, God willing, I have several more years to go before I hang it up for good. I continue to add bricks to the wall and probably will right up until the final bell. Firefighting is the best career ever, but like any other career, it comes with its down sides. I unfortunately came across a particularly rough one of those potholes but was able to make it out.

This brief recap of my story has many more parts and detailed events, but this is how I rose then fell and am rising again. Not everyone will emerge from this type of crossroads like I did, but I hope anyone that has the opportunity to overcome such an obstacle gives it a shot. My advice would be to get help early, disregard the stigmas, and take it at your own pace. Just don't go it alone.

Andy experienced the freeze response that night in the attic fire. He says, "I froze in place, went to my knees, and embraced my certain demise. How could I have so many things going for me only to have it all end like this?"

His inability to move was a physiological response to his circumstances. When an emergency arises, the cortisol (the "stress" hormone) level shoots up, the heart races, the muscles tense, and energy surges through the body.[17] This is all for the sake of giving the individual the physiological strength to escape the circumstance. But when the trauma is insurmountable and the individual thinks, "I'm a goner," the brain dissociates the individual from the trauma being experienced, and the body freezes.

Why does this happen? It is similar to when an animal "plays dead." Animals often play dead when facing a severe threat. This inability to move in the face of overwhelming trauma is called *tonic immobility*. The mind shuts down mental output to the body, so the body cannot move. This reflex is involuntary; in hope of survival, the body freezes.[18] No matter how the individual thinks he will respond to a trauma beforehand, tonic immobility can take over because it is a reflex. You cannot control it.

People who experience tonic immobility and survive often freeze more easily in future traumatic events. This is because the brain recorded the first freeze response as a success, as survival. So in future traumatic events, the individual may freeze very readily in response to emergency experiences.

What Is the Result of the Freeze Response?

The problem with the freeze response is that it is debilitating. Those who experience tonic immobility are 3 times more likely to experience PTSD. They are 3.5 times more likely to develop severe depression than those who did not experience tonic immobility.

Researchers found that when people do not have the chance to "let go" or "thaw out" after they freeze in the face of danger, they can develop PTSD, phobias, panic attacks, obsessive-compulsive behaviors, and various anxieties.[19]

Being unable to move during severe danger is never a good thing. The freeze response is not your friend. Without movement, all of the stress chemicals just sit in your body. They do not get worked out of your body through your lymphatic system via muscle contractions and heavy breathing, so they begin wreaking havoc on the body and the psyche. Movement is essential for releasing cortisol.

What if movement is not possible during the situation? If you are in an emergency situation where you are forced to stay still, immediately begin slow, intentional breathing. You can downshift your body's stress hormones by inhaling slowly and intentionally. For example, you could count to four as you inhale, and exhale slowly and intentionally, possibly for six counts. If possible, add light movement. Walking is great, as is yoga, if you have the opportunity to do either activity soon after the event.

Mind-Body Intervention

In a study designed to enable people to physically work the excess cortisol out of their systems, researchers introduced mind-body intervention (MBX) to nurses who had symptoms of PTSD. Twice a week for eight weeks, the nurses participated in 60-minute sessions of stretching and breathing, and at the conclusion of the study, it was determined that all of the MBX students showed a significant reduction in cortisol serum and in PTSD symptoms. They also experienced improvements in sleep, stress resilience, energy levels, emotional regulation under stress, and resumption of pleasurable activities they had previously discontinued. In bringing the body back to its baseline, movement and breathing had significant impact on removing excess stress chemicals from the body.[20]

Disconnecting After Trauma

When an individual experiences trauma, his brain dissociates from the experience—his mind checks out as a defense mechanism. This disconnection often remains even after the event is over. So the individual goes on feeling disconnected from himself and numb to the world around him. Some psychologists refer to this as "the missing roommate," or the idea that although the person is there physically, he no longer "lives there" anymore. This can be unsettling for the individual's loved ones, and if the individual remains disconnected from the world for too long, it can lead to PTSD.[21]

In order to reconnect, studies show that individuals should intentionally spend time doing things that make them feel connected or things that previously have made them feel connected. Connecting can reduce the symptoms of PTSD. Some people feel connected to nature. Others feel connected when they are with trusted friends or family members. Some feel connected when they pet dogs or ride horses. Other connections include spirituality/self/intuition, body/sensory experience, community/groups, art, God/universe/higher power/bigger picture, and meaning and purpose.[22]

Trauma and the Nervous System

Now you know how the brain responds to trauma, but how does the body respond to trauma?

For starters, there are two components to the nervous system in your body—the central nervous system and the peripheral nervous system. Your central nervous system is your brain and spinal cord. Your peripheral nervous system is your autonomic nervous system (ANS)—which controls your body's involuntary processes, such as pulse and breathing, homeostasis, and your metabolism—and your somatic nervous system—which controls the voluntary things your body does, such as moving muscles, sensing motion, touch, balance, and senses of your body's movement and stress.

The key duality for those who experience trauma is the duality of the autonomic nervous system, which includes the sympathetic nervous system (SNS) and the parasympathetic nervous system (PNS) (fig. 7–2).

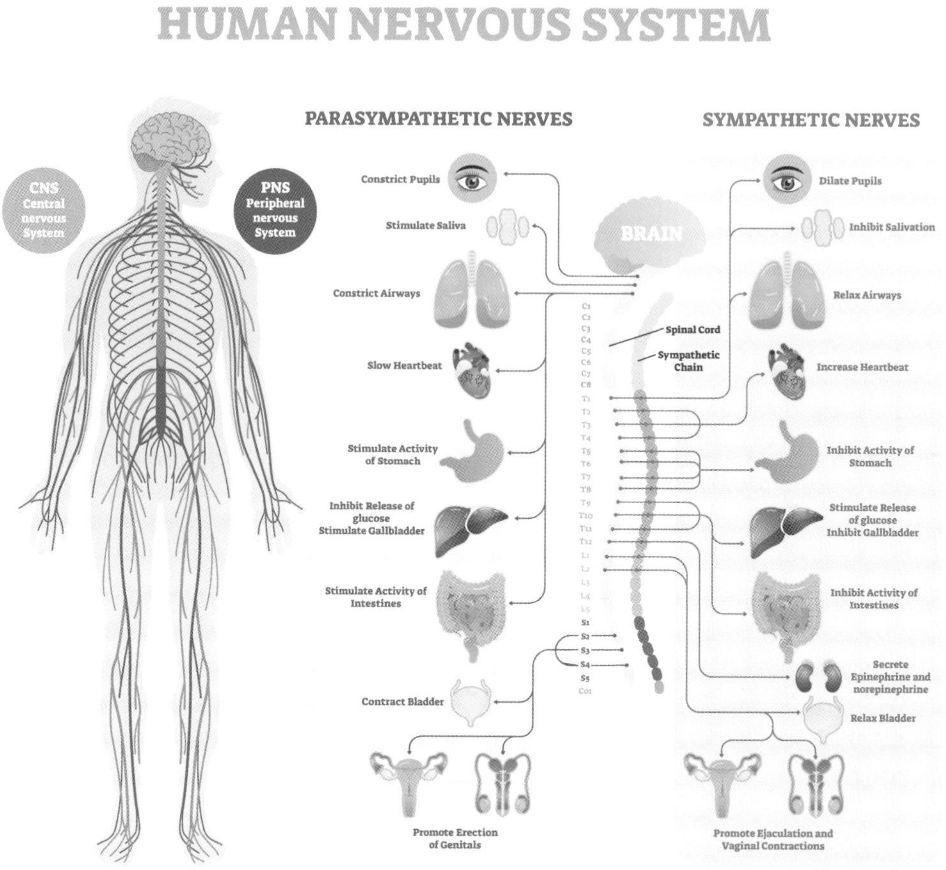

HUMAN NERVOUS SYSTEM

Figure 7–2. The peripheral nervous system includes all the nerves in the body outside of the spinal cord and the brain. These nerves carry information to and from the central nervous system to provide complex body functions.

The sympathetic nervous system (SNS) is your fight-or-flight nervous system. According to research conducted by Sang Hwan Kim, Nursing Research and Translational Science, National Institutes of Health, and others, it turns on in stressful or emergency situations and is responsible for the following physiological responses:

- Increased heart rate
- Increased blood pressure
- Increased respiration
- Dilation of bronchial tubes
- Contraction of the muscles (especially in the arms, legs, hands, and feet to mobilize for fighting or fleeing)
- Dilation of pupils
- Decrease in stomach movement and secretion
- Decrease in saliva production
- Release of adrenaline

- Increase in glycogen to glucose conversion
- Decrease in urination output[23]

The parasympathetic nervous system (PNS) is your rest and digest nervous system. It is where you operate when you are in homeostasis. Individuals should spend most of their lives with their parasympathetic nervous systems governing their body's function. But as I mentioned before, our society is overwhelmed with stress, which pushes people to live with their sympathetic nervous system turned on. The problem with this is that it puts the heart on overdrive and the digestive system on pause. Many people experience strong imbalances as a result of living in such a heightened state of stress.

First Responders Misdiagnosed with "Bipolar Disorder"

Stress responses can affect every single area of life for first responders if they are not intentionally kept in check. Even when a first responder is aware of feeling "revved up," his body's stress response may be more complicated than simply downshifting his mind and body.

Biologist Robert Sapolsky, Stanford University, identifies the areas of the human body commonly affected by stress, which include the following:

1. Functioning of glands, hormones, and neurotransmitters
2. Heart, blood pressure, cholesterol, and breathing
3. Metabolism, appetite, digestion, stomach, and gut functioning
4. Growth and development
5. Sex and reproduction
6. Immune system and vulnerability to disease
7. Pain
8. Memory
9. Sleep
10. Aging and death
11. Mental health and well-being
12. "Depression," motivation, and ability to experience pleasure
13. Personality and temperament
14. Vulnerability to addiction[24]

The human experience is essentially driven by responses to stress and the body's ability to ramp up for action and then deescalate afterward for well-being, balance, and longevity. According to author Sarah Knutson, when we are fearful, our bodies pump adrenaline, cortisol, and glucocorticoids to empower our bodies for a fight-or-flight emergency response, directed by our sympathetic nervous systems. Knutson notes that this response can mimic the similar physiological response of a manic episode of bipolar disorder.[25]

The following are the *DSM-5* criteria for a *manic episode*:

(A) A distinct period of abnormally and persistently elevated, expansive, or irritable mood and abnormally and persistently increased goal-directed activity or energy, lasting at least one week and present most of the day, nearly every day (or any duration if hospitalization is necessary).

(B) During the period of mood disturbance and increased energy or activity, three (or more) of the following symptoms (four if the mood is only irritable) are present to a significant degree and represent a noticeable change from usual behavior:

1. Inflated self-esteem or grandiosity.
2. Decreased need for sleep (e.g., feels rested after only 3 hours of sleep).
3. More talkative than usual or pressure to keep talking
4. Flights of ideas or subjective experience that thoughts are racing.
5. Distractibility (i.e., attention too easily drawn to unimportant or irrelevant external stimuli), as reported or observed.
6. Increase in goal-directed activity (either socially, at work or school, or sexually) or psychomotor agitation (i.e., purposeless non-goal-directed activity).
7. Excessive involvement in activities that have a high potential for painful consequences (e.g., engaging in unrestrained buying sprees, sexual indiscretions, or foolish business investments).[26]

Consider the scenario of a first responder going out on a call. His body will enter a period of "elevated, expansive, or irritable mood and abnormally and persistently increased goal-directed activity or energy," as the *DSM-5* categorizes mania. Is the first responder's mood elevated? Of course it is! Is he irritable? Only if someone slows him down or performs in a way that prevents him from being effective in the situation. Is he goal-directed? He absolutely is. He is on a call, and his goal is to save someone. Knutson points out that, unlike mania, a stress response does not usually last a week.[27] In fact, it usually only lasts the duration of the emergency call. But it could last a week if the situation required more of him, or if he failed to decelerate his emotions after the call.

A review of the *DSM-5* criteria for mania[28] reveals that the stress response in a first responder's emergency may show up looking like many, if not all, of the second criterion for a manic episode.[29] The first responder may seem to have inflated self-esteem or grandiosity, even after the call, because of the enormous surge of strengthening hormones coursing through his body. He may have a decreased need for sleep, again because his adrenaline levels are quite high. He may be extra chatty before, during, or after the call because he wants everyone around him to understand the situation and make sure it is carried out perfectly. He may be distracted easily as he takes in all the stimuli needed to respond to the details of the call. He may be goal-directed to the extreme in response to the situation. And with his endorphins raging, he may be unphased in response to danger and prone to making dangerous moves without thinking.

Clearly the stress response very strongly resembles a manic episode of someone with bipolar disorder, but categorically, it is different. A manic episode is a part of a larger, more persistent mental health problem. In contrast, the stress response is a part of the body's natural physiological response to emergency. The SNS is governing all action, and the person feels powerful, laser-focused, and unstoppable, at least for a time.

Once the crisis is over, however, the responder is absolutely exhausted. All of the survival hormones that his body has been pumping for the emergency and even in the days following the emergency have to come from somewhere. His body has been on overdrive, and the power he has been using is borrowed from tomorrow and the next day and the next.[30] This is why it is absolutely critical for first responders to be intentional to activate their parasympathetic nervous systems (PNS). In order to bring the body back out of the emergency state and restart rest, digestion, and well-being, the first responder has to recognize when he is being governed by his SNS.

Your SNS is in charge when your blood pressure is high, blood sugar and oxygen levels are out of whack, blood and saliva tests indicate a presence of adrenaline and glucocorticoids, skin conductivity tests are firing "hot," muscle tension and twitching are present, movement and fidgeting are common, pupils are dilated, perception of things seen is focused toward the big picture rather than details, and brain scans show neural pathways are "hot."[31] Some of these signs are noticeable without a medical test from the doctor. Other signs may need to be determined by a doctor and medical testing. The bottom line is this—when you are governed by your SNS, every part of your body is affected. So if you are a first responder, and you are diagnosed with bipolar disorder, consider getting a second opinion from a professional clinician who is experienced with first responders and the stress response.

After the body has come out of the freeze response, the SNS is primarily associated with emergencies, and the PNS is normally associated with homeostasis. So working to activate the PNS is a critical part of helping the body come out of trauma and back to balance. Self-care and mindfulness are two ways people can switch from being governed by their SNS to being governed by their PNS.

Whenever possible, observe yourself. Note when you feel your body ramping up, note what is causing you to ramp up. If you are not on the job, be intentional to activate your parasympathetic nervous system. Some simple ways to do this are to bring your attention to the physical sensations of your five senses, breathe deeply, relax your muscles, focus on positive emotions as you breathe, engage in regular meditation practice, intentionally put on an optimistic attitude, and practice yoga regularly.[32] More information about ways to regain control of your mind, body, and nervous systems is presented in chapter 12.

What will it feel like when you activate your PNS? Knutson describes the feeling, saying, "Waiting out this transition can feel really uncomfortable. The more activated I am, generally the harder it is to sit tight. But if I'm able to keep trusting in the process (instead of panicking and ratcheting myself back into stress reactivity), my body gets progressively more comfortable."[33]

Reflection Questions

1. What is one event that you have experienced that you would label as "traumatic"?
2. What traumatic symptoms did you experience?
3. What surprised you about the description of the freeze response?
4. Have you ever experienced the freeze response? If so, have you had the chance to tell someone about it?
5. If you could go back and tell yourself one truth in the middle of your traumatic event, what would you tell yourself?
6. Read chapter 12 for information on caring for yourself to maintain emotional wellness.

Notes

1. Mind Tools, "What Is Stress?" Mindtools.com (accessed August 9, 2021), https://www.mindtools.com/pages/article/newTCS_00.htm.
2. J. Fay, "Resiliency for First Responders," First Responder Support Network, 2018, https://www.frsn.org.
3. F. Figueroa-Fankhanel, "Measurement of Stress," *Psychiatric Clinics of North America* 37, no. 4 (December 2014): 455–487, https://doi.org/10.1016/j.psc.2014.08.001.
4. Fay, "Resiliency for First Responders."
5. C. Ashton, "Yoga to Transform Trauma: Leadership Training and Intensive with Catherine Ashton," Illumine Chicago (2014), https://illuminechicago.com/events/yoga-to-transform-trauma-leadership-training-intensive-with-catherine-ashton/.
6. E. Wall, "Self-Care Practices and Attitudes Toward CISD and Seeking Mental Health Services among Firefighters: A Close Look at a Mid-Sized Midwestern Urban City," UST Research Online (2012), https://ir.stthomas.edu/ssw_mstrp/123.
7. Ashton, "Yoga to Transform Trauma."
8. T. Hanks and the Northern Illinois Critical Stress Management Team, "Critical Incident/Traumatic Events Information" (2007), https://cdn.ymaws.com/www.nena.org/resource/collection/BAB1E806-C074-4D0F-8374-7B08E8D976FB/Information_And_Symptoms_2011.pdf.
9. D. Lisak, "The Neurobiology of Trauma," Arkansas Coalition Against Sexual Assault, February 5, 2013, YouTube video, 34:30, https://youtu.be/pyomVt2Z7nc.
10. Lisak, "The Neurobiology of Trauma."
11. Lisak, "The Neurobiology of Trauma."
12. S. McCracken (Crown Family School of Social Work, Policy, and Practice, University of Chicago), in discussion with the author, October 22, 2019.
13. R. A. Bryant, M. L. Moulds, and R. M. Guthrie, "Acute Stress Disorder Scale: A Self-Report Measure of Acute Stress Disorder," *Psychological Assessment* 12, no. 1 (March 2000): 61–68, https://doi-org.du.idm.oclc.org/10.1037/1040-3590.12.1.61.
14. Lisak, "The Neurobiology of Trauma."
15. Lisak, "The Neurobiology of Trauma."

16. A. Perry (captain of the Peoria [IL] Fire Department and peer support team member), in discussion with the author at peer support team training session.
17. S. Kim et al., "PTSD Symptom Reduction with Mindfulness-Based Stretching and Deep Breathing Exercise: Randomized Controlled Clinical Trial of Efficacy," *Journal of Clinical Endocrinology & Metabolism* 98, no. 7 (2013): 2,984–2,992, https://doi.org/10.1210/jc.2012-3742.
18. B. P. Marx et al., "Tonic Immobility as an Evolved Predator Defense: Implications for Sexual Assault Survivors," *Clinical Psychology: Science and Practice* 15, no. 1 (2008): 74–90, https://doi.org/10.1111/j.1468-2850.2008.00112.x.
19. Kim et al., "PTSD Symptom Reduction with Mindfulness-Based Stretching and Deep Breathing Exercise."
20. Kim et al., "PTSD Symptom Reduction with Mindfulness-Based Stretching and Deep Breathing Exercise."
21. B. O'Hanlon, "Resolving Trauma Without Drama," *Possibilities*, October 2016, https://www.billohanlon.com.
22. O'Hanlon, "Resolving Trauma Without Drama."
23. Kim et al., "PTSD Symptom Reduction with Mindfulness-Based Stretching and Deep Breathing Exercise."
24. R. M. Sapolsky, *Why Zebras Don't Get Ulcers: The Acclaimed Guide to Stress, Stress-Related Diseases, and Coping* (Holt, 2004).
25. S. Knutson, "How the Human Stress Response Explains Away 'Bipolar Disorder.'" Mad in America, April 1, 2018, https://www.madinamerica.com/2018/04/how-the-human-stress-response-explains-away-bipolar-disorder/#fn-154832-2/.
26. Substance Abuse and Mental Health Services Administration (SAMHSA), "Table 3.19, *DSM-IV* to *DSM-5* Adjustment Disorders Comparison," 2016, https://www.ncbi.nlm.nih.gov/books/NBK519704/table/ch3.t19/.
27. Knutson, "How the Human Stress Response Explains Away 'Bipolar Disorder.'"
28. SAMHSA, "Table 3.19, *DSM-IV* to *DSM-5* Adjustment Disorders Comparison."
29. Knutson, "How the Human Stress Response Explains Away 'Bipolar Disorder.'"
30. Knutson, "How the Human Stress Response Explains Away 'Bipolar Disorder.'"
31. Knutson, "How the Human Stress Response Explains Away 'Bipolar Disorder.'"
32. N. Earl, "Chill 101: How to Activate the PNS," HealthVibed.com (2017), http://healthvibed.com/relaxation-101-how-to-activate-the-pns/.
33. Knutson, "How the Human Stress Response Explains Away 'Bipolar Disorder.'"

Chapter 8

POST-TRAUMATIC STRESS DISORDER (PTSD)

My father passed away in 2017. He lived a long life, and his death caused me to reflect on his life, our family, and what could have been different. For the majority of my life, my dad was controlling. I did not understand it until right before he died, but now I believe this was because he was trying to keep his family safe.

My dad had grown up in an unstable situation, and sadly, our family had our own share of trauma at home. On top of that, he had seen a great deal of trauma as a first responder. Pain and hardship began at a young age for my father. His childhood was filled with abuse, poverty, and bullying. He was raised during a time when the men buffered much of the stress for the family, keeping depression and fear away from the women of the family. Since my father was the only boy in his family, he would often work odd jobs to provide for the others. I believe this stressful childhood drove my dad to be guarded and somewhat short with people. His word choice was abrupt and sometimes harsh, but I realize now that it was almost as if he never felt he had enough time to get the words out.

Perhaps it was his traumatic childhood or his first responder training that taught him to communicate quickly, but perhaps it is also what caused us to waste so much time on misunderstandings. My father communicated as if everything was a crisis. We lived in emergency mode, and it created a great deal of anxiety in the people around him.

My dad used to play a game with me when we stopped at a gas station. He would buy me a piece of candy and ask me to guess which hand was holding the candy before he would give it to me. He had grown up in a keep-you-on-your-toes type of environment, always shifting from apartment to apartment, experiencing little stability and even less money. He frequently changed his mind. It was as if he was teaching me, "You don't know which hand you're going to be dealt in life."

As explained in chapter 5, one of my brothers, Jimmy, committed suicide when I was 12. In response, my dad distanced himself from our family and turned to alcoholism to cope with his pain. My dad's life consisted of layer upon layer of pain and hardship. I believe he was primarily disconnected from himself and emotional wellness because he carried around the weight of undiagnosed post-traumatic stress disorder (PTSD).

What Is Post-Traumatic Stress Disorder?

Post-traumatic stress disorder is an illness caused by experiencing a significant traumatic event. According to the American Psychiatric Association's *Diagnostic and Statistical Manual of Mental Disorders,* Fifth Edition (*DSM-5*), in order for someone to be diagnosed with PTSD, five criteria must be present:

1. *Life-threatening incident.* The individual or someone else was exposed to threatened or actual death, serious physical injury, sexual violence, or other extreme danger. Or the individual was exposed to repeated adverse details of such events.
2. *Intrusion.* Intrusion occurs when "what the firefighter was exposed to is persistently re-experienced by recurrent, involuntary, and intrusive memories, nightmares, flashbacks, prolonged distress, or physical reactivity."
3. *Avoidance.* The individual intentionally avoids triggers such as memories, places, people, and conversations that remind him of the traumatic event.
4. *Negative thoughts and feelings.* These negative thoughts and feelings began or became worse after the traumatic event. These emotions can include "fear, horror, guilt, shame, or anger." Individuals may lose interest in things they previously enjoyed and may isolate themselves.
5. *Hyperarousal.* The individual may begin to behave differently after the traumatic event, including showing irritable or aggressive behavior, self-destructive or reckless behavior, hypervigilance, an exaggerated startle response, problems concentrating, or sleep disturbance.[1]

I often wonder what life would have been like if my dad had seen a counselor experienced in treating trauma and PTSD. I believe the symptoms he was experiencing could have been healed, his relationships could have been transformed, and his joy in life could have been restored.

Symptoms of PTSD

The symptoms of PTSD include physical, cognitive/emotional, and behavioral outcomes. Physically, PTSD can cause difficulty breathing, profuse sweating, rapid heart rate, elevated blood pressure, migraines, exaggerated startle response, and difficulty sleeping. PTSD affects the mind and emotions by causing someone to become easily agitated, have trouble concentrating, have negative expectations of self or distorted blame, be unable to experience positive emotions, have nightmares or flashbacks of the event with strong emotional response, and/or feel overwhelmed. As listed above, individuals experiencing PTSD will avoid feelings, thoughts, people, places, or events related to the traumatic event. They will be hyperalert, detached, and withdrawn. They may consume alcohol or use drugs, change activities or lose interest in hobbies, or have disciplinary issues.

PTSD literally changes the way the brain works. According to researchers Rajdip Barman, Department of Behavioral Medicine and Psychiatry, West Virginia University, and Mark Detweiler, staff psychiatrist, Salem Veterans Affairs Medical Center:

> The brain areas thought to be the most important in PTSD are the hippocampus, amygdala, and medial prefrontal cortex. In several neuroimaging studies, PTSD patients were found to have increased function in posterior cingulate, motor cortex and amygdala. There was also decreased prefrontal, parietal, hippocampal, and temporal cortical function. The PTSD patients had increased left amygdala activation with fear acquisition, and decreased anterior cingulate function during extinction, relative to controls.[2]

This means that the brains of those with PTSD are structurally and functionally different than those without PTSD, note Barman and Detweiler. These changes, along with changes in neurotransmitter balances in the brain, can show up as personality changes, weakened short-term memory and remote memory, and learning challenges.[3]

Ben's Story

I frequently see people with these symptoms come through the doors in my counseling office. One of my clients, Ben, came to me with PTSD symptoms a few years ago. Stan McCracken shared with me once that the military are trained to kill from behind whenever possible, so they never have to look into the eyes of death.[4] First responders are always looking into the eyes of death. In the case of Ben, there were five very traumatic calls, and every single one that bothered him involved interaction with families on scene. Looking into the eyes of the grieving families is what I believe was so traumatizing for him. It flipped a switch, taking him out of that disconnected, doing-my-job mode into having empathy and realizing that this was somebody's child, spouse, or loved one. It took him into a relational place, which causes a great deal of trauma.

In addition to those calls, one incident that was extremely traumatic for him occurred when he saved a young person who was paralyzed as a quadriplegic as a result of the accident. The father of that child yelled at Ben for saving his son, telling Ben, "Now he has to live his life as a quadriplegic." That traumatized Ben because he obviously had done the right thing by saving a person's life. That is what he is supposed to do. Firefighters are rescuers, helpers, and heroes who save the day. When someone is unhappy with them for doing the right thing, it is hard to process. Somebody was angry at him for saving a life! It outrages me to think about it, but it happens, and it can cause PTSD in firefighters.

PTSD Assessment

PTSD has five qualifiers that make up its psychological diagnosis: duration, functional significance, exclusion, dissociation, and delayed expression.[5] When assessing an individual for PTSD, I ask the following questions:

- Has this individual experienced symptoms for more than a month?
- Is this event coming up in his mind every day or every week, even though it happened a long time ago?
- Has this incident affected the firefighter's personal relationships or occupational behavior?
- Have the individual's sleeping habits, including dreams/nightmares, changed as a result of the event?
- Does this individual intentionally avoid people, places, or situations that remind him of a previous call?
- Is the firefighter's personality, demeanor, or behavior different now?
- Has the individual felt guilty about the call, a death, or his actions?
- Is the individual experiencing disturbances that are not related to substances, medicine, or another diagnosis?
- Is the firefighter experiencing high levels of reaction to trauma?
- Is he feeling detached from himself? Does he feel like an outsider watching his life happen?
- Is the firefighter feeling *derealization*, or a distance from his life like things are not real?
- Did the individual seem fine after the event, but then his symptoms began to worsen six months after the event? (*Note:* Many people with delayed onset PTSD have symptoms but do not quite meet the criteria for PTSD. Their existing symptoms may get more severe, and they may experience more symptoms as well.)

When clients come to me and begin to share their stories, I watch for these things in their lives. If you suspect that you or someone in your life has PTSD, you may be able to answer yes to a number of these questions. If so, I recommend talking to someone, especially someone trained in first responder counseling, trauma counseling, and/or accelerated resolution therapy (ART). (More information about treatments for trauma is presented in chapter 12.)

Post-traumatic stress disorder is a normal stress response to an event that disrupts a person's world. But often society expects first responders to be emotionless, unshakable, and strong in the face of unbelievably difficult calls. In fact, this massive gap between what firefighters are *supposed* to feel—unshaken, no emotion—and what firefighters *actually* feel in response to a traumatic event—fearful, overwhelmed, and shocked—can make symptoms of PTSD worse.[6]

Complex PTSD

Mental health professional Pete Walker explains that some individuals cope with a different form of PTSD, known as *complex PTSD (CPTSD)*. Unlike PTSD, CPTSD arises from constant or ongoing exposure to trauma, common in first responder fields. Many firefighters are relieved to discover that there is a name for the symptoms they are experiencing. CPTSD is a normal, instinctual response to ongoing exposure to trauma. Walker points out that these first responders are not "hopelessly oversensitive, and/or incurably defective" but instead are responding normally to the things they have seen.[7]

If you are experiencing complex PTSD, some of these things may be present in your life:

- Do you have trouble regulating/controlling your emotions and finding healthy ways to respond to situations that make you mad?
- Do you feel negatively about yourself, like you are worthless or defective, and suffer from shame or self-criticism?
- Do you have interpersonal problems, such as feeling disconnected from people around you?[8]

If so, you might be struggling with complex PTSD. The onset of complex PTSD is not as immediate as standard post-traumatic stress disorder, but it is nonetheless destructive. Complex PTSD may not have one event to which it is linked, but it can be equally debilitating for firefighters. If you believe you may have complex PTSD, please talk to a first responder counselor.

Moral Injury

One phenomenon that can create deep psychological wounds for first responders is *moral injury*. Moral injury occurs when a firefighter does or observes something that he knows, deep inside him, is not right. The things a first responder sees, actions he takes, or betrayals by leaders can violate the individual's core beliefs and shatter the reality on which he has built his worldview.[9] Moral injuries are situations that break our trust and disrupt our moral values, ethical values, core beliefs, and expectations.[10] Because they dig into the deepest foundations of our thoughts and beliefs, they wound people in excruciating, unseen ways.

Moral injury and PTSD are not the same thing, but moral injury frequently arises in the middle of a traumatic situation, and it often gets grouped into the traumatic memory.[11] I have seen that moral injury can be the most triggering part of the memory. The symptoms of moral injury include the following:

- Shame, guilt, irritability, anger, embarrassment, anxiety, and feelings of worthlessness
- Thinking negatively about oneself or others
- Avoiding, withdrawing, isolating, and emotional numbness
- Feeling emotionally distant from others
- Increased substance use
- Less empathy
- Distress, anxiety, fear, or demoralized feelings
- Doubting God, feeling abandoned by God, or questioning one's beliefs/purpose[12]

People who have experienced moral injury may begin to experience "depression, post-traumatic stress disorder (PTSD), suicidal ideation, interpersonal conflict, prolonged recovery, difficulty resuming daily activities or returning to employment, substance use disorder, and self-isolation," according to PsychCentral's Katy Kamkar.[13] Those who have experienced both PTSD and moral injury are more likely to experience suicidal ideation and behaviors.[14]

McCracken notes that moral injury for firefighters might look different than for military or police. For firefighters, moral injury typically involves the terrible results of acts that others have committed, including murders, domestic violence, sexual abuse, rape, and child sex crimes. Moral injury could also arise in situations that involve understaffing, which could hinder firefighters from doing their jobs effectively. Moral injury could even result from leaders betraying subordinates or leaders committing immoral acts.[15]

One of my clients, Ben, experienced moral injury when he attended the scene of a horrific car accident. The mother showed up on the scene, and she asked him where her child was. He had seen the child in a terrible, dismembered state, and lied to the mother, telling her that the child was in the ambulance. He did this to spare her having to see her baby that way, but he was haunted by the situation, and I believe the fact that he lied stuck with him.

During the 2020 COVID-19 pandemic, I was in touch with a number of first responders, and I believe many of them may have moral injuries from the hospital situations surrounding the unknown and highly contagious coronavirus. In the height of the pandemic, some responders were told not to take life-saving measures on certain patients to prevent the spread of the disease. For example, when a patient was having trouble, responders were often not permitted to do compressions and nebulizer treatments. In addition, hospitals were short-staffed, and resources were extremely limited, which hindered first responders from giving the best care possible. Factors like this often contribute to moral injury.

In chapter 14, I share the story of a client of mine named Brad. He experienced a training incident in which he almost died. The response of his leaders to this event conveyed the message that his life did not matter to them. His feeling of being unprotected by his leaders was a moral injury for him because their betrayal disrupted his moral belief that leaders care about and protect their subordinates. Leaders are supposed to be like fathers to their subordinates, so when they are not, it creates a moral injury for the subordinates. This moral injury contributed greatly to Brad's traumatic symptoms.

The reason firefighters and other first responders encounter moral injury so frequently is because of the proximity they have to violence, situations involving children, and events to which they can relate personally.[16] This repeated exposure to mortality, sickness, and the unthinkable can cause deep psychological injuries that can stay hidden and become worse unless they are brought to light. Talking about what happened can help first responders identify moral injuries and bring truth to them, so that they can begin healing and can reconstruct their belief system in a healthy way.

Reflection Questions

1. Based on the lists of symptoms above, do you believe you or someone you know may have post-traumatic stress disorder, complex PTSD, or moral injury? If so, which symptoms are you seeing?
2. What surprised you about the complex PTSD information?
3. What events can you remember that could build into complex PTSD in your life?
4. What surprised you about the idea of moral injury?
5. Have you or someone you know experienced an event that may have caused a moral injury?
6. What steps do you personally need to take now that you have read this chapter?

Notes

1. American Psychiatric Association (APA), *Diagnostic and Statistical Manual of Mental Disorders*, 5th ed. (Washington DC: American Psychiatric Publishing, 2013).
2. R. Barman and M. B. Detweiler, "Late Onset Stress Symptomatology, Subclinical PTSD or Mixed Etiologies in Previously Symptom Free Aging Combat Veterans," *Journal of Traumatic Stress Disorders and Treatment* 3, no. 4 (July 28, 2014), https://doi.org/10.4172/2324-8947.1000132.
3. Barman and Detweiler, "Late Onset Stress Symptomatology."
4. S. McCracken (Crown Family School of Social Work, Policy, and Practice, University of Chicago), in discussion with the author, October 22, 2019.
5. American Psychiatric Association (APA), *Diagnostic and Statistical Manual of Mental Disorders*, 5th ed.
6. E. C. Nielson et al., "Traditional Masculinity Ideology, Posttraumatic Stress Disorder (PTSD) Symptom Severity, and Treatment in Service Members and Veterans: A Systematic Review," *Psychology of Men and Masculinities* 21, no. 4 (January 27, 2020): 578–592, https://doi.org/10.1037/men0000257.
7. P. Walker, "Emotional Flashback Management in the Treatment of Complex PTSD," Psychotherapy.net, September 2009, http://pete-walker.com/pdf/emotionalFlashbackManagement.pdf.
8. M. Cloitre et al., "Distinguishing PTSD, Complex PTSD, and Borderline Personality Disorder: A Latent Class Analysis," *European Journal of Psychotraumatology* 5, no. 1 (September 2014): 25,097, https://doi.org/10.3402/ejpt.v5.25097.
9. B. J. Griffin et al., "Moral Injury: An Integrative Review," *Journal of Traumatic Stress* 32, no. 3 (January 28, 2019): 350–362, https://doi.org/10.1002/jts.22362.
10. A. Honneth and J. Farrell, "Recognition and Moral Obligation," Johns Hopkins University Press, *Social Research* 64, no. 1 (Spring 1997): 16–35, https://www.jstor.org/stable/40971157?seq=1#metadata_info_tab_contents.
11. McCracken, discussion with the author.
12. Griffin et al., "Moral Injury"; and Honneth and Farrell, "Recognition and Moral Obligation."

13. K. Kamkar, "First Responders Suffering from 'Moral Injury,'" TheSafetyMag.com, January 7, 2019, https://www.thesafetymag.com/ca/news/opinion/first-responders-suffering-from-moral-injury/187489/.
14. Griffin et al., "Moral Injury."
15. McCracken, discussion with the author.
16. K. Papazoglou and B. Chopko, "The Role of Moral Suffering (Moral Distress and Moral Injury) in Police Compassion Fatigue and PTSD: An Unexplored Topic," *Frontiers in Psychology* (November 15, 2017): 8, https://doi.org/10.3389/fpsyg.2017.01999.

CHAPTER 9

RESILIENCY

Challenges are what make life interesting. Overcoming them is what makes life meaningful.
—Joel Fay

Between stimulus and response, there is a space. In that space is our power to choose our response. In our response lies our growth and our freedom.
—Viktor E. Frankl

Whether you have been through a formal resiliency training or you are just beginning to learn about how to become resilient, know this: *anyone can develop resiliency because it all comes down to your thoughts.*

What Is Resiliency?

The American Psychological Association (APA) defines *resilience* as "the process of adapting well in the face of adversity, trauma, tragedy, threats or significant sources of stress—such as family and relationship problems, serious health problems, or workplace and financial stressors." The APA notes that resilience is as much about your ability to rebound from hardship as it is about your personal growth through hardship.[1] Daryl Conner, organizational leadership consultant, explains that "highly resilient people are positive, focused, flexible, organized, and proactive in their approach to change."[2]

Psychologist Joel Fay observes that *resiliency* is the ability to rebound from hardship, trauma, adversity, physical or emotional pain, loss, or overwhelming circumstances that seem impossible to overcome. Resiliency requires acting when necessary to find healing and a new vision for the future. Resiliency is the choice to move forward and find new meaning in life after the unthinkable happens. Resiliency is moving from victim to survivor, according to Fay.[3] Resiliency is making lemonade from lemons. Resiliency is finding beauty from the ashes and molding a new piece of pottery when everything crumbles to dust around you.

What Is the Difference between Coping and Resiliency?

Coping is taking action to deal with a big or small life situation. Researchers Valerie Rice, US Army Research Laboratory, San Antonio, and Baoxia Liu, DCS Corp., note that in coping, the demands of the situation exceed the person's internal and external resources, and thus the individual is left shifting tactics to manage, reduce, or tolerate the demands of the situation. Resiliency, on the other hand, is a measure of how you gather resources to surmount the situation you are facing. Resiliency is more than adapting strategies to manage a tough situation. It is about building and leveraging new internal and external resources so that the demands of the situation no longer exceed your ability to thrive in the face of them. Rice and Liu define *resiliency* as "the ability of a person, who has been exposed to traumatic event, to maintain or quickly return to, a healthy and stable state, physically, cognitively, emotionally, and behaviorally."[4]

Politics and Resiliency—Blake's Story

Sometimes people come to a crossroads where they have to choose between merely coping or thriving. This is what happened to Blake. Blake came to me about three years ago because he was struggling with depression and isolation within his department. At that point, Blake was 15 years into his career. He was experienced and comfortable with his responsibilities, but he was getting a little tired of all the new initiatives his department kept putting forward, and he showed it through his body language.

That is when his department's leadership launched a new regime to clean house. While many people felt targeted, Blake felt singled out almost daily as his leaders made things difficult for him. He was written up on multiple occasions, although he never actively disobeyed an order or broke protocol.

Blake was exhausted, depressed, and hopeless. He felt like there was nothing he could do to work his way back into the brotherhood, so he began looking for other career options. He explored construction work, along with a few other interesting, movement-oriented type of jobs and found that nothing could pay what his family needed as well as the fire service could. He felt stuck. He felt alienated. He did not feel like any brotherhood existed in his department.

For three years, every time Blake came to me, the repeated theme was "one more shift." Together, we talked through his frustrations and isolation, and each time he left with just enough determination to make it through one more shift. That was all. He was ready to get out and did not see any acceptable alternatives. This was the point at which Blake had the choice to cope or to become resilient. Coping would mean showing up and doing the minimum necessary to survive. Resilience would mean digging in and leveraging this challenge in his department as an opportunity to grow as a firefighter and as a person.

Shift by shift, Blake decided to jump through one hoop at a time. There was an assignment he was quite good at and that he enjoyed, but his leaders took it away from him. Blake decided to attend a training specifically for that assignment, even though he already had extensive experience doing this task. Nevertheless, he went. He decided not to consider himself entitled to anything. This season of having to prove himself was a challenge, but he knew he could overcome it with some humility and hard work.

Determined to earn respect, rank, and a position on the team, he decided to "see himself as a probie." Blake was determined to view his situation as a challenge instead of a threat. He explained, "If you're under the microscope, treat every shift as if you're on probation." Every shift was an opportunity to learn something new. If he did not know how to do something, he asked. When he was at home, he studied training materials to get better and better at his job. He read any manual he had not read yet, and he went to new trainings. He researched new ways to perform fireground operations. And through this, he never let his leaders see him struggling. They had no idea he was depressed. They just saw a guy who acted with a positive attitude and was more determined than ever.

Appraising a situation as a challenge actually enables your body and mind to respond with more effective actions. Alternately, viewing situations as threats to your well-being can overwhelm the system and actually cause your mind and body to shut down (more about this is presented in chapter 14). What happened in Blake's situation is that he chose to view his situation as a challenge, and his body and mind responded by empowering him to step up his game.

Looking back, he thinks his body language initially brought on this scapegoating. Formerly, he would roll his eyes and sigh audibly to show his disapproval when his leadership would bring on new initiatives. Though he was not directly disrespectful, his body language clearly revealed his attitude. But about three years into this daily determination to change his outlook, Blake noticed something. No one was making him jump through hoops anymore. Blake's attitude was different, yes, but his leaders' demeanor toward Blake had completely turned around. All of the sudden he was respected, left alone, and treated like a brother again.

Blake was invigorated. Instead of depressed and exhausted, he felt ready to go to work. We no longer had to talk about "one more shift." He had found resilience of his own in the middle of a storm of isolation, and his leaders had learned to respect him. This was a total turnaround.

Blake shared his advice for those who go through something similar:

- Put yourself in the spotlight to show what you are good at.
- Take criticism with acknowledgement and respect, even if it might be unfounded.
- Be someone your department wants. Be someone your department is proud to have represent them.
- Always remember the end game. The shift will end, this month will end, and another year will go by, and you will get closer to the end of your career and closer to the retirement you want on your own terms.

After 15 years of work, it was as if Blake found himself in a newly plowed field. Though he felt like the work he had put in until that point would be enough, his field was plowed bare by the impossible expectations of his leaders. So Blake sowed seeds of hard work, humility, determination, and willingness to learn, and three years later, the harvest was abundant.

Resiliency is deciding to roll up your sleeves in an empty, plowed field. Resiliency means picking up the seeds and planting again, even when it seems like the harvest should already be ripe. And when you do replant, you will find that when the harvest comes, it is that much sweeter for all the sweat and tears you put into it. Keep planting, my friends.

If It Does Not Kill You, Does It Make You Stronger?

Is Friedrich Nietzche's famous phrase, "What does not kill me makes me stronger,"[5] actually true? The short answer is that hardship *can* make you stronger, but it is up to you how you respond.

Often, one of two things can happen when people experience adversity:

1. *They struggle.* They find themselves haunted by the memories of the horrible event or season of their lives. These memories consume their daily thoughts and make it difficult to experience life after the hardship.
2. *They grow.* They process the events or circumstances that they walked through, opening up to others about them, and sorting the memories into their rightful places. They find meaning because of how they have grown as a result of the hardship.

Let us explore these two paths more closely.

Struggle: What Does Not Kill You Makes You Weaker

Individuals who experience hardship and choose to "stuff it" by closing themselves off from the world become less able to cope with challenges and become angry or irritable more quickly. After trauma, it is common for people to develop anxiety, acute stress, and even post-traumatic stress disorder (PTSD). Some people try to keep moving because busyness is a great distractor from pain. Others turn to alcohol or other substances as an escape. Some people begin considering suicide because of their pain. But healing is possible!

Those who choose to open up about their pain, talking to a trusted friend, spouse, or counselor about the situation, often find healing. They process what happened and how they felt about it, and they begin to see a light at the end of the tunnel. Emotional support, self-care, and openness can make a world of difference in how people weather hardship.

Growth: What Does Not Kill You Makes You Stronger

As many as 90% of trauma survivors report having a renewed appreciation for life.[6] After enduring a trauma or hardship, grieving what happened, and facing changes and loss, they find the most beautiful kind of healing. People discover that they have made positive psychological changes in response to adversity. They experience warmer, more intimate relationships. They live more in the present, with gratitude for each day. They find that their thinking has changed significantly. They have a greater sense of purpose and clarity concerning what they want out of life. They say things like, "I have a new lease on life" or "I'm glad to be alive." Furthermore, those who have walked through trauma or adversity and have come out the other side with healing and renewed strength find that they can more easily navigate future traumatic events or seasons of life. They actually do become stronger. (This is called *post-traumatic growth*, which is discussed in greater detail in chapter 10.)

If you are walking through a valley right now, or if you have experienced a traumatic event in the past, you can choose how you weather this storm. You can isolate and struggle, or you can open up and grow. Take a risk and talk to someone you trust, so you can come out the other side better and more alive than you were before.

Daily Thinking: Optimism Versus Pessimism

Resiliency begins with your perspective. When facing a potentially traumatic event, resiliency begins with whether you assess the event as a challenge or as a threat. In your daily life, your mindset about whether something is either a problem or an opportunity will also influence your ability to ride it out.

Viktor E. Frankl once said, "Everything can be taken from a man but one thing: the last of the human freedoms—to choose one's own attitude in any given set of circumstances."[7]

The circumstances optimists and pessimists face might be the same, but perspective changes everything. Optimists look at the world around them and believe that bad events are about isolated issues. They blame bad events on causes outside of themselves. Conversely, pessimists look at the world around them and believe that bad things always happen to them. They blame bad events on causes inside of themselves. Optimists anticipate a better future and take steps necessary to create it. Pessimists are skeptical that their own actions can lead to good results, and they tend to overlook positive outcomes when they do occur.

So why does the pessimist feel like he always loses? He tells himself that his life is bad, blames the bad on himself, and does not expect himself to succeed. Interestingly enough, self-doubt is actually universal. It keeps us from becoming narcissists. Self-doubt challenges us to rise up and surprise ourselves with what we can accomplish.

A few years ago, I had the honor of attending the first four days of the California Highway Patrol (CHP) Academy, which is all about building resiliency. During the training, we went around the table hearing from each cadet about why he or she came to the academy and why each intended to make it to graduation. One cadet shared that his father's dream was to become an officer. Though the father was not able to realize this dream, his son pursued the dream with new fervor so that he could send money back to support his father financially. As I heard him share, I could see that he had embraced his circumstances as a challenge to overcome. He chose optimism, and it motivated him.

Is the situation a challenge or a threat? Perspective can decide. Threats jeopardize our well-being. Challenges strengthen us. When a child is challenged, he grows by experimenting and developing his own resources and strengths. His flexibility allows him to see a new challenge as an exciting opportunity to learn. Even as adults, seeing a threat may make us shrink back. Seeing a challenge invites us to look ahead and actively engage in changing our circumstances.

Psychologist and resiliency expert Joel Fay once said, "What is very clear is that we can make ourselves more or less vulnerable by how we think about things."[8]

Positive Versus Negative Thoughts

Positive	Negative
I like a challenge.	I am incompetent.
I can get it done.	I must do this perfectly.
Relax, I'm doing what I can.	I can't handle this.

Positive Versus Negative Behaviors

Positive	Negative
Assertive	Withdrawal
Productive	Avoiding situations
Task-oriented	Overreacting

Source: American Counseling Education, *Stressors and Stress Response*, 1994.

Perspective is so powerful that using proactive language can actually increase the likelihood of completing the challenge. Saying things like, "I'll do it," and "When I complete this . . ." will make the reality come alive.

Optimistic thinking occurs when you take control of your perspective. Two tools to help you accomplish this are *self-efficacy* and *gratitude*. Self-efficacy is your belief in your own ability to succeed in a specific situation or accomplish a task. Take a second to look back at the good you have already done. Think of a time you overcame something hard in the past. You had it in you to win that challenge. You can definitely find that strength in you again this time.

Gratitude is being thankful for what you have. Did you know that thinking and expressing gratitude changes the brain's activity and chemistry? According to Rita Carter and coauthors of *The Brain Book*, those who expressed gratitude raised the hypothalamus activity level in their brains. The hypothalamus is responsible for body temperature, hunger, thirst, sleeping, metabolic activity, and managing stress. Carter and others note that not only is this area of the brain more responsive and effective when exposed to gratitude, but the limbic system is also activated by gratitude. The limbic system releases dopamine—the "feel-good" neurotransmitter—and oxytocin—the "cuddle and calm" neurotransmitter—in greater quantities.[9] This positive stimulus can then send the person into a positive loop of thinking that makes them more productive, positive, and creative.

Consider starting your day by thinking of three things you are thankful for. Maybe it is your home, your family, or your health. If you can, think of more. Then end your day by recounting three good things that happened and write them down. Challenge yourself, at least once a week, to write down 20 things for which you are thankful. The more you practice gratitude, the more emotionally well you will become. In fact, those who kept track once a week of things they were grateful for were more upbeat and had fewer physical complaints than those who did not take time to remember what they were thankful for.[10]

Your perspective is powerful. So much can be determined by your mind. In fact, the mind is wired with a *confirmation bias*, which means that your brain searches for cues in the world around you to confirm the ideas you have in your head.[11] So when you express gratitude, the brain will look for more ways to confirm your notion that life is good. Conversely, people can get stuck in a negative loop of thinking everything is against them, and their brains will work to take note of the ways this is evidenced in the world around them.

Your brain wants you to be grateful, and it can take a bit of intentionality to cultivate gratitude if you have gotten out of the habit of being grateful. But when you begin to recognize things you are thankful for, your brain will become more productive, creative, focused, prone to personal growth, and emotionally well.[12] Gratitude is abounding in benefits.

How to Stay on Top of Your Thoughts

In the heat of an intense moment, it can be easy to ramp up so quickly that you get overwhelmed. When you get overwhelmed, it can be easy to make mistakes and experience tonic immobility (the freeze response, as discussed in chapter 7). Stress hormones then remain in your body without being used for effective action. This will shift your mind from your frontal lobe into your amygdala, which will impair your logical thinking. In order to stay on top of your thoughts and stay sharp for your job, your first action should be to take a deep breath.

Deep, intentional breathing helps you keep a clear head, so your frontal lobe—the part of your brain responsible for logical, sequential thinking—can stay in control. One way to do this is to take a drink of water. Taking a drink will force you to breathe, hold your breath, swallow, then breathe again.[13] It is a natural action that forces intentional breathing. After you have taken a drink, continue by counting your inhales and your exhales, trying to regulate them: inhale for six counts, exhale for eight counts.

Breathing deeply will help you have clarity of thinking to problem solve. Police call this *tactical breathing*. Staying on top of your breathing will help you solve your problem with your best thinking abilities. Then you can take a minute to do a personal review of the situation, identifying the problem and the goal and generating multiple solutions. Make sure you continue breathing intentionally as you begin to implement your solution.[14]

There is a phrase in the Marine Corps, "Slow is smooth, and smooth is fast." This applies here. In a crisis situation, moving slowly with deep breathing actually enables you to be smooth, accurate, and effective, which is far faster in the long run than trying to make decisions with jumbled thinking and potentially flawed plans.

Slow down. Focus on intentional choices and effective actions. Finish the task at hand. Then give yourself some mental recovery time.

After the intense situation is over, allow your mind to reset by giving yourself a mental rest. The American Counseling Education suggests several actions to help you recover mentally: Divert your attention to something else, reflect on your expectations, and think back to the words you used—"have to," "must," and "perfect" should be avoided.[15] Take a minute to see if there is another way to look at the situation. If the situation was overwhelming, try to put it in perspective with the rest of your life and even with your spirituality. Identify erroneous thinking and replace it with positive thinking. Learn to say "good-bye" to a thought, feeling, or emotion that is not serving you. Learn to accept what you cannot change about the situation.

Taking some time to unwind physically after an intense event will also contribute to your resiliency.[16] Doing relaxation exercises, taking a walk, meditating, and doing yoga are great ways to bring yourself back to your baseline. Some cardiovascular intensity can help, too. Experiment with doing cardio exercise, gardening, playing tennis, or engaging in another positive activity that can help you work off excess cortisol. You might even consider "draining off" excess cortisol by taking a hot bath, spending some time in a sauna, or getting a massage. Taking personal time out to do something else for a short period of time or talking to a trusted friend or your spouse also can help you recover and find resiliency after an intense situation.

The Seven Resiliencies

Sybil Wolin and Steven Wolin, authors with the Resiliency Project, note that in order to get back up every time you get knocked down, there are seven strengths you can leverage: insight, independence, relationships, initiative, creativity, humor, and morality.[17] Using these strengths will help you thrive in the midst of a highly stressful career. We all have these traits, and we all rely on some more than others. Here is how they play out.

Insight

One way to develop resiliency is to ask difficult questions of yourself.[18] Being honest with yourself will help you understand what needs to change. It will help you have different perspectives on change. Then it is up to you to make that change. Maybe you are in the habit of turning to alcohol in the evening. Maybe it is not an addiction yet, but you can see that it is becoming a pattern in your life to have a little too much to drink after a stressful day at work. Insight would say, "Why am I actually drinking? Is there something else I actually need instead of alcohol?" Insight can lead you to important conclusions. Perhaps you need to talk to someone about something that happened at work. Maybe connecting with your spouse and going out on a date to relax together would be a more effective means of dealing with stress. After you identify these potential solutions, implement them. This builds resiliency.

Independence

In building resiliency, independence means choosing to go forward and do something that you have never done before, even if no one is going with you.[19] Maybe you are not a person who likes to think about your feelings. Maybe you go to your AA meeting alone. Maybe independence means you take some time to write down your thoughts or talk to a peer supporter or counselor, even if no one else around you seems to be responding that way. It is okay to be the only one drinking water at a barbecue. You are choosing to be flexible to meet the needs you have at the moment. That is resiliency.

Relationships

Resiliency thrives on connections with others. Friends and people we love are as vital as oxygen to our lives.[20] They help us grow. When we face adverse situations, it is often relationships with trusted family members and friends that help us have the strength to go forward. Ask yourself who those people are in your life. If your list is small, what can you do to change that?

Surrounding yourself with people builds resiliency. A 2006 study from the University of Wisconsin looked at how spouses play a role in reducing the fear responses of the brain when a person is exposed to danger or pain. The study analyzed 16 happily married couples by putting the wife in an MRI machine and giving her a signal before providing her with a shock or no shock. In one session, each woman had her husband holding her right hand. In a second session, each woman had a stranger holding her right hand. In the third session, the women had no one holding their hand. While the study subjects were inside the MRI machine, researchers analyzed the areas of the brain connected with fear reactivity. The study found that those with someone holding their hand did much better in response to threat cues, but those whose spouse held their hand had even more regulatory benefits in the face of threat.[21] The results of this study suggest that first responders who are surrounded by social support, specifically by the support of a spouse, physiologically should have much lower reactions to threat and fear in the brain. Social support changes our outlook on the world and our ability to thrive amidst adversity.

Initiative

One way to build resiliency is to take control of a situation when others have not. Taking control of a situation is something you do every day. When you step up and offer a solution to a workflow problem, you are taking initiative.[22] When you look at something that needs to be done at home and jump into it, you are taking initiative. When you take initiative, it boosts the dopamine in your brain, and it helps you feel accomplished. Choosing to thrive rather than merely cope empowers you even more to take charge of your thoughts, feelings, and actions. Taking charge of your life builds resiliency. One simple example might be to plan a date night with your partner. This simple act not only can boost your confidence, but it can build your relationship immensely.

Creativity

Using your imagination is a fabulous way to become resilient.[23] Creativity is more than just art or music. It is also outside-the-box thinking and problem solving that can help you come up with solutions to challenges throughout your life. When you get creative to find solutions, express your thoughts or feelings, or dig into hobbies that give you time to think, you are leveraging your creative brain to help you become resilient.

Humor

The ability to find things funny when they are bad or tense is an amazing way to be resilient.[24] If you are stuck in an uncomfortable training exercise, laugh about it. The ability to laugh at yourself or your circumstances is a great way to keep perspective and teach yourself to be resilient. This is the primary resiliency most first responders use, and I encourage you to keep it up. When something is not going well, try to see it as funny. It will help you keep your perspective.

Morality

According to Wolin and Wolin, knowing the difference between right and wrong helps to build resiliency. Showing up to do the right thing, even when it is unpopular, builds your sense of significance and justice and helps you thrive in the middle of hardship. Is there something you need to stand up for in your life? Speaking up will build resiliency.[25]

Marks of Resilient People

In my years as a first responder counselor, I have become attuned to the traits of a person who is struggling in this career. I have also become attuned to what resiliency looks like. The people who are resilient unknowingly apply the seven resiliencies to their lives on a daily basis. You can watch for these traits in your own life as you monitor your progress in becoming more resilient.

Speaking up for Their Needs

Looking at a person, you will notice resiliency when they speak up for their own needs. This requires three of the seven resiliencies: insight, independence, and initiative. In order to advocate for yourself, you are choosing to identify what you are actually feeling and what you need in any given moment. Then you are choosing not to let the demands of other people supersede your needs. This can be difficult if you are a person who needs to be needed (see chapter 3 for more information). Finally, you are choosing to take initiative to meet your needs in a healthy way.

Speaking up for your needs is healthy, and it affords you the self-respect of having boundaries, especially boundaries that allow you to heal. Resilient people speak up and take time off work when they have a rough call. They give themselves the dignity of time to recover, so they can be well. Resilient people speak up for their need for rest when they turn down a late night with a friend. They allow themselves to say no so they can be the emotionally well person they want to be.

Selecting Quality Friends

You will notice resiliency in a person who chooses to confide in trusted friends. People you spend time with can help you to be balanced in thinking and living. Trusted friends offer a safe place to talk about the hard parts of a first responder job. When you can confide in someone or simply hang out and laugh with a friend, you build trust and develop a connection that anchors you in the midst of the tumultuous seas of a first responder career.

Did you know that there are different types of friendships? Aristotle proposed that there are three types of friendships.[26] The shallowest friendships are *friendships of utility*, in which both people benefit, but these can be short-lived. The middle level of friendships are *friendships of pleasure*, which can last longer but are more about the fun of the friendship than they are about longevity. The third, deepest level of friendships are *friendships of the good* (or *virtue*, as Aristotle called it). Friendships of the good are based on unconditional love. They take a long time to build and are more meaningful because they are the friendships that stick around through thick and thin. Though the investment is big, the payout for friendships of the good are huge. These are the types of friendships I encourage firefighters to build.

According to Thomas Joiner, men especially have difficulty maintaining long-term friendships and starting new friendships as they age because they tend to be busy and focused on their jobs.[27] Subsequently, they find themselves lonely and depressed in retirement. Your emotional wellness is heavily influenced by your friendships. As you invest in friendships at any stage of life, keep in mind that you are building a more balanced, connected future for yourself. It is true that hanging out with friends takes time, but the benefits are innumerable for those who are willing to work for it.

As you are choosing friends to spend time with, make sure you include both first responder friends and friends who are not first responders. Your friends outside the sphere of first response are great for keeping perspective on a world that is not always filled with crisis, tragedy, loss, and emergency. First responder friends, peer supporters, and counselors are also important because they can listen and understand what it is like to see and experience the things common to first responders.

Resilient people build a support network of people who can help them live life to the fullest and remain emotionally well because they are surrounded by support on all fronts. Having outside support helps you separate your painful memories from your own sense of self. First responders do not have to be defined by what happens to them on the job, and trusted friends can help remind them of this truth.

A key thing to remember as a firefighter is that people do not succeed in isolation. You may shine for a time because of your skills, but your longevity depends to some degree on the social support and interconnectedness you develop with the people around you. You need your family, peers, friends, and leaders. There are areas where they are strong and can help you. There will be areas where you are strong and can help others. Letting others help you is not weakness; it is balance. So surround yourself with a strong support network if you want to be resilient in this career.

Comfortable with Themselves

You can recognize people who are resilient by how they are able to spend time alone doing what they love, without noise or distraction.[28] When people are constantly busy or constantly listening to music or having a show on in the background, it could be a cause for concern. Are they running away from feelings they do not want to deal with? Resilient people are not running. They are comfortable talking, thinking, and using healthy practices to help them move forward after trauma or adversity.

Using Physical Activity and Other Self-Care Practices Regularly

Resilient people are intentional in taking care of themselves. Specifically, regular physical activity is key to overall health. It builds physiological toughness so the body responds with the right concoction of adrenaline and cortisol to empower effective action and avoid PTSD during emergency situations. (See chapters 11 and 14 for more about physiological toughness.)

As I have shared before, regular physical activity facilitates information processing and improves memory. Activity burns off excess cortisol and the hormones that are left in the body after an intense situation, hormones that function to shut off the prefrontal cortex. Some types of physical activity can activate the parasympathetic nervous system and bring you back to your baseline. Furthermore, exercise trains your body to relax in healthy ways without the need for alcohol. All of these reasons contribute to a first responder's ability to be resilient amid highly stressful careers. (More about self-care is presented in chapter 11.)

Talking About Things with Perspective

Resilient people speak about adversity with a hint of hope or positivity. When they talk about a negative experience, they may say things like, "At least I . . ." or "I really learned . . ." or "I'm glad I talked about it with . . ."

One sign of a resilient person is how he or she relates the current hardship to the context of his or her life. When people face hardship but understand it to be only a piece of their wonderful, hard, beautiful, good, and sometimes trying life, they can deal with the momentary hardship in a healthy way. In fact, keeping a healthy perspective on your current trial will help you deal with your current difficulty in whatever way is best for you *right now*. We call this *coping flexibility*. Sometimes exercise helps. Sometimes reading helps. Sometimes friendship helps. People who are resilient have perspective that helps them recognize the uniqueness of their current difficulty, so they can address it with whatever self-care practice is most effective at the moment.

Recharging Your Batteries

I love the analogy of a battery when it comes to resiliency. It offers a more realistic picture of emotional energy than the idea of a person gritting it out as things get harder around him. Authors Sam Achor and Michelle Gielan explain resiliency by noting that you do not gain resiliency by merely toughing things out.[29] They explain that resiliency is about how you recharge.

Like a battery, you cannot keep going indefinitely without taking time to recharge. Eventually your battery will die, and you will need a more significant mental or physical health rest period. You need regular, intentional times to disconnect from your work and plug yourself into things that energize you.

Just because you are off work does not mean you are recovering from the stress or pain of the job. If you are bringing work home, checking your phone, thinking about solutions to work-related problems as you fall asleep at night, talking about work over dinner, or waking up in the middle of the night thinking about work, you are not recharging for recovery.[30] Sadly, it is quite easy to be away from work without it being away from us, thanks to our cell phones and laptops.

If you pretend you are recharging but are constantly sneaking work into your personal life, you will eventually burn out. Our minds and bodies will devote an extreme number of resources to come back to their baselines when our bodies are dictated by such a high level of stress.

Achor and Gielan draw a distinction between internal and external recovery. *Internal recovery* is taking short breaks throughout the workday and shifting tasks when your mental or physical resources for that task are maxed out. *External recovery* is time away from work in evenings, weekends, and vacations.[31] It might also mean not taking on second and third jobs. Allowing your mind to rest rather than spending mental energy on politics, house projects, and the like is how you recharge.

Ironically, productivity rises when people take time to recharge. Think of your phone being on power-saving mode. It is not as bright. It is not as effective at doing its job when it has a low battery. But when it is freshly charged, it is bright and powerful.

People who choose to stop, intentionally giving themselves a break from work for rest, find themselves not only more emotionally well but also more effective at their jobs and in their relationships.

What Can You Control?

Everyone has a sphere of influence. Outside that circle are all the things that are "not about me."[32] This includes your circumstances, family, coworkers, and friends. Inside the circle is everything you can control—your thoughts, emotions, beliefs, and actions. Your sense of control dictates whether you will be able to be resilient; exposure to stress with the ability to control it has no adverse effects. Exposure to stress without the ability to control it activates all sorts of biological systems and impairs parts of the immune system. If you believe that the stressors exceed your ability to cope, then that becomes your reality.

One of the biggest items in your circle of control is yourself. Do you take care of yourself or neglect yourself? When you take time to care for yourself, you build your physical and emotional resiliency.

Resiliency Through Transition

I personally love yoga. I regularly attend hot yoga classes that challenge me both physically and mentally. In the process of enjoying yoga, I have discovered that it can be used as a metaphor for life transitions.

About a year ago, an instructor of mine said something that perfectly captured the truth about first responders and life transitions. She said, "Be slow and easy on yourself during the transitions." In the lifespan of a first responder, there are many transitions. In restorative yoga, individuals spend an extended period of time in each pose. When moving from one pose to the next pose, the greatest risk of injury comes from moving too quickly.

The phrase struck me because of the way I have seen so many first responders suffer after transitions because of their choice not to slow down and adjust properly. No matter what the situation, change is hard. It can be unsettling and disorienting, whatever your transition. For some it is the transition to a new shift schedule or the transition to a new boss. For others it may be a divorce, the death of a parent, or kids moving away to college. It can even be a personal injury. Transitions come in many forms, as they do in yoga, and they put you into new and different situations. This can be great, but it can also be exhausting.

In every life transition, the safest, most effective way to adjust is by slowing down. It is true that some transitions are more extensive than others and will require more time and more intentionality. But all transitions require slow, deliberate expenditure of your energy and thoughts.

Though it may feel like slowing down is unproductive, transitions are filled with unseen productivity. This is where growth takes place. Transitions are packed with deep, beneath-the-surface change, healing, and realignment. Just as in yoga, life's transitions allow you to return to alignment, clarify your thinking, process the past, and gear up for the new. Rest is balancing and nourishing for the soul.

What to Do When You Are in Transition

One common time of transition for people occurs when their children move away. They love their children are attached to them. They miss their kids, and daily communication is not what it used to be. If you are approaching a similar transition in your life, keep in mind that it is important for you to slow down and take the time to remember. Your thoughts during this time should center on sorting out and cultivating gratitude for the time you had with them under your roof. As you remember, you are placing memories in their rightful places, so they are not painful as you remember them.

All change is stressful. So your secondary objective during times of transition in your life is to rest. Especially for introverts, but also for all people, rest is a part of your mental adjustment and physical healing. When you rest, your body repairs itself and your mind organizes and consolidates memories. Slowing down in the midst of life transitions means processing and rest.

Perhaps your transition is your age. Perhaps you are finding that you are more winded or more easily fatigued than before. In all transitions, especially those in your body, your job is to slow down and find a new rhythm. This does not mean you call yourself "old" and forego the fun and exciting activities you enjoyed when you were younger. It means you play along as you want at your own pace. It may mean you warm up first before playing sports so you do not get injured.

If your interests change, it may mean that you begin to explore new activities that are kind to your body but still challenge it to be strong and capable. Think about trying yoga, tennis, swimming, isometric-based workouts, or even taking breaks during the day to walk with a coworker. Taking the transition slow means being kind to yourself and celebrating things you enjoy and can still do.

Assistance Can Help You Hold Steady

Many yoga instructors provide yoga props, like foam blocks, bands, and balls to help people maintain proper form in a pose. Just as people need props to hold steady in a yoga pose, people need assistance and support to hold steady in a life transition.

Like a prop in a yoga position, helpful friends, secure and thoughtful leaders, and open communication with a spouse or counselor can help bolster an individual during some of the most tumultuous changes of life. Humans are not meant to transition alone. Relationships with healthy, open communication and safety remind the individual that the world is a safe place and that he is not alone. They form a support structure around an individual that, like yoga props, can hold him up and help him remain well in a new phase of life.

I encountered this useful similarity in yoga class. As my instructor encouraged us to use props, I realized that I would actually accomplish my goal better if I allowed myself some form of help.

As you face the transitions of your life, do not be afraid to reach out for help. Sometimes asking a friend, spouse, or leader, "Do you have a minute to talk?" can start a conversation that will help you grapple with the transition you are facing and find new stability and strength to go forward. Perhaps for you it means talking to a counselor to help resolve any anxiety or trouble you are having with the transition and to find new possibilities and vision for the next phase you are facing. Other people are your greatest source of help when it comes to weathering a transition in a healthy way. They are one of the greatest means by which you will recognize growth through a transition.

Adjustment Disorder in First Responders

Adjustments are stressful and difficult for everyone, but there is a deeper level of difficulty I sometimes see first responders experience when encountering life transitions. It can be easy for first responders to turn to their occupational training in times of change. This training has taught them to keep their heads down, remain focused on tasks, disregard emotion, and move forward with action. The problem with treating life transitions this way is that it does not allow people the slowness and kindness they need to transition well. As a result, many first responders develop what is called *adjustment disorder*.

Adjustment disorder is "the development of emotional or behavioral symptoms in response to an identifiable stressor(s) occurring within 3 months of the onset of the stressor(s)," according to the *DSM-IV* and *DSM-5*.[33] Basically, with adjustment disorder you are having a hard time coping with the stressor. It can affect the way you interact with people at work or in your personal life. In this disorder, individuals can experience an intense spike in stress, mood depression, anxiety, disturbing conduct, and strong emotions.

First responders are trained to focus on the task at hand, disregard emotion, and move forward quickly. During life transitions, however, you need to go directly against that training to avoid getting hurt. As in yoga, if you transition too quickly, you can put yourself in a bad position and get injured. So it is in life.

That is why what my yoga instructor said resonated with me. First responders who wish to remain emotionally well through life's transitions need to slow down, be kind to themselves, ask for support when needed, and seek to find meaning and excitement in the new phase of life.

Transitions are important. They are defining moments and set you up to find beauty and life in your next phase, but only if you slow down and take the time to do it right and be kind to yourself.

Searching for More of Yourself

Just after the death of my father, I felt compelled to take a daily yoga class. This intense craving for yoga perplexed me. What was I really craving? As I attended class daily, I realized that I had been pushing through the grief, staying busy, and not connecting at all with what was happening. Yoga gave me an appointment every day to connect with my grief. What I was craving, hidden underneath my busy schedule and my insatiable desire to go to yoga class, was space to think, be alone with my thoughts, and cry. And I did. I often cried on the mat.

I was busy planning my dad's funeral, taking care of my mom, and continuing on in my work and volunteer work of traveling to do peer support trainings, but I was missing out on time for me.

I yearned to have time to meditate and reflect and gain insight that showed me I was moving forward with my pain and processing it well. I found myself irritable when I did not get to attend a yoga class because it was more than just yoga exercise—it was space to mourn.

Though I really do believe yoga is a great self-care practice to develop, I also believe there was something more going on in my life. In my difficult transition, I needed to slow myself down and find who I was, to simplify my thinking and take a break from the world around me to realize that the world is not resting on my shoulders—though sometimes it feels like it is. As I was adjusting to the loss of my dad, I chose to slow down. In my slowness, I found peace, and I reconnected and discovered more about who I truly am.

Maybe You Need to Slow Down

Maybe you need to give yourself permission to slow down. Maybe slowing down for you means going to yoga class. Or maybe it means turning down a project that would crowd out precious reflection time. Maybe it means choosing to take a daily walk by yourself to allow time to think. Whatever slowing down looks like for you, I believe that if you slow things down, it will help you find yourself. You will gain insight. You will anchor yourself and become stronger for the season ahead. If you skip the slowdown period, however, you will miss the chance to get into the right position for the next step.

Reflection Questions

1. What is one resiliency that you use on a regular basis?
2. What is one resiliency you learned about in this chapter that you think could help you? Give an example of a situation when you could use that resiliency.
3. Who is someone in your life that exhibits resiliency?
4. What are some marks of resiliency you see in that person?
5. What do you think you can learn from having that person as an example in your life?
6. Have you been recharging or just resting? What is one thing you would like to do differently to truly be able to recharge your batteries?

Notes

1. Palmiter et al., "Building Your Resilience," American Psychological Association (2012), https://www.apa.org/topics/resilience.
2. D. R. Conner, *Managing at the Speed of Change* (New York: Villard Books, 1992).
3. J. Fay, "Resiliency for First Responders," First Responder Support Network (2018), https://www.frsn.org.
4. V. Rice and B. Liu, "Personal Resilience and Coping Part II: Identifying Resilience and Coping among U.S. Military Service Members and Veterans with Implications for Work," *Work* 54, no. 2 (July 5, 2016): 335–350, https://doi.org/10.3233/wor-162301.
5. F. Nietzche, *Twilight of the Idols*, trans. Duncan Large (Oxford: Oxford University Press, 1998), 6, https://www.google.com/books/edition/_/iSNeybYAgNkC?hl=en&gbpv=0.
6. L. G. Calhoun and R. G. Tedeschi, eds., *Handbook of Posttraumatic Growth: Research and Practice* (Routledge, 2014).
7. V. E. Frankl, *Man's Search for Meaning* (Simon and Schuster, 1985).
8. J. Fay, "With Help, Life Gets Better. PTSD," *The Squad Room*, episode 7, audio podcast, June 29, 2015, http://www.thesquadroom.net/episode7/.
9. R. Carter et al., *The Human Brain Book: An Illustrated Guide to Its Structure, Functions, and Disorders* (London: Dorling Kindersley Ltd, 2019).
10. H. Cloud, *The Law of Happiness: How Ancient Wisdom and Modern Science Can Change Your Life* (Simon and Schuster, 2011).
11. Carter et al., *The Human Brain Book*.
12. Carter et al., *The Human Brain Book*.
13. Fay, "Resiliency for First Responders."
14. C. Ashton, "Yoga to Transform Trauma: Leadership Training and Intensive with Catherine Ashton," IllumineChicago.com (2014), https://illuminechicago.com/events/yoga-to-transform-trauma-leadership-training-intensive-with-catherine-ashton/; and Fay, "Resiliency for First Responders."
15. American Counseling Education, *Stressors & Stress Response* (American Counseling Education, 1994).
16. American Counseling Education, *Stressors & Stress Response*.
17. S. Wolin and S. Wolin, "Resilience as Paradox, " Project Resilience (1999), http://projectresilience.com/framesconcepts.htm.
18. Wolin and Wolin, "Resilience as Paradox."
19. Wolin and Wolin, "Resilience as Paradox."
20. Wolin and Wolin, "Resilience as Paradox."
21. J. A. Coan, H. S. Schaefer, and R. J. Davidson, "Lending a Hand: Social Recognition of the Neural Response to Threat," *Psychological Science* 17, no. 12 (December 1, 2006): 1,032–1,039, https://doi.org/10.1111/j.1467-9280.2006.01832.x.
22. Wolin and Wolin, "Resilience as Paradox."
23. Wolin and Wolin, "Resilience as Paradox."
24. Wolin and Wolin, "Resilience as Paradox."
25. Wolin and Wolin, "Resilience as Paradox."

26. B. Helm, s.v. "Friendship," *Stanford Encyclopedia of Philosophy*, August 7, 2017, https://stanford.library.sydney.edu.au/entries/friendship/.
27. T. E. Joiner, *Lonely at the Top: The High Cost of Men's Success* (New York: Palgrave Macmillan, 2011).
28. B. Waters, "10 Traits of Emotionally Resilient People," *Psychology Today*, May 21, 2013, https://www.psychologytoday.com/us/blog/design-your-path/201305/10-traits-emotionally-resilient-people.
29. S. Achor and M. Gielan, "Resilience Is About How You Recharge, Not How You Endure," *Harvard Business Review* (June 24, 2016), https://hbr.org/2016/06/resilience-is-about-how-you-recharge-not-how-you-endure.
30. Achor and Gielan, "Resilience Is about How You Recharge, Not How You Endure."
31. Achor and Gielan, "Resilience Is about How You Recharge, Not How You Endure."
32. Fay, "Resiliency for First Responders."
33. Substance Abuse and Mental Health Services Administration (SAMHSA), "Table 3.19, *DSM-IV* to *DSM-5* Adjustment Disorders Comparison" (June 2016), https://www.ncbi.nlm.nih.gov/books/NBK519704/table/ch3.t19/.

CHAPTER 10

POST-TRAUMATIC GROWTH

Emotional pain cannot kill you, but running from it can. Allow. Embrace. Let yourself feel. Let yourself heal.
—Vironika Tugaleva

The most beautiful people we have known are those who have known defeat, known suffering, known struggle, known loss, and have found their way out of the depths. These persons have an appreciation, a sensitivity, and an understanding of life that fills them with compassion, gentleness, and a deep, loving concern. Beautiful people do not just happen.
—Elisabeth Kübler-Ross

Philosopher and psychologist William James believed that the happiest people are those who have gone through a crisis or season of significant suffering, followed by significant personal growth. He called these individuals "twice born."[1] I meet twice-born individuals all the time in my work with first responders.

While serving as the clinical director of program development for the Illinois Firefighter Peer Support (ILFFPS), I had the privilege of hearing the stories of countless firefighters who have weathered traumatizing experiences, devastating childhood losses, and emotional wellness issues that nearly killed them, only to come out on the other side shining like gold.

One of the most powerful stories I witnessed was that of Matt Olson, the cofounder and former executive director of ILFFPS. Matt Olson served as a lieutenant for the Bolingbrook Fire Department in Bolingbrook, IL until he recently retired. He is the vice president of the Associated Fire Fighters of Illinois. Matt truly is a beacon of light to many men and women who are wrestling with darkness, disconnection, and pain in their lives. But Matt was not always like this. As with many firefighters, Matt had seen his share of trauma on the job. (*Note:* Matt's story is also available as a podcast,[2] where you can listen to the interview in Matt's own words.)

For Matt Olson, 2012 was the toughest year of his life. His emotional pain began on January 1, 2012, when his mother was found dead in a bathtub. Her suicide was hard for him to process, and he dealt with his grief by staying busy. He helped with all the logistics of closing her estate. He sold her house, helped the family sort through her things, and stayed extremely busy at work. "Busy sort of got me through it," he explains.

Two months later, his grandmother ended up in the local emergency room, and he knew immediately that it was serious. She died within the next couple of days. Then while on duty, Matt attended a call in which an 11-year-old girl died. This specific call rocked him to his core. Rather than processing these events, he continued coping by staying busy and drinking more and more. This gradual progression was never intentional. He will tell you that he was just doing the best he could at any given moment.

A couple of months later, his wife's best friend died while he and his wife were on a trip. Later that year, his grandfather died, and it was not until his grandfather's death that Matt realized what a profound impact his grandfather had had on Matt's fire career and life. Matt's grandfather had been a firefighter in the Chicago Fire Department for many years. This influenced Matt's career choice, and from a young age all Matt ever wanted to do was become a firefighter.

The layer upon layer of pain made Matt shrivel inside. He was not okay, and it was starting to show. Next he had a close friend lose a four-year-old son. Matt had no answers. He had no sense of hope in the world, and in desperation, he became severely depressed and suicidal. When Matt reflects upon this time of life, he will say, "Things pile up. If we don't process them, it becomes a problem."

Matt would go to work and "try to be functional," but he felt fragile, damaged, and weak. He secretly wondered what was wrong with him. Although he felt strengthened by his department's perception of his strength, his pain was still there.

One evening, he and his wife took a walk near a peaceful part of a nearby town. As they walked, Matt just could not stop crying. For an hour, tears rolled down his face. He had spent the last few months upset, and he would drink, cry, feel sad, and not sleep. Eventually, his wife encouraged him to see a counselor.

He began to realize that his wife was right. He needed help. He had been feeling like nothing could cure the pain he had. He felt permanently stuck. Though he could easily reach out and hug his precious children, he began to think that was not enough to cure the pain he had. Well aware of how much it would hurt his children and his wife if he took his own life, he sought help and healing and went to a counselor.

His counselor helped him identify that he was depressed, and that depression is not forever. He explains, "Depression is a mood. It's a tough place to be *for now*." He continued going to see his counselor. Sometimes he would just sit and cry in her office, but often he would recognize truth and hope that had been hiding behind his pain. One of the main things he realized from going to counseling was that he may never forget the traumatic things he has seen, but he can process them so they become less invasive in his life. He can live his life intentionally making new and better memories by being present and enjoying the beauty of his kids, his wife, and the moments of life all around him.

Through counseling he became more connected with himself, and after about six months of counseling, he started to feel okay. He wanted all the firehouses that he visited to understand that "it's okay not to be okay." He made it his personal mission to "make it safe" for first responders to talk about their pain.

Matt began to recognize how his pain could become a beautiful part of his story, remaking him into someone better and somehow more fully alive than he had been before. The more he continued to seek healing, find connection with his wife and kids, and discover new hope in his life, the more passionate he became about others finding the same kind of healing. That is when his desire to start Illinois Firefighter Peer Support dawned.

If you rewound his story to the middle of the despair, however, and looked him squarely in the eye in the midst of his pain, he would not yet have been able to picture this chapter of his story. In the middle of the storm, it is easy to think that the only way to survive is to get out of it. But perhaps it is the process of choosing to walk forward in the midst of the storm, even while getting rained on, that makes you able to burst forth as bright as day when the storm is over. You can stand and say, "I'm here, and I'm not going down without a fight!"

Growing from the Pain

When you feel emotional pain, it is going to be uncomfortable. But even as you are crying and sadness settles in your core, you are healing. The only answer is to feel this uncomfortable feeling, all the while knowing and believing that it will not last forever. Find hope in between sobs because your pain has meaning, and you will grow from it. Someday you will see that this despair will change you in the most positive ways. This uncomfortable feeling has a purpose, and the purpose is learning, growing, and healing. When you choose to stay still and not distract yourself from your losses, allowing yourself to feel whatever feelings stir inside you, you are growing into a truly remarkable, compassionate, and emotionally intelligent person.

What if instead of seeing pain as a breaking process, we viewed it as a tunnel through which we must walk to find healing? Instead of trying not to talk about it, what if we made a point to talk about it openly within trusted relationships? What if we gave ourselves the dignity of a personal high-five every time we let ourselves cry about it?

Life can be full of momentary setbacks, traumatic experiences, disappointments, breakups, job losses, and other devastating misfortunes. But running away from them will never bring healing from them.

Though our natural reaction as human beings is to flee when we experience negative feelings, they are critical for shaping you into the future person you are going to be. In fact, according to Roy Baumeister, psychology professor at Florida State University, and others, many of life's most meaningful, most transformative experiences are actually quite painful.[3] This is not to say that we should allow ourselves to remain in depression or intentionally suffer long-term. But it does give us permission to allow ourselves to feel pain, fear, sadness, worry, and even loss without running away or solving our problem right away. In fact, perhaps the most powerful part of your story is how you learned to face these feelings in healthy ways and then stood in wonder as new life, hope, and resilience bloomed out of your dust.

Baumeister and others note that in contrast, when we avoid negative feelings, we forfeit the richness and depth of character formed in us as a result of experiencing them.[4] When negative feelings come, it takes emotional energy to push them aside and focus on something more positive. Instead of this avoidance, what if we took that opportunity to process, write down our story, or talk to a trusted friend? Baumeister and others have observed that facing negative feelings as they come results in greater mental clarity and focus for the individual. The mind is not divided by suppressed feelings that spontaneously surface. The biggest outcome of taking the time to feel and to think is meaningfulness. Happiness is based on current circumstances, but meaningfulness integrates past, present, and future experiences into the stable foundation of who you are and what your life is all about.

I think there would be many more first responders alive down the road who steered clear of suicidal thinking and action if they felt permission to walk through the tunnel of sadness and find healing, instead of sweeping emotion under the rug and letting it silently decay their souls. Meaning brings lifelong satisfaction and a strong sense of self, but it only comes from letting yourself think and feel.

A New Perspective on Life

The beautiful thing about trauma is that those who are forced to walk through it often gain a renewed perspective on life when they come out the other end of it. Psychologists call this *post-traumatic growth*. When an individual endures adversity or trauma, he often develops an entirely new perspective on life, his relationships, his possessions, and his future. These positive psychological changes significantly transform the way he thinks. From that point forward, his brain is reoriented to equip him to recover faster from future trauma or adversity, and he gains a greater appreciation for life.

According to retired psychotherapist and speaker Bill O'Hanlon, post-traumatic growth usually propels an individual's connection, compassion, and contribution ahead to live more fully than he ever did before the trauma. Post-traumatic growth (PTG) has a way of accelerating connection. O'Hanlon also notes that people who experience post-traumatic growth feel a new and refreshing sense of connection with themselves, others, or something beyond. They may say things such as, "I have a greater sense of closeness with others," or "I changed my priorities about what's important in life." One may even hear the occasional, "I love you man," slip from lips that were previously reticent of praise.[5]

Post-traumatic growth often awakens compassion for oneself and others. O'Hanlon observes that the individual may seek forgiveness for wrongs done, or he may finally decide to forgive someone who hurt him in the past. You will hear him say, "I realized how silly that was," or "It was finally time for me to let that go." He may even notice growth in himself and say something like, "I know better that I can handle difficulties," or "I'm stronger for it," according to O'Hanlon.[6] He may say, "I discovered that I'm stronger than I thought I was," or "I've really grown through this." He may even feel a sense of contented pride, thinking, "I'm not the same as I was before. This has broken pieces out of me that needed to go, and I'm a better person than I was before this happened."

In post-traumatic growth, the person who has experienced wounds or trauma begins to desire to help others or change the world in a positive way.[7] You will see this in individuals who start organizations to help others endure the same hardship they have been through—men and women who start support groups at church for cancer patients and their families, families of homicide victims who start victim support organizations to provide legal and emotional support through the process, or people who write hope-inspiring books as they come out of the shadows of grief and loss. You will see people like Matt Olson start peer support organizations or people like Ryan Elwood's family start organizations to help with suicide prevention because they want others to have powerful hope in their darkest moments.

People who grow through trauma find that the world is abounding with life, connection, beauty, and meaning. But it does not happen all at once. It starts with talking to someone about the pain. While looking at the pain without distracting yourself with being busy, you can choose to endure unmedicated until a glimmer of gratitude dawns. Gratitude will grow for all the good things that have come into this life because of the bad things that happened. Then as gratitude starts to shed some light in the darkness, life will become brighter and brighter until you are absolutely, overwhelmingly blinded by the radiance of the beautiful pieces of this new life you have—this unexpected life, this life after the storm.

As a mountain that was once consumed by wildfire sprouts, grows, and teems with life again, this individual has let the ashes of his past become the fertilizer for his new and abundant life.

Take heart, my friend. If you have been through a storm, keep walking. Keep talking about it. Keep pressing on toward healing because life after the storm is richer than you could ever dream.

Reflection Questions

1. What is a traumatic event that you have experienced?
2. Think about how far you have come since that event.
3. What are some of the ways you are different as a result of weathering that storm?
4. What is one thing you are proud of now that you are on the other side of that event?
5. If you have not reached a point where you are proud of being on the other side of a traumatic event, or if you are currently dealing with a traumatic memory, would you consider seeking professional help?

Notes

1. J. H. Evans, "Race, Religion, and the Pursuit of Happiness," in *Theological Perspectives for Life, Liberty, and the Pursuit of Happiness: Public Intellectuals for the Twenty-First Century*, A. M. Isasi-Díaz, M. McClintock Fulkerson, and R. Carbine, eds. (New York: Palgrave Macmillan, 2013): 13–22.
2. J. Sanders, "Interview with Lieutenant Matt Olson," *The Fire Inside*, episode 3, audio podcast, https://thefireinsidepodcast.com/003/.
3. R. F. Baumeister et al., "Some Key Differences between a Happy Life and a Meaningful Life," *Journal of Positive Psychology* 8, no. 6 (August 20, 2013): 505–516, https://doi.org/10.1080/17439760.2013.830764.
4. Baumeister et al., "Some Key Differences between a Happy Life and a Meaningful Life."
5. B. O'Hanlon, "Resolving Trauma without Drama," *Possibilities*, October 2016, https://www.billohanlon.com.
6. O'Hanlon, "Resolving Trauma without Drama."
7. O'Hanlon, "Resolving Trauma without Drama."

Chapter 11

First Responder Self-Care

Almost anything will work again if you unplug it for a few minutes, including you.

—Anne Lamott

Whether on or off the clock, first responders are conditioned to help others. Their days are filled with projects and tasks that benefit everyone else around them. While helping others is great, it is important that first responders understand the value of self-care and learn how to incorporate it into their lives.

Anthony's Story

Anthony was a client of mine who learned the value of self-care in his personal life. In his desire to help other first responders become emotionally well, he agreed to share his story in this book:

> Like many others in the fire service, I have struggled with maintaining a healthy mental state. Over the past five or so years, I have found that the quality of sleep that I get each night is the foundation for each day that follows. Both the quantity and quality of sleep were always a problem for me throughout most of my career. Whether it was the issue of not being able to fall asleep, waking easily, not feeling rested when I did wake, or simply not sleeping at all, it has been a long road to creating good habits for myself.
>
> The lack of sleep led to a multitude of issues ranging from being irritable, short-fused, or outright mean to my family, friends, and coworkers. The majority of my issues came at home. I would find myself counting the hours to go back to the fire house. The way that I treated my family was degrading at best, and neither my wife nor my children deserved to be treated that way. I knew how to be a fireman, and a good one at that. What I didn't know was how to deal and cope with the stress and emotions from work and being a husband or a father. It took an

accumulation of lack of sleep, being overstressed at home and work, and leaning on alcohol as a crutch, at times, to drive me to make good changes in my life.

At the nucleus of this change was sleep. The simple fact is that without quality sleep, I didn't have a solid foundation to build from. I needed it, and I didn't know how to get it. Thanks to a good support network of friends and a counselor, I was put on a path for developing good habits that would make lasting change in both my physical and mental health. Through experimenting with a multitude of different things, I found what has worked for me. They are all very simple changes that I added into my daily life.

First, I need to work out a minimum of four times per week. I simply classify this as getting my heart rate up and being active. It can be going for a hike, hitting the gym, playing hockey, or simply labor-intensive yard work. Staying active is a key in reducing the levels of cortisol that build up in my body over time. I notice a difference in my sleep on the days I am active compared to the days that I am not.

Second, I try to do something that is close to meditation. I have found the greatest impact is from doing yoga a minimum of once and as many as three times per week. Most fireman would find this to not be macho enough; however, after doing research on the health concerns associated with sleep loss, I was game to try anything. Restorative yoga was what I found to have the greatest impact. It is one that is not so much of a workout as it is a calm and quiet atmosphere where you hold stretches for three to five minutes at a time. It gives me a space where I can actively try to clear my head and focus on my breath. It sounds easy, but with the world we live in, it is more of a challenge than you would think. If I do not have time for that, my default is to go for a walk. I try to follow the same process of clearing my head and focusing on my diaphragmatic breathing. Same as with the workouts, I can see a noticeable difference in my patience and temper between when I stick to it and when I don't.

Third, my bedtime routine consists of writing in a journal, reading, and stretching or foam rolling. This idea came from my own life, recognizing how my children go to bed faster when we follow a simple routine every night, compared to letting them watch TV. Their simple routine consisted of no TV one to two hours before bed, reading multiple books, and lately we have added in kids' yoga, on occasion. I use writing as an outlet to get my thoughts from the day out, and it helps me stay mindful of my actions and choices. Reading books, rather than [watching] TV, helps prepare my body and mind for bed. There have been countless studies about how the light emitted from our phones and TVs impacts our brains' ability to turn off. Finally, I use foam rolling for 5 to 10 minutes as the last thing I do before bed. The use of a foam roller helps to release hormones that prepare the body to relax. It is like a poor man's massage, and the effects are very similar. This entire routine can take as little as one hour, and if I add in the yoga at the end of the night, it can go as long as I want.

This was a long road to get to a point where I am sleeping through the night on a regular basis. With every step, I saw a noticeable difference in my sleep and my mental state when I am awake. Each time I found something that helped, it was a small victory, and it gave me more motivation to continue. In contrast, if I was having a rough couple of days or a week, I knew what worked, and I would implement them slowly to get back on track. It wasn't uncommon for those ruts to last a month or longer in the beginning. As time went on, the ruts were less and less. I am to a point where, if I go more than three days in a rut, that is a long time. In all honesty this process took me five years to figure out, and it still isn't perfect, but my family is my why. Everything that is worth something is worth the effort.

First responders have a responsibility to care for themselves as the foundation of serving others. You cannot rescue and serve civilians effectively if you are depleted physically and emotionally. First you have to recharge yourself so you can be powered up and ready to serve others.

Even for those who understand its benefits, however, self-care is not always easy to achieve. There is a reason it is considered a discipline. Tami Forman, contributor to ForbesWomen, identifies that self-care requires tough-mindedness, a deep understanding of your priorities, respect for yourself, and respect for the people with whom you choose to spend your life.[1]

What Is Self-Care?

Self-care is about recognizing what you need, choosing to do what is good for your mind and your body, and being consistent and committed to helping yourself stay balanced in the midst of the demands of a busy career and life. One of the key pieces of self-care is intentionally recognizing the stressors that weigh on your mind and taking time to disengage and downshift from an amplified, adrenaline-rushed state to find balance and peace again.[2]

Self-care is about doing things that are actually good for you instead of following your momentary whims and doing what feels good in the moment, according to Forman. This requires first responders to give up the responsibility for the emotional well-being of other adults. Other adults are responsible for their own well-being, and you are responsible for yourself. Forman further explains that taking complete responsibility for your own emotional well-being means you choose to engage in life-giving activities and rest to care for your soul.[3] Self-care is not something you indulge in just to restore equilibrium in your life. Self-care is a core part of your vitality of life, and regular, habitual self-care is a central responsibility of every first responder.

Where Does Self-Care Start?

Self-care begins by paying attention to what rejuvenates your mind, body, and soul and to what depletes them. According to author Jill Metzler Patton, identifying what actually makes you happy and healthy helps you shift your schedule accordingly. By taking a critical look at the people with whom you spend your time, and the activities and responsibilities to which you are committed, you will begin to set priorities based on what enhances your life and what heaps stressors on your shoulders, observes Patton.[4]

Identifying what makes you feel good helps you shift your nutrition, exercise, and rest accordingly. According to Patton, as you pay attention to which foods make you feel better and which foods make you feel worse, you attune your awareness to be able to recognize what is good for you and what is bad for you. Patton notes that by paying attention to how you feel after a nap, you increase your ability to recognize how much rest you need.[5] In increasing awareness of yourself and your needs, you become better able to meet those needs in healthy ways.

In fact, I recommend going a step further and writing down what you have observed about yourself. Writing engages your brain on a deeper level than just thinking or even typing. Writing is incredibly therapeutic, and I frequently use it in my practice to help people sort through their thoughts. Patton believes asking yourself a few good questions while you are writing out your thoughts can help you break past the surface-level thoughts to go deeper into recognizing what does good for your soul.[6] This will help you recognize and implement ways to care for yourself.

For you, self-care could look like turning off the television instead of watching another episode of your favorite show, so you can get up and get to the gym in the morning. Self-care could look like declining the second drink (or maybe a first drink) at your buddy's get-together. Self-care could mean saying no to something you do not want to do, even if someone is going to be upset with you for declining. Self-care also means letting others take care of themselves.[7]

The Duty to Care for Yourself

Let us re-envision what self-care looks like in your mind. Erase your mental picture of a woman sitting in a nail salon getting a mani-pedi when you think of self-care. Instead, see self-care as a disciplined, courageous effort to help yourself become emotionally well and balanced. Acting on this new understanding will result in the best version of you for yourself, your family, your department, and your community. That is what Anthony did with his sleep habits, which spilled over into his other habits and breathed new life into his weary mind, body, family life, and other relationships.

Taking care of yourself is not selfish or vain. Taking care of yourself is your duty as a first responder so you can be equipped to take care of others who need you in some of the most trying moments of their lives. People need you, but you must be well physically, emotionally, and physiologically to be able to help others to the best of your ability.

Self-Care Practices for First Responders

The following steps are a good place to start when creating a personal self-care routine:

1. *Create a good work-life balance.* This involves taking time off to recharge and avoiding working long hours or even a second job.
2. *Exercise to relieve stress.* The benefits of exercise are infinite, especially for a first responder. Try a yoga class, lifting weights, taking a hike, high-intensity interval training (HIIT), running or jogging, or going for a walk. Having a regular exercise routine will increase feel-good endorphins and help combat the physical and emotional stress that comes with a first responder career. Try to exercise at least three to five times a week.
3. *Meditate daily.* Meditation is a great way to bring more mindfulness and peacefulness into your life. Quietly sitting for just 10 minutes a day can help control anxiety, reduce stress, and improve your overall mental health. It can aid in fighting addictions. Not great at sitting still? Meditate while you take a walk. Walking as moving meditation and restorative yoga are also effective.
4. *Surround yourself with healthy relationships.* First responders can truly benefit from a good support system. It is important to make meaningful connections with family, friends, neighbors, and coworkers.
5. *Laugh more.* Humor is good medicine when it comes to relieving stress and improving your mood. Watch a funny video, play with a pet, read a funny book—whatever helps you relax. Getting together with your favorite buddy who makes you laugh relieves stress.
6. *Connect with nature.* It is important to remember to utilize nature for its therapeutic benefits. Going for a hike, sitting by a fire, walking along a waterfront, or spending the day fishing can be helpful ways to reconnect with yourself and your thoughts.
7. *Get involved outside of work.* Take your mind off of work by investing your time in a hobby, such as woodworking, target shooting, camping, or hitting some balls at the driving range.
8. *Meet with a counselor.* Unfortunately, first responders are very susceptible to depression, addictions, and PTSD. A counselor can help you put things in perspective and identify coping skills. Seeing a counselor is beneficial for your mental wellness and your personal relationships, and it can improve your performance at work. You do not have to be in a terrible spot to go see a counselor. Everyone could use a guide sometimes.

Although all helping professionals are in danger of developing compassion fatigue, having a good self-care plan in place can reduce the emotional wellness risks that many first responders face.

Exercise for Mental Health

The importance of exercise in first responder fields cannot be overstated. Even though you are a busy first responder and would like to have a personal life, I would argue that the *best* version of yourself will be available to you, your family, and colleagues, *if you work out*. Making time can be difficult, but it matters. Exercise develops you physiologically, physically, and emotionally, and it breaks the power of stress in your life.

Chris Marella, full-time firefighter/paramedic and owner of 4th Shift Fitness, explains how stress affects first responders:

> When we begin to define stress as it truly is, we can start to handle it differently. Stress is quite simply a challenge to react and adapt to a changing situation. When we view it as such, we start to realize that everything is stress. Tying your shoes, walking to your car, getting fired, it's all stress! Stress in the fire service has the same spectrum—from discovering a dirty tool, to expecting to do one sign-off at a car accident and ending up with four, to any high-risk fireground rescue operation.
>
> Now the level of reaction to that stress can certainly affect us mentally, emotionally, and physically. When you encounter stressors, chemicals start flying. Stress receptors in your brain are shooting signals, activating all kinds of hormones controlling heart rate, respiratory rate, blood pressure and blood sugar, just to name a few. All these things culminate in the face of elevated stress as the "fight-or-flight" reaction. So, you either grab your spear and kill the lion or you throw your hands up and run away screaming, except that neither of those reactions is appropriate in today's society. So, we end up suppressing one of the most powerful and primal natural reactions we have in an effort to remain composed and useful in the back of the ambulance or on the fireground. Now, you have all sorts of crazy corrosive and high-power chemicals pumping through your system, and all you can usually do is sit there and continue to think about how stressed you feel.[8]

The problem for firefighters is that you are often called into situations where you can only do so much. Your body's fight-or-flight chemicals are firing, but you may be prevented in some way from being able to fully help the patient or even yourself. When this happens, those survival chemicals just sit in your muscles, and it mimics the freeze response. This becomes one of the most harmful physical situations when it comes to your mental state. (See chapter 7 for more information on what happens in your body and specifically in your brain during trauma, including the freeze response.) Movement is the core piece of emotional wellness because you need to use your stress hormones, not let them sit in your muscles.

Psychologists have begun researching how stress hormones and PTSD are linked. The benefits of mind-body intervention (MBX), which is a program similar to yoga and deep breathing, as discussed previously in chapter 7, have been studied with nurses employed by the University of New Mexico Hospital. The MBX group of nurses in the study who had reported struggling with PTSD symptoms showed reduced cortisol serum levels in their bodies *and* reduced symptoms of PTSD after eight weeks of twice-weekly MBX sessions.[9]

Clearly, stress is both mental *and* physical. Our bodies are vessels of stress hormones. We can either use them up as we fight to rescue people and fight fires, or we can get thwarted and saturate our bodies with them. Either way, our bodies require physical movement to remain emotionally and physically healthy.

In addition to removing the power of stress in your life, there are three ways exercise improves your performance as a first responder:

1. It develops physiological toughness.
2. It extends your physical health and longevity.
3. It establishes your emotional wellness.

Physiological Toughness

When you exercise, you build what psychologists call *physiological toughness*. Basically, your body is better able to ramp up its fight-or-flight responses, pump blood more efficiently, breathe more effectively, and make the best use of your strength to tackle a physically demanding call. Exercise is the means by which you train your body for appropriate physiological responses to stress. It teaches you to regulate yourself by powering up effectively and powering down effectively when needed.

Your body responds to an intense situation by ramping up the chemicals in your brain and muscles to empower you for action. Specifically, when your body gets ramped up, your brain and body release noradrenaline (norepinephrine), your body releases adrenaline (epinephrine), and your brain releases dopamine. All of these actions work together simultaneously to help you act to the best of your ability to resolve the situation. Likewise, your body releases cortisol to mobilize your energy stores, trigger your central nervous system, and inhibit your immune system. An excess of cortisol actually inhibits your performance. But did you know that you can train your body to respond with maximum adrenaline, noradrenaline, and dopamine and only a limited amount of cortisol?

Exercise builds physiological toughness in specific ways. Studies show that individuals who willingly undertake stressful situations in which they have control, such as swimming in cold water, show rapid and intense spikes in adrenaline in response to stressful tasks.[10] These adrenaline spikes tend to enhance performance. They also show a rapid rise in noradrenaline and a limited increase in cortisol, so their bodies experience a rise in heart rate, sharpened mental activity, relatively constant blood pressure, and efficient conversion of fat to energy, which increases muscular activity. All of this occurs without the inhibiting effects of an oversupply of cortisol.

Physiological toughness can be built through three factors, according to psychology researcher and lecturer Simon Moss:

1. *Exposure to a stressful event in a controlled environment.* For example, one study monitored individuals exercising in a cold pool.
2. *A coinciding sense of control over the stressor.* One example would be if the individual is allowed to choose when to jump into the pool and to determine how long to stay in the pool.
3. *Adequate recovery time.* A sufficient period of time is allowed for recovery before the individual has to repeat the activity.[11]

These three factors, when repeated, establish a pattern of physiological response to subsequent challenging events, which enhance a person's emotional stability and performance.

Human bodies are amazingly adaptive. Your mind and body can learn to be better at dealing with intense events based on the things you expose yourself to, including positive training experiences. If you experience stressful events on the job, but you have some degree of control over them, and you have sufficient recovery time afterward, this same development of physiological toughness can occur, which is known as *passive toughening*. More likely, however, you will have control over your own exposure to stressful/challenging things when it comes to the things you intentionally push yourself to do, like exercise, which is *active toughening*. Childhood experiences that individuals have adequately recovered from can also create physiological toughness, according to research conducted in 1980 by Marianne Frankenhaeuser, Ulf Lundberg, and Lennart Forsman, University of Stockholm, and in 1989 by R. A. Dienstbier, University of Nebraska–Lincoln.[12]

I would argue that building physiological toughness goes beyond swimming in a cold pool to include any sort of challenging exercise that you have control over and adequate recovery time. As you take yourself outside your comfort zone and give yourself adequate recovery time, you will actually become better at your job. According to Chris Marella, firefighter paramedic:

> By giving yourself a physical outlet, we are able to activate your stress response, take it to the limit, and follow through with action! We get to use all those gnarly chemicals in physical display. By putting your stress response through its paces, you also build a little bit of "stress armor" every time. Imagine that each workout is a brick. Every time you hit that level of intensity, you add a brick to the wall that protects you from being "stressed out."[13]

Researchers Jim Blascovich, University of California–Santa Barbara, and Joe Tomaka, New Mexico State University, note that individuals with physiological toughness usually show less fear or avoidance in stressful contexts in the future, and they usually perform better. Emotionally, as they think about the event they are about to face, they tend to conceptualize these events as challenges rather than threats. According to Blascovich and Tomaka, the elevated levels of adrenaline somehow prevent fearful or avoidant responses.[14]

If you were thinking that exercise does not make that big of a difference, think again. You will be a better firefighter and physiologically better able to meet the demands of the job if you push yourself to exercise. On top of that, you will be less likely to experience harmful cortisol dumps, which are usually linked to PTSD and negative emotional outcomes.

Physical Health and Longevity

Not only does exercise increase your physiological toughness, but it builds your physical longevity. Your family will thank you for this.

Some forms of exercise and deep breathing help you activate your parasympathetic nervous system (PNS), which is responsible for rest, digestion, and a general sense of well-being. When you are constantly in a state of being stressed or aroused for action, your sympathetic nervous system (SNS) takes over. After emergencies are over, the SNS can continue to be activated by flashbacks, memories, or a continual state of arousal. Chronic activation of the SNS can create mental, emotional, and physical health problems. Symptoms of PTSD are associated with overactivation of SNS.

When the SNS is overactivated, it can result in cardiovascular problems. These problems can take many forms, such as hardening of arteries and heart attacks, gastrointestinal issues of all sorts, including IBS, chronic diarrhea, and constipation, immune system weakening, and endocrine system issues such as type 2 diabetes, sexual impotence, and decreased longevity.[15]

By contrast, when you do low-intensity exercise like walking and yoga, you transition yourself back to letting your PNS run the show. Some forms of exercise, like running or high-intensity interval training, activate the SNS, while low-intensity exercise activates the PNS. This is the state that your body should be in most of the time. On top of that, your exercise decreases your risk of heart disease. Certainly you see accidents and crisis situations all the time as a first responder, but did you know that actually the most common killer in America is heart disease? When you exercise, your body pumps more blood through your heart and strengthens the heart muscle. It also decreases your blood pressure by empowering your heart to dilate its blood vessels more effectively, making your heart less likely to become congested and your arteries clogged.

Exercise builds your physical strength, restores your body to the right balance of the nervous system, and helps you live in physical and emotional wellness.

Emotional Wellness

Exercise helps your body get rid of excess stress hormones that can cause negative thoughts and anger, so working out helps you become emotionally well.

In first responder careers, many first responders deal with chronic stress. This means that their bodies are saturated with the stress hormone cortisol, which begins to wreak havoc on the functioning of their bodies and affects their mental state. One of the most important things individuals can do to overcome the emotionally taxing side of being a first responder is to physically work the cortisol out of their bodies through exercise, stretching, and breathing.

Studies have proven the mood-boosting benefits of exercise, as well. Even 20 minutes of moderate to intense exercise will stimulate your brain to release endorphins and dopamine, which make you feel good. It also works the cortisol out of your muscles to destress you. According to Gary Cooney, Division of Psychiatry at Royal Edinburgh Hospital, and others, numerous studies and meta-analyses of the studies have demonstrated that exercise has long-term benefits for those who struggle with depression and anxiety. Cooney and others also noted that many researchers see exercise as an alternative to antidepressant medication.[16] In older adults, one study compared antidepressant drugs to exercise and found exercise to be equally effective in mitigating the symptoms of depression.[17]

Exercise in Overcoming Addictions

As I mentioned in chapter 4, when firefighters experience trauma or uncomfortable feelings, many of them may begin to experience flashbacks or triggers that drive them to substances to numb their pain. Rather than taking care of themselves and talking about their pain, they stuff it down and neglect self-care. This often leads to addictions. According to Jonice Webb and Christine Musello, authors of *Running on Empty*, individuals struggling with addiction are "masters of self-neglect," and in order to recover, they must learn "consistent and effective self-care."[18] For example, exercise can reduce the incidence of relapse. Learning to successfully live with uncomfortable feelings and making healthy choices that support one's physical and emotional well-being can help people overcome addictions and deal with their pain because exercise releases dopamine, which makes the mind feel better.

Self-Care for Anger

Often our focus is on the negative aspects of anger and the need to control our anger. What if we focused instead on ways we could leverage anger for good?

Anger is often an indication that something deeper is going on. Often something we have seen has struck a chord with something that happened to us earlier in the past. But what if, instead of letting that make us defensive, depressed, or anxious, we let it motivate us to make a positive change? If we view anger as a state of energy in response to an issue in our lives that needs to be examined, we can start to leverage those feelings to help us speak up against injustices or take action to make improvements.[19] The key here, however, is making sure the anger is directed at the right thing, at the right time, and in the right way so that we do not let our anger propel us into overpowering and brutal displays toward the wrong person or thing.

Therapist Laurie Ure believes that "chronically suppressing anger contributes to a variety of health problems, including constipation, headaches, and high blood pressure. It also often underlies conditions like depression and anxiety, or behaviors like substance abuse."[20] So how do we redirect anger in the right way?

The first step is finding safe, nondestructive ways to discharge that energy physically. For first responders, I recommend exercise first and foremost. If your muscles are filled with adrenaline and ready to fight, you may find it helpful to lift weights, do high-intensity intervals, go for a run, or even take a walk. Movement will help you use some of that adrenaline to bring your brain from fight mode back into the prefrontal cortex, where you can use logical, controlled thought.

Another alternative for discharging your energy so you can leverage anger for good is finding other outlets for expressing your anger. These could include playing an instrument, writing down your thoughts on paper—especially any angry sentiments—or even playing a game of basketball or golf.

Your mind and body are intricately and powerfully connected. If you have been experiencing crippling anger, perhaps safe, nondestructive physical movement could help you work it through your muscles toward a place where you are better able to express what is going on that is frustrating you. Then you can make a change in your life that is much needed.

Sometimes you will find that you are angry about something that is appropriate for you to try to influence. That is a great opportunity for you to have a constructive conversation to change what you can control, but only after you have discharged some of that angry energy. At other times you will find that something is beyond your control, in which case letting it go may be your only option. This can be frustrating, but your anger does not need to sit in your muscles. Every time you find yourself bothered by the situation, take action—for example, work out, write, or make music. Then maybe you can find the right words to use in the right way and at the right time to make a positive change in response to your anger.

Anger is energy. Instead of letting that energy burn you from the inside out, burn off that energy and then come up with a constructive way to use it for good.

Nutrition for Mental Health

Eating nourishing food is a key part of self-care, and it is a core component of having a healthy mind. Many of today's mental health issues are exacerbated by poor nutrition, resulting from the surplus of processed foods in today's standard American diet. Thankfully, psychiatrists are beginning to incorporate nutrition and supplements into their approach to treat a number of mental health disorders. Perhaps the best news about this approach is that nourishing foods, vitamins, and minerals can lead to lasting changes in mental health from the inside out. Instead of turning to antidepressants, people can restore their body's balance naturally, without the negative side effects, stigma, and potential for dependency on prescription medications.[21]

Though the field of nutrition and mental health is still young, researchers are certain that nutrition impacts brain health. UK researchers Lauren Owen, University of Central Lancashire, and Bernard Corfe, University of Sheffield Medical School, explain:

> Both cross-sectional and longitudinal studies have shown that the more one eats a western or highly processed diet, the more one is at risk for developing psychiatric symptoms, such as depression and anxiety. Conversely, the more one eats a Mediterranean-style diet, the more one is protected from developing a mental disorder.[22]

Owen and Corfe note that this is partly because the hippocampus in the brain, which is the location of learning, memory, and mood, is where new neurons are formed. As more new neurons formed, through a process called *neurogenesis*, the memory and mood generally improve. They also point to evidence that suggests that *gut microbiota* (the microorganisms in the gut) have a strong impact on the neurotransmitter activity in the hippocampus of the brain.[23] This is one primary reason why pursuing good nutrition, specifically good gut health, can help balance the brain's mental health.

The first step in reducing inflammation in the body is to eat real food, avoiding foods that are made in factories and focusing on foods that are grown on farms. According to physician and biochemist Catherine Shannahan and coauthor Luke Shannahan, two of the biggest contributors to harmful inflammation in the gut are bad fats and processed foods. Bad fats include canola oil, soy oil, sunflower oil, cottonseed oil, corn oil, grapeseed oil, safflower oil, and nonbutter spreads (including margarine and the so-called trans-fat free spreads) because they easily change molecular structure in the presence of heat, forming free radicals and toxic compounds. Shannahan and Shannahan note that processed foods that promote gut inflammation include margarine and spreads, store-bought salad dressing, rice milk, soy milk, soy cheese, soy-based meat, breakfast cereals, french fries, crackers and chips, granola, soft breads, buns, and store-bought muffins. These often are high in unhealthy oils, sugar, and bad fats, which can impair vitamin absorption and promote disease.[24]

In addition to avoiding foods that increase inflammation in the gut, you should also avoid foods that create hormonal imbalances in your body. When you consume sugar, caffeine, excessive alcohol, cigarettes, or other drugs, your body ramps up its fight-or-flight response, sending adrenaline and cortisol into your bloodstream to empower you for action. The problem is that you now have stressed out your system, and your mind is working overtime not to become anxious or depressed.

If you are a big coffee drinker, you may want to shift to decaf or other options. This may be a hard shift to make, but you should consider it a shift in the direction of self-care. Here are some ideas. Instead of having coffee first thing in the morning, consider having green or black tea or even decaf coffee, all of which have less caffeine. Or at least wait an hour when you wake up in the morning to let your body ramp up its own natural cortisol and adrenaline for the day before you stress it out with added caffeine. Then try to limit your caffeine intake to one cup of coffee. Consider the same idea for sugar. When there are treats at work or home, decide beforehand to limit yourself to one. Excessive sugar intake stresses out your system and can create imbalance in your mind, driving you toward anxiety and depression. Nutrition is a core component of taking care of your body and mind.

Many nutritionists believe that historical eating is more beneficial than our modern way of eating. This is because historically, food sources included a wider variety of fruits and vegetables, which people ate seasonally, and included a better balance of high-fat seasons and low-fat seasons without the fad diets.[25] Shannahan and Shannahan note that foods our grandparents ate tended to have higher concentrations of vitamins, minerals, healthy fats, and probiotics that naturally helped our bodies and minds function optimally. Historically, some of the healthiest societies ate fresh meat, often slow-cooked on the bone, broths and stocks cooked with bones, fermented and sprouted foods, offal meats, which are extremely high in nutrients that are missing in today's diet, and fresh vegetables and fruits.[26] A general rule of thumb for improving your nutrition is to try to eat the way your grandparents ate and to shift toward a more Mediterranean diet.

Vitamins and supplements can also help with restoring mental health. One study on women in their 20s and 30s found that the following supplements reduced the likelihood of developing major depressive disorder:

- *Omega 3*. These healthy fats increase the reception of neurotransmitters in the brain. Not getting enough essential fatty acids can contribute to low mood, cognitive decline, and poor comprehension.
- *Folate and vitamin B_{12}*. These are both needed for the body to be able to metabolize serotonin and other monoamine neurotransmitters and catecholamines. Getting enough of these means a healthier presence of feel-good neurotransmitters and better ability to calm back down after an influx of stress.
- *Vitamin C, vitamin E, and carotenoids*. Because the brain consumes oxygen more than most parts of the body, it is susceptible to peroxidation, which can damage neurotransmitter function and central nervous system function. Antioxidants including vitamin C (ascorbic acid), vitamin E, and carotenoids stop or slow the oxidation, promoting healthy neurotransmitter function.
- *Selenium*. Selenium is an important regulator of mood. Low selenium is related to higher occurrence of depression and hostility.
- *Iron*. Not having enough iron decreases synthesis and communication of neurotransmitters. Low iron is associated with declined memory, concentration, and brain function as well as fatigue, irritability, and apathy.
- *Zinc*. Zinc can regulate neurotransmitter activity and may even act as a neurotransmitter, carrying messages within the brain, which can improve healthy communication within the brain.[27]

This is a general list for young women, but these nutrients can also have positive impacts for other populations. I suggest talking to your health-care provider about adding a food-based multivitamin to your daily habits to help ensure that you are getting the nutritional building blocks needed for healthy cognitive function.

As you can see, self-care is about layer upon layer of giving yourself the respect of meeting your own needs. When you exercise and nourish your body, so many of your physical and mental systems come back into order. When you give yourself the respect of taking time to unwind, you restore your sense of well-being and your body's ability to function optimally.

If I were standing next to you each day in your department, I would tell you, "You matter. Give yourself what you need, so you can be at your best." That might mean changing what you eat for lunch every day. Going from a burger and fries to a sandwich and salad can make a big difference in your physical well-being, your body's inflammation, and your brain health. It might mean giving yourself a break from drinking alcohol for a while, so you can get better REM sleep and begin to restore more mental balance. Or it might mean setting firmer boundaries and saying no when your friends are going out because you just need some time to rest. I respect a first responder who puts his needs on the schedule instead of pushing them off to the side. Your needs matter. What is one thing you can do today to take better care of yourself?

Nature for Mental Health

In chapter 1, I discussed the power of nature in building environmental wellness, which is a necessary part of being a connected, whole individual. Connection with nature is a key part of self-care because nature shifts us away from the stressors of a concrete society. It physically, mentally, and spiritually refocuses us on life, gratitude, and a higher power.

Intentional Actions for Happiness

Your body is wired to help itself come back to balance. If you have been experiencing depression, anxiety, or other imbalanced thinking, you can intentionally take actions to help your brain bring you back to balance. Author Emily Swaim offers the following 10 ways to boost your dopamine and serotonin naturally:[28]

1. *Exercise.* Working out for 30 minutes or more per day of aerobic exercise boosts dopamine levels in the brain. For even more of a boost, add a novel twist and do a "zombie run" or "color run."
2. *Time in nature.* Spending time in nature for a minimum of two hours per week can bear amazing fruit in your life. Even just five minutes a day in nature can boost your mood, motivation, and self-esteem.

3. *Nutrition.* As we mentioned before, good nutrition, including omega-3s and tryptophan can boost serotonin.
4. *Meditation.* Practicing meditation or prayer every day boosts dopamine and develops focus, relaxation, and controlled breathing.
5. *Gratitude.* Writing down three things you are thankful for every day of the week will improve dopamine production, which will boost happiness and lessen depressive symptoms.
6. *Essential oils.* Your sense of smell can trigger serotonin and dopamine to release, especially with plants like bergamot, lavender, and lemon. So if you have some oils, consider diffusing a few drops.
7. *Happy memories.* Dwelling on happy memories rather than sad memories increases serotonin. Think about one of your proudest moments, and your brain will release serotonin as if you were living that moment all over again.
8. *Novelty/new things.* By seeking out something you have not done before, you are triggering dopamine to release. The newer the experience, the more dopamine. Traveling or touring someplace new—whether far away or close to home—can also trigger this dopamine release.
9. *Therapy.* Talking to a trusted friend or counselor can improve your mood and your body's ability to create serotonin.
10. *Flow.* Flow is a concept coined by Mihaly Csikszentmihalyi, University of Chicago, to describe the happiness we feel when we are so mentally satisfied and focused on our activity that we lose track of time.[29] Flow often occurs during a favorite hobby, while exercising, or even at work.

You can also find happiness by using your unique strengths in your daily life. According to researchers Ed Diener and Martin Seligman at the University of Pennsylvania, people can find greater happiness in life by working within their unique signature strengths.[30] (See chapter 13 to read more about incorporating your strengths into your life.)

Many of the self-care practices listed earlier in this chapter benefit the brain, the body, and many areas of our lives. That is why I continually recommend them to my clients, and I encourage you to try them also.

Self-Care Actions That Are Also Treatments

In my counseling practice in Sugar Grove, IL, I invite first responders to try some of the following activities both in our sessions and at home to begin healing. Feel free to incorporate them into your daily/weekly routine as much as possible.

Moving Meditation

One of the practices I recommend to clients in my practice is *moving meditation*. Basically, it means you go for a walk to give yourself some space to think through a problem/

concern or intentionally not think about anything at all. Sitting and meditating can be difficult for first responders, who are naturally driven and active people. Walking gives your body something to do while your mind begins to sort things out. You give yourself the opportunity to notice the world around you and listen to the sounds while you process subconscious thoughts.

When you walk, your body releases serotonin, a neurotransmitter that helps you feel better. This can immediately begin to combat depression. Walking activates your parasympathetic nervous system, the nervous system in your body that helps you be calm, digest well, and rest, as explained previously. This is different than going for a run or taking on an aggressive workout. Working out aggressively has its own benefits, but when it comes to meditation, high-intensity exercise can actually activate the sympathetic nervous system. Working out at a higher speed is great for releasing dopamine and endorphins, which are other types of neurotransmitters associated with feeling better. I do recommend 20 minutes of cardio as a way of de-stressing, but that is not what I am talking about here.

Moving meditation is conscious rest. When your arms and legs move, it triggers bilateral stimulation of your brain. This means you are using both sides of your brain, so you can use logic and creativity to make sense of a situation. This can help you find better clarity on the situation. If you add intentional breathing—for example, inhaling for five counts, holding your breath for five counts, and exhaling for five counts—the cortisol in your body will begin to be worked out through your lymphatic system. Mindful meditation gives you space to think, coupled with physiological healing processes that your body naturally does while you are walking.

So your assignment is to go for a walk today and start thinking through some of your circumstances and concerns. Or better yet, do not think at all and focus on nature and turn the experience into moving meditation. Your mind will be working even if you are not actively trying to think. This kind of thinking is less painful than you might imagine. The serotonin released in your brain will help you work through your problems, making them less overwhelming.

Slowing Down Your Breathing

Both on and off the job, how you breathe sends a sequence of messages to your body, which can either ramp up your stress responses or calm them down. Your goal is always to shift yourself back into your parasympathetic nervous system whenever possible. That is where you should live most of your life. For firefighters, whose sympathetic nervous systems get activated regularly on the job, practicing mindful breathing should become a part of your daily habits to help you be emotionally well.

To start breathing intentionally, one easy pattern is to make your exhalations longer than your inhalations. Try this: inhale to a count of four, then exhale to a count of six. If you are worked up, you may notice that four is about as long as you can go. As you calm, gradually go longer, and try to inhale for six and exhale for eight counts, and so on.[31]

The reason intentional breathing is so important is because when you exhale, your vagus nerve—the largest nerve in your body, which runs from your neck to your pelvis—

releases a chemical to slow your heart rate. This neurotransmitter by-product is called *acetylcholine* or ACh. When you breathe out, ACh signals for your heartbeat to slow down.[32]

Your heart rate is very much determined by your sympathetic and parasympathetic nervous system. This balance is what keeps your heart rate in a healthy range. Controlling your breathing increases your *heart rate variability*, a reflection of your physical and physiological health.[33] Being intentional with your breathing is thus a key part of recovering from an intense situation. It is a simple but critical part of daily self-care for a first responder.

Yoga for PTSD

Yoga offers innumerable benefits for those sorting through trauma and PTSD. I personally have fallen in love with the practice of yoga because of all the amazing stress-reducing, mental refocusing work it provides to me. Beyond my personal life, I have seen what it can do for struggling first responders. Now some might think, "Yoga is for girls," but yoga has actually been around for thousands of years and is enjoyed by both men and women. I have witnessed firsthand how it can benefit first responders.

A few years ago, I attended a conference, "Yoga to Treat Trauma," with Catherine Ashton and fell in love with the benefits of yoga for those struggling with PTSD.[34] Since then, the practice of using yoga and meditation to help first responders and veterans with PTSD has grown exponentially, and it is now becoming a normal part of the way that counselors, social workers, yoga instructors, doctors, the VA, and other practitioners treat PTSD and other emotional wellness issues. It is a holistic and effective avenue of treatment.

According to the National Center for Complementary and Alternative Medicine, yoga is "a mind and body practice with origins in ancient Indian philosophy. The various types of yoga typically combine physical postures, breathing techniques, and meditation or relaxation."[35] For the sake of simplicity, I will refer to yoga classes that incorporate breathing exercises, postures, meditation, and guided imagery as *restorative yoga classes*. Meditation is often intricately linked to the practice of yoga:

> The National Center for Complementary and Integrative Health (NCCIH) classifies meditation and yoga as complementary mind and body health approaches. There are many forms of meditation, some of which teach practitioners to observe thoughts, feelings, and sensations in a non-judgmental manner. Mindfulness meditation, for example, teaches participants to orient their attention to the present with curiosity, openness, and acceptance. Experiencing the present moment non-judgmentally and openly may encourage practitioners to approach rather than avoid distressing thoughts and feelings, which may reduce cognitive distortions and avoidance.[36]

Autumn Gallegos, University of Rochester Medical Center, and others have noted that *mantra meditation*—repeating a phrase or belief to reinforce it mentally—helps reprogram the mind and helps individuals with PTSD move forward. Gallegos and others also have recognized that yoga and meditation work hand-in-hand to help the mind and body experience and express the emotions of PTSD in healing ways by addressing the various facets of PTSD.[37] When individuals incorporate these restorative yoga practices into daily life, they can begin to see a more complete resolution to the physical and mental symptoms of PTSD.

In addition to restorative yoga, the two other types of yoga that have been shown to be maximally effective for helping those with trauma and PTSD are *trauma-sensitive yoga (TSY)* and *yoga nidra*. Caitlin R. Nolan, researcher and instructor at Florida State University, points to research conducted by David Emerson and his colleagues at the Trauma Center of the Justice Resource Institute in Brookline, MA. Emerson and colleagues developed trauma-sensitive yoga techniques that are designed to reduce triggers that may be experienced in a typical yoga class by those with trauma and PTSD. For example, instructors do not do hands-on corrections in a TSY class. Also, they guide students through a series of poses that end with a traditional rest (*Savasana*), during which students can rest with their eyes open or even sit with their eyes open or closed to avoid feeling vulnerable and unsafe. Nolan notes that for trauma survivors, this emphasis on only doing things they feel comfortable doing with their bodies is immensely powerful for giving them control over their bodies and reconnecting with themselves.[38]

Neuroscientist S. Mayanil, who has 45 years of practical experience with regular yoga and *dhyana* (meditation), believes that the specific practice of yoga nidra provides the transformational mind work necessary to reverse cortisol's negative effects and mitigate some of the symptoms of PTSD.[39] Yoga nidra is becoming increasingly integrated into the practices of social workers and counselors because it creates a REM-like state in the brain, although the individual is still fully conscious.[40] Yoga nidra has been used effectively on veterans with PTSD and on a range of other emotional wellness issues like anxiety, depression, stress, addiction, grief, and insomnia.[41]

These three types of yoga—restorative yoga, trauma-sensitive yoga, and yoga nidra—are uniquely powerful for first responders struggling with anxiety, depression, trauma, or PTSD. If you have experienced a traumatic event and are struggling, I highly encourage you to try one of these three types of classes. I wholeheartedly believe you will experience, as I and many of my clients have, that yoga can accelerate the healing work you are trying to do more powerfully than just psychotherapy alone. There are a number of studies that have demonstrated the efficacy of yoga for those with traumatic experiences.

Science Supporting Yoga and Meditation for Trauma and PTSD

Researchers believe in the power of yoga and meditation for mitigating symptoms of anxiety, depression, and PTSD.[42] One study conducted by Bessel van der Kolk, Boston University, and others on 64 women in the US suffering from chronic PTSD resistant to standard psychological treatments found that "trauma-informed yoga" significantly reduced PTSD symptomatology. The study divided participants into two randomized groups. The first group was assigned to a weekly hour-long session of "trauma-informed yoga," while the control group attended "supportive women's health education" classes. At the end of the study, half of the trauma-informed yoga participants no longer met the qualifications for having PTSD. Of those in supportive women's health education, only 6 of the 29 saw enough change to qualify as no longer having PTSD. Even though both groups showed significant decreases in PTSD about halfway through the study, the trauma-informed yoga group was the one that maintained this permanent reduction in PTSD symptoms over time. The researchers concluded that yoga significantly reduces PTSD symptomatology in the same degree as other psychopharmacological treatments.[43]

Citing studies by other researchers, van der Kolk and colleagues noted, "The extensive research on the effects of mindfulness meditation [shows that it] has been demonstrated to positively influence numerous psychiatric, psychosomatic, and stress-related symptoms, including anxiety, depression, chronic pain, immune function, blood pressure, cortisol levels, and telomerase activity."[44]

As mentioned in chapter 7, a study on ER nurses with PTSD symptoms conducted by Sang Hwan Kim, Clinical Center, National Institutes of Health, and others described a similar outcome. One group of the nurses attended eight weeks of mind-body intervention (MBX) classes, which were similar to yoga and meditation and included stretching, balance, and breathing, with a focus on mindfulness. Those who participated in the classes saw a significant reduction in PTSD symptoms, as well as a reduction in serum cortisol in their bodies. Kim and others also noted that the participants reported improvements in sleep, stress resilience, energy levels, emotional regulation under stress, and resumption of pleasurable activities they had previously discontinued.[45]

Why does yoga and other exercise help with PTSD? Remember that when you are under stress, your body releases cortisol. When cortisol sits in the body, it can wreak havoc on the mind, as is often the case with those with PTSD symptoms. According to Mayanil, cortisol can also destroy the immune system and make the body vulnerable to foreign pathogens. It literally knocks you down from the inside.[46] When people can work cortisol out of their bodies through stretching, breathing, and mindfulness meditation, it resets their mental and physical wellness.

A clinical trial was performed by Troels Kjaer, John F. Kennedy Institute, Denmark, and others to analyze the impacts of yoga nidra on the brain. They found that the brain released endogenous dopamine during yoga nidra meditation, as a result of the brain shifting from everyday thinking to active meditation.[47] The results of the study by Kjaer and others suggest that this specific form of relaxation and meditation has the ability to help you regulate your conscious state, impacting your brain all the way down to the synaptic level. That is why it is a powerful tool for first responders who are looking to reduce stress and regain control of traumatic memories.

Another study performed on 16 male combat veterans from the Vietnam and Iraq Wars found that one yoga nidra class per week reduced rage, anxiety, and emotional reactivity and increased self-awareness, self-control, self-efficacy, peace, and relaxation.[48] As you can see, trauma-informed yoga, mindfulness meditation, and breathing, as found in restorative yoga and yoga nidra, are all effective in reducing cortisol and helping individuals with traumatic memories reduce PTSD symptoms.

What Occurs in the Body During Yoga and Meditation?

Physiologically, yoga, deep breathing, meditation, and relaxation, as found in restorative yoga, trauma-sensitive yoga, and yoga nidra practices, help decrease the physiological arousal experienced by those with PTSD.[49] In addition to calming the stress response, the brain and gut secrete beneficial neurotransmitters in response to yoga and meditation. Those struggling with addiction and trauma benefit from restorative yoga because it increases production of the neurotransmitter gamma-aminobutyric acid (GABA),[50] which is statistically low in people with anxiety, substance abuse, and depression.[51] Yoga nidra has been shown to increase the release of endogenous dopamine levels up to 65%.[52] Boosting dopamine through restorative yoga helps first responders gain a natural sense of well-being. Yoga and meditation help people come back to physiological balance.

Physically, restorative yoga rebalances the nervous system. In an emergency situation, the sympathetic nervous system takes over, increasing the heart rate, dilating bronchial tubes, and contracting muscles to mobilize for fight or flight. It also dilates pupils; decreases digestion, saliva production, and urine output; releases adrenaline; and increases glycogen and glucose conversion, all to empower the body to fight.[53] First responders often live in a constant state of emergency, living with the sympathetic nervous system governing the body, which can cause cardiovascular problems, digestive issues, a weakened immune system, and endocrine issues, in addition to emotional wellness issues like anxiety, depression, and addiction.[54]

In order to bring the body back into balance after a stressful event, the parasympathetic nervous system needs to kick in as the primary executor of the body's functions. The PNS is the "rest and digest" nervous system; it lowers blood pressure, stabilizes blood sugar, increases energy, and improves sleep. Knowing how to activate the PNS is a critical part of emotional wellness for first responders. The stretching, meditation, and deep breathing used in yoga tell the body that it is not in an emergency situation, engaging the PNS.[55] Nolan explains,

> Breathing is essential to survival; thus, information the brain receives regarding respiration is given top priority. Slow, rhythmic, and controlled breathing sends information to the brain that positively impacts perception, cognition, emotion regulation, and behavior through the stimulation of the vagal nerves, which act as the primary pathway of the PNS.[56]

As mentioned previously, the vagus nerve is one of the major nerves connected with your body's sense of wellness. It runs from the brainstem to the major organs of your body and regulates heart rate.[57] Doing restorative yoga, which includes intentional breathing, helps improve function of the vagus nerve so that it can "work like a well-played instrument [to] calm your heart rate and lower your blood pressure."[58]

What Occurs in the Brain During Yoga and Meditation?

In the brain, similar healing is taking place during yoga and meditation. Chronic pain triggers the brain to change, decreasing the volume of gray matter and decreasing the connectivity of white matter.[59] When grey matter decreases, people often experience depression, anxiety, and impaired cognitive function.[60] Yoga has the opposite effect on the brain, increasing gray matter through neurogenesis and increasing the connectivity of white matter through an ability of the brain called *neuroplasticity*.[61] According to brain expert and NIH scientific director Catherine Bushnell, "Brain anatomy changes may contribute to mood disorders and other affective and cognitive comorbidities of chronic pain. The encouraging news for people with chronic pain is mind-body practices seem to *exert* a protective effect on brain gray matter that counteracts the neuroanatomical effects of chronic pain." In discussing the results, Bushnell notes that there is a correlation between increases in gray matter "and duration of yoga practice, which suggests there is a causative link between yoga and gray matter increases."[62]

Yoga and meditation, specifically yoga nidra, can also help the brain to move memories around so that traumatic memories are no longer intrusive. Memories are stored in the limbic system of the brain (comprised of the hippocampus, amygdala, thalamus, and hypothalamus). Traumatic memories get organized by the amygdala and are stored subcortically, which can mean that they are recalled often. This is a protective action taken by the brain to make life-and-death information easy to access. As a result, these memories can become intrusive.[63] Research shows that people who meditated for 30 minutes a day for eight weeks had reduced gray matter in the amygdala and increased gray matter in the hippocampus. These memories became less intrusive as a result of meditation.[64]

Mayanil believes that yoga nidra is the most effective form of yoga for restructuring traumatic memories in the brain because this ancient meditation technique moves the person into a restorative "conscious deep-sleep state," where post-traumatic responses can heal.[65] Additionally, yoga nidra increases dopamine levels in the brain, which promotes neuroplasticity.[66] This is helpful in brain rewiring and to erase some of the traumatic memories that make life difficult. Mayanil believes in the power of yoga because he has seen how a person's outlook can transform through regular yoga practice. He states:

> Yoga has been a blessing to me, and I want to encourage all first responders to try it. Although yoga has been an advanced science since thousand plus years, I am working fiercely to educate people by making yoga a part of the modern neuroscience, called yogic neuroscience. The brain is the biggest pharmacy there is, and yoga and pranayama are the only way to access that benevolent pharmacy at will.[67]

Yoga also helps strengthen communication within the brain. The prefrontal cortex of the brain is responsible for reason, discrimination, abstract thinking, creativity, personality, and behavior. Studies using EEG have identified that those who meditate regularly had stronger communication between the prefrontal cortex and other parts of the brain.[68] The value of this increased activity is the ability to better identify emotions and to better regulate uncomfortable or distressing emotions and sensations.[69]

These are merely a few of the ways that yoga improves the equilibrium of your body and your brain. There is so much more information, but to avoid overwhelming you with the details, let me say this: Your goal in healing from trauma is to come back to homeostasis, where traumatic memories are no longer invasive. This needs to be both a physical and a mental rebalancing, and restorative yoga, trauma-sensitive yoga, and yoga nidra incorporate both the physical and the mental components needed for rebalancing the body and restructuring the brain. That is why I believe yoga and meditation should be a key part of every first responder's strategy to heal from traumatic memories. The more you incorporate yoga and meditation into your daily life, the sooner your mind and body can begin working together to heal trauma and find wellness again.

Reflection Questions

1. Can you identify regular self-care practices in your life that are helping keep you well and resilient? What are they? How do you think they are helping you?
2. What is one thing you need to add to your self-care regimen that would help you grow in wellness?
3. How do you think that would help you?
4. What is something that stood out to you from this chapter? How do you want to change your thinking/behavior as a result?

Notes

1. T. Forman, "Self-Care Is Not an Indulgence. It's a Discipline," *Forbes*, December 13, 2017, https://www.forbes.com/sites/tamiforman/2017/12/13/self-care-is-not-an-indulgence-its-a-discipline/#30e901c8fee0.
2. J. M. Patton, "5 Truths About Self-Care," *Experience Life*, August 23, 2019, https://experiencelife-com.cdn.ampproject.org/c/s/experiencelife.com/article/5-truths-about-self-care/amp/.
3. Forman, "Self-Care Is Not an Indulgence. It's a Discipline."
4. Patton, "5 Truths About Self-Care."
5. Patton, "5 Truths About Self-Care."
6. Patton, "5 Truths About Self-Care."
7. Forman, "Self-Care Is Not an Indulgence. It's a Discipline."

8. J. Hudson, "Exercise Does More Than You Think," Wellness Supplement, *Fire Engineering*, September 22, 2018, https://digital.fireengineering.com/fireengineering/201809/MobilePagedArticle.action?articleId=1423595#articleId1423595.
9. S. H. Kim et al., "PTSD Symptom Reduction with Mindfulness-Based Stretching and Deep Breathing Exercise: Randomized Controlled Clinical Trial of Efficacy," *Journal of Clinical Endocrinology & Metabolism* 98, no. 7 (2013): 2,984–2,992, https://doi.org/10.1210/jc.2012-3742.
10. S. Moss, "The Distinction Between Challenge and Threat Appraisals," SicoTests.com, June 28, 2016, https://www.sicotests.com/psyarticle.asp?id=281.
11. Moss, "The Distinction Between Challenge and Threat Appraisals."
12. M. Frankenhaeuser, U. Lundberg, and L. Forsman, "Dissociation Between Sympathetic-Adrenal and Pituitary-Adrenal Responses to an Achievement Situation Characterized by High Controllability: Comparison between Type A and Type B Males and Females," *Biological Psychology* 10, no. 2 (March 1980): 79–91, https://doi.org/10.1016/0301-0511(80)90029-0; and R. A. Dienstbier, "Arousal and Physiological Toughness: Implications for Mental and Physical Health," *Psychological Review* 96, no. 1 (1989): 84–100, https://doi.org/10.1037/0033-295X.96.1.84.
13. J. Hudson, "Exercise Does More Than You Think."
14. J. Blascovich and J. Tomaka, "The Biopsychosocial Model of Arousal Regulation," *Advances in Experimental Social Psychology Advances in Experimental Social Psychology* 28 (1996): 1–51, https://doi.org/10.1016/s0065-2601(08)60235-x.
15. C. Ashton, "Yoga to Transform Trauma: Leadership Training and Intensive with Catherine Ashton," *Illumine Chicago*, 2014, https://illuminechicago.com/events/yoga-to-transform-trauma-leadership-training-intensive-with-catherine-ashton/.
16. G. M. Cooney et al., "Exercise for Depression," *Cochrane Database of Systematic Reviews* 9 (September 12, 2013), https://doi.org/10.1002/14651858.CD004366.pub6.
17. M. Babyak, J. A. Blumenthal, and S. Herman, "Exercise Was More Effective in the Long Term than Sertraline or Exercise Plus Sertraline for Major Depression in Older Adults. (Therapeutics)," *Evidence-Based Mental Health* 4, no. 4 (November 2001): 105–106.
18. J. Webb and C. Musello, *Running on Empty: Overcome Your Childhood Emotional Neglect* (New York: Morgan James Publishing, 2019).
19. L. Ure, "Rethinking Anger," *Psychotherapy Networker,* October 1, 2020, https://www.psychotherapynetworker.org/blog/details/1824/rethinking-anger?utm_source=linkedin&utm_medium=social.
20. Ure, "Rethinking Anger."
21. L. Owen and B. Corfe, "The Role of Diet and Nutrition on Mental Health and Wellbeing," *Proceedings of the Nutrition Society* 76, no. 4 (July 14, 2017): 425–426, https://doi.org/10.1017/s0029665117001057.
22. Owen and Corfe, "The Role of Diet and Nutrition on Mental Health and Wellbeing."
23. Owen and Corfe, "The Role of Diet and Nutrition on Mental Health and Wellbeing."
24. C. Shannahan and L. Shannahan, *Deep Nutrition: Why Your Genes Need Traditional Food* (Flatiron Books, 2008).
25. J. Douillard, *The 3-Season Diet: Solving the Mysteries of Food Cravings, Weight-Loss, and Exercise* (New York: Harmony Books, 2000).

26. Shannahan and L. Shannahan, *Deep Nutrition*.
27. L. M. Bodnar and K. L. Wisner, "Nutrition and Depression: Implications for Improving Mental Health among Childbearing-Aged Women. *Biological Psychiatry* 58, no. 9 (July 27, 2005): 679–685, https://doi.org/10.1016/j.biopsych.2005.05.009.
28. E. Swaim, "10 Ways to Boost Dopamine and Serotonin Naturally," *Good Therapy*, December 12, 2017, https://www.goodtherapy.org/blog/10-ways-to-boost-dopamine-and-serotonin-naturally-1212177.
29. M. Csikszentmihalyi, *Finding Flow: The Psychology of Engagement with Everyday Life* (New York: Basic Books, 1997).
30. E. Diener and M. E. Seligman, "Very Happy People," *Psychological Science* 13, no. 1 (January 1, 2002): 81–84, https://doi.org/10.1111/1467-9280.00415.
31. C. Bergland, "Longer Exhalations Are an Easy Way to Hack Your Vagus Nerve," *Psychology Today*, May 9, 2019, https://www.psychologytoday.com/us/blog/the-athletes-way/201905/longer-exhalations-are-easy-way-hack-your-vagus-nerve#:~:text=A%20myriad%20of%20breathing%20patterns,and%20calm%20one's%20nervous%20system.
32. Bergland, "Longer Exhalations Are an Easy Way to Hack Your Vagus Nerve."
33. Bergland, "Longer Exhalations Are an Easy Way to Hack Your Vagus Nerve."
34. Ashton, "Yoga to Transform Trauma."
35. National Center for Complementary and Alternative Medicine, "Yoga: What You Need to Know," US Department of Health and Human Services, National Institutes of Health (2015); latest version (updated April 21, 2021) may be found at https://nccih.nih.gov/health/yoga.
36. A. M. Gallegos, W. Cross, and W. R. Pigeon, "Mindfulness-Based Stress Reduction for Veterans Exposed to Military Sexual Trauma: Rationale and Implementation Considerations," *Military Medicine* 180, no. 6 (June 2015): 684–689, https://doi.org/10.7205/MILMED-D-14-00448.
37. A. M. Gallegos et al., "Meditation and Yoga for Posttraumatic Stress Disorder: A Meta-Analytic Review of Randomized Controlled Trials," *Clinical Psychology Review* 58 (December 2017): 115–124, https://doi.org/10.1016/j.cpr.2017.10.004.
38. C. R. Nolan, "Bending Without Breaking: A Narrative Review of Trauma-Sensitive Yoga for Women with PTSD," *Complementary Therapies in Clinical Practice* 24 (August 2016): 32–40, https://doi.org/10.1016/j.ctcp.2016.05.006.
39. S. Mayanil, "Yoga and PTSD," in discussion with the author (October 10, 2020).
40. S. Parker, S. V. Bharati, and M. Fernandez, "Defining Yoga-Nidra: Traditional Accounts, Physiological Research, and Future Directions," *International Journal of Yoga Therapy* 23, no. 1 (January 1, 2013): 11–16, https://doi.org/10.17761/ijyt.23.1.t636651v22018148.
41. R. Miller, "About iRest," iRest Institute (accessed June 25, 2021), https://www.irest.org/about-irest-institute.
42. R. J. Macy et al., "Yoga for Trauma and Related Mental Health Problems: A Meta-Review with Clinical and Service Recommendations," *Trauma, Violence, & Abuse* 19, no. 1 (December 9, 2015): 35–57, https://doi.org/10.1177/1524838015620834.
43. B. A. van der Kolk et al., "Yoga as an Adjunctive Treatment for Posttraumatic Stress Disorder: A Randomized Controlled Trial," *Journal of Clinical Psychiatry* 75, no. 6 (June 2014): 559–565, https://doi.org/10.4088/jcp.13m08561.
44. van der Kolk et al., "Yoga as an Adjunctive Treatment for Posttraumatic Stress Disorder."

45. S. H. Kim et al., "PTSD Symptom Reduction with Mindfulness-Based Stretching and Deep Breathing Exercise."
46. Mayanil, "Yoga and PTSD."
47. T. W. Kjaer et al., "Increased Dopamine Tone during Meditation-Induced Change of Consciousness," *Cognitive Brain Research* 13, no. 2 (April 2002): 255–259, https://doi.org/10.1016/s0926-6410(01)00106-9.
48. L. Stankovic, "Transforming Trauma: A Qualitative Feasibility Study of Integrative Restoration (iRest) Yoga Nidra on Combat-Related Post-Traumatic Stress Disorder," *International Journal of Yoga Therapy* 21, no. 1 (October 1, 2011): 23–37, https://doi.org/10.17761/ijyt.21.1.v823454h5v57n160.
49. van der Kolk et al., "Yoga as an Adjunctive Treatment for Posttraumatic Stress Disorder."
50. Nolan, "Bending Without Breaking."
51. C. C. Streeter et al., "Effects of Yoga on Thalamic Gamma-Aminobutyric Acid, Mood and Depression: Analysis of Two Randomized Controlled Trials," *Neuropsychiatry (London)* 8, no. 6 (2018): 1,923–1,939, https://www.jneuropsychiatry.org/peer-review/effects-of-yoga-on-thalamic-gammaaminobutyric-acid-mood-and-depression-analysis-of-two-randomized-controlled-trials.pdf.
52. Kjaer et al., "Increased Dopamine Tone during Meditation-Induced Change of Consciousness."
53. Ashton, "Yoga to Transform Trauma."
54. C. C. Streeter et al., "Effects of Yoga on the Autonomic Nervous System, Gamma-Aminobutyric-Acid, and Allostasis in Epilepsy, Depression, and Post-Traumatic Stress Disorder," *Medical Hypotheses* 78, no. 5 (May 2012): 571–579, https://doi.org/10.1016/j.mehy.2012.01.021.
55. B. A. van der Kolk, "Clinical Implications of Neuroscience Research in PTSD," *Annals of the New York Academy of Sciences* 1,071, no. 1 (July 26, 2006): 277–293, https://doi.org/10.1196/annals.1364.022; Kim et al., 2013; and Y. Tang et al., "Central and Autonomic Nervous System Interaction Is Altered by Short-Term Meditation," *Proceedings of the National Academy of Sciences (US)* 106, no. 22 (June 2, 2009): 8,865–8,870, https://doi.org/10.1073/pnas.0904031106.
56. Nolan, "Bending Without Breaking."
57. R. Adelson, "Stimulating the Vagus Nerve: Memories Are Made of This," *PsycEXTRA Dataset* 35, no. 4 (2004), https://doi.org/10.1037/e362742004-023.
58. D. Asprey, "Stimulate Your Vagus Nerve to Improve Memory, Says New Study," *Dave Asprey* (blog), June 26, 2018, https://blog.daveasprey.com/vagus-nerve-affects-memory/.
59. R. Rodriguez-Raecke et al., "Brain Gray Matter Decrease in Chronic Pain Is the Consequence and Not the Cause of Pain," *Journal of Neuroscience* 29, no. 44 (November 4, 2009): 13,746–13,750, https://doi.org/10.1523/jneurosci.3687-09.2009.
60. S. M. Grieve et al., "Widespread Reductions in Gray Matter Volume in Depression," *NeuroImage: Clinical* 3 (2013): 332–339, https://doi.org/10.1016/j.nicl.2013.08.016.
61. B. Froeliger, E. L. Garland, and F. J. McClernon, "Yoga Meditation Practitioners Exhibit Greater Gray Matter Volume and Fewer Reported Cognitive Failures: Results of a Preliminary Voxel-Based Morphometric Analysis," *Evidence-Based Complementary Alternative Medicine* (December 5, 2012): 1–8, https://doi.org/10.1155/2012/821307.

62. American Pain Society, "Yoga and Chronic Pain Have Opposite Effects on Brain Gray Matter," ScienceDaily.com, May 15, 2015, http://www.sciencedaily.com/releases/2015/05/150515083223.htm.
63. D. Hernandez (licensed clinical psychologist), email communication with the author about yoga and the brain, October 30, 2020.
64. J. D. Bremner et al., "A Pilot Study of the Effects of Mindfulness-Based Stress Reduction on Post-Traumatic Stress Disorder Symptoms and Brain Response to Traumatic Reminders of Combat in Operation Enduring Freedom/Operation Iraqi Freedom Combat Veterans with Post-traumatic Stress Disorder," *Frontiers in Psychiatry* 8 (August 25, 2017), https://doi.org/10.3389/fpsyt.2017.00157.
65. Mayanil, "Yoga and PTSD."
66. D. J. Surmeier, J. Plotkin, and W. Shen, "Dopamine and Synaptic Plasticity in Dorsal Striatal Circuits Controlling Action Selection," *Current Opinion in Neurobiology* 19, no. 6 (December 2009): 621–628, https://doi.org/10.1016/j.conb.2009.10.003.
67. Mayanil, "Yoga and PTSD."
68. Bremner et al., "A Pilot Study of the Effects of Mindfulness-Based Stress Reduction."
69. D. Hernandez, email communication with the author.

Chapter 12

HOW I TREAT TRAUMA IN MY PRACTICE

I have seen amazing benefits for first responders from the treatments I have listed in this chapter, and I encourage you to find a first responder counselor in your area who provides some of these options.

Accelerated Resolution Therapy

Accelerated resolution therapy (ART) is a brief, safe, and effective treatment for combat-related symptoms of post-traumatic stress disorder (PTSD) among veterans and US service members, researchers at University of South Florida College of Nursing reported in 2014. The researchers noted that this newer treatment—a combination of psychotherapies and use of eye movements—was shorter and more likely to be completed than conventional therapies formally endorsed by the US Department of Defense and the Veterans Administration.[1]

As reported by the American Psychological Association (APA), E. C. Nielson and others studied male veterans with PTSD and found that "masculine norms can create barriers to getting necessary treatment."[2] The study noted that a first responder who does seek treatment may still not feel free to express emotion fully, which can hinder progress during treatment. This is why I appreciate ART, which does not require the individual to speak at all. It is a silent treatment that invites mental processing as a way of refiling memories in the brain to remove the intrusion they have been creating. ART is a fabulous treatment for those who are perhaps unable or unwilling to talk about what happened.

My friend and colleague, Diego Hernandez, is a licensed clinical psychologist, a master ART therapist, and a lead ART trainer. He has been working in trauma and mental health for active-duty military and veterans for many years. He explains ART, saying:

> ART bypasses social monitoring and focuses on the experience as recalled. Its structure provides a context and langue for images, feelings, and sensations that make them the focus of attention. But not the focus of discussion. This context is helpful for not only resolving a difficult or traumatic memory. It is also helpful in

giving expression to thoughts, feelings, and sensations that may be typically challenging to process in traditional therapeutic approaches due to internalized conceptualizations of masculinity which may devaluate them.[3]

I appreciate this perspective and the passion Hernandez has for ART.

How Does ART Work?

ART builds on two existing PTSD therapies—*imaginal exposure* (reliving) and *imagery rescripting*. Traumatic memories are often intense sensory images that are not well integrated with other positive imagery. When an individual has a fear-provoking memory, it is first stored (consolidated). When it is later retrieved, it must be reconsolidated. That leaves the memory open for modification.

In order to change the placement of these memories and their emotional associations, a clinician will oscillate his or her hand at eye level while the patient recalls the traumatic event. In sets of 40 oscillations at a time, the patient moves his eyes back and forth following the counselor's hand, while recalling the traumatic memory. This involves interhemispheric communication in the brain and contributes to problem-solving similar to that in REM sleep. Since sleep is often disrupted in PTSD patients, their brains do not have as much time to process emotional memories efficiently. These eye movements are also believed to have a calming effect as they shift the individual from being dominated by his sympathetic nervous system, which is ramped up for fight-or-flight response, to his parasympathetic nervous system. This REM-like processing ultimately results in reduced sleep interference.[4]

After having the client recall the memory while following the counselor's hand with his eyes, the counselor will invite the individual to "rescript" the images in his mind by asking the individual to think of a solution or "replace" the distressing images with positive images. This is called *voluntary image replacement (VIR)*, and individuals find that it changes the recall of memories from negative to positive. This VIR involves sensation processing, which removes the psychological sensations invoked by remembering the event. The memory then becomes a reconsolidated hybrid memory that is rescripted and less traumatic to recall.

Is It Effective?

Clinical studies of ART on PTSD patients showed a significant decrease in patients re-experiencing events (intrusion), avoidance, numbing, and hyperarousal. Of those studied, 71.4% showed a meaningful change in their symptoms of depression by their four-month follow-up.[5]

One of the greatest strengths of ART is that it has a short treatment protocol—typically just one to four sessions. This increases the likelihood of people completing the treatment and finding healing. Other studies found that all of the participants who completed treatment showed a reduction in PTSD, depression, anxiety, and aggression, and many of them also showed a significant decrease in trauma-related distress and an increase in self-compassion.[6]

ART is endorsed by the US government, the Substance Abuse and Mental Health Services Association (SAMHSA), the National Registry of Evidence Based Programs and Practices (NREPP), and by military leaders. The technique has been used on Green Berets, special operations commanders, and hundreds of veterans and active duty troops to treat PTSD. I have found the protocol to be of tremendous benefit to first responders since incorporating it in my practice in 2017.

Mike's Story Using ART

I shared the story of a client of mine named Mike in chapter 6. Mike had experienced two calls involving the deaths of infants, which became intrusive flashbacks during his wife's pregnancy. Early on, when Mike noticed the anxiety starting to control his everyday life, he contacted his doctor and asked about medication. His doctor prescribed Zoloft, which can take a few months to reach full effectiveness. Mike desperately wanted to feel "normal" again.

In the midst of his anxiety, Mike began meeting with me regularly for ART. He explains his experience with ART, saying:

> When Jada first brought up ART, I was intrigued but skeptical at the same time. She told me to go on her website and watch a video she had posted about ART. The video is about a veteran named Brian Anderson, a special forces operator, and his struggle with PTSD and how ART changed his life. I can remember watching this and feeling extremely emotional, and after it, feeling hope. Hope that this would help me through this dark time in my life.
>
> Jada and I have done several of these sessions. Never in a million years would I have thought someone moving their hand back and forth in front of my face, mimicking eye movement of REM sleep, would help me through anxiety. Brian said in his video, "ART changed my life," and like him I know ART changed mine as well. The calls that I was dwelling on faded away and I was able to talk about them without breaking down in tears. I physically felt changes in my body. The constant tense and on edge feelings went away. ART is an amazing treatment, and like Brian, I want every one of my brothers and sisters to know about it.
>
> I was recently catching up with a very close friend from high school. It had been a while since we last spoke to or saw each other. We did the typical, "How's life? How's work?" At one point, I told her about being diagnosed with PTSD, and

how I was in a dark place for a while. Her response was, "Honey, I knew you had PTSD. You weren't yourself."

Hearing that made me realize that the people we are closest to will see the pain, see the destruction, [and] see the dark place you're in. I know guys at the fire house saw the pain I was in but never really had the balls to ask me if I was ok. They would say things to other people but never directly to my face. There were a few people that I confided in and still do. They know who they are, and I can't thank them enough. I guess where I'm going with this is if you know a brother, sister, friend, loved one is struggling, say something! Don't try and fix them, just be there for them, love them, and help them get the help they need.

If it wasn't for the support of my wife, dad, and close friends, I don't know if I'd still be here today. That saying "God puts people in our lives exactly when they need to be" hits home for me. One of those people, for me, is Jada. We don't leave the job the same person as we started, but thank God for Jada and her dedication to helping first responders. I know someone right now is reading this and is in a dark place like I was. Have hope, as dark as it is right now, there will be light again. I promise!

Narrative Therapy: Writing Your Way to Healing

Healing from trauma can seem like a massive undertaking, but some of the most profound healing can come from some of the simplest activities, like writing with a pen and paper. So whether you have experienced a trauma in the past that you need to sort through, or you are mentally preparing yourself in case you experience a trauma in the future in your career, here is a tool that is simple, inexpensive, and powerful. I frequently recommend this to my counseling clients.

The most important thing you can do to begin healing from a trauma is to give yourself a 24-hour-period in which you allow yourself a full sleep cycle to begin processing. Then begin to deal with it right away by thinking through what exactly happened and how it made you feel. In fact, one of the most helpful ways to begin to think through a traumatic event is by writing about it. There are so many subconscious thoughts that human beings hold internally—for example, dreams, ambitions, fears, habits, and desires—and when people begin to write down their reflections on a situation, it engages their thinking on a deeper level. By using a different part of the brain, reflective writing can help you think about a traumatic situation differently, shedding new light on it and helping you be able to wrap your mind around it. You may be surprised. The part of the brain that is engaged during reflective writing may produce even more thoughtful insights than talking aloud can produce.

This is not to minimize the importance of talking through a situation with a peer supporter or a counselor. Having someone to ask insightful questions and relate to what you have been through is invaluable. But there is also power in taking a pen to paper and pouring your heart out. Think of this as translating your feelings into a story. There was an event, framed by all of your other life events, that occurred, and at the moment it makes no sense. But as you fill in the words on the page, the story starts to take shape. You are the author and the main character, and you can begin to see how the event fits.

James Pennebaker, research psychologist at the University of Texas, conducted a number of studies that found that writing about a stressful or traumatic event helped people come to terms with the emotional aftermath of such an event. Not only did writing help mentally and emotionally, but it also helped physically. Pennebaker noted that not only were those in his study less prone to symptoms of post-traumatic stress disorder (PTSD), but they were also less likely to suffer from asthma, arthritis, and even chronic fatigue syndrome later. Pennebaker was able to quantify his results. He found that when people write for about 20 minutes a day—ideally at the end of the day—for three or four consecutive days, they were likely to have *half* the number of medical visits.[7]

Writing goes beyond just processing trauma. Writing is also powerful for reducing anxiety in your life. Your worries take up mental energy, so that your brain is constantly multitasking to juggle your worries alongside your daily endeavors. When you take those worries and write them out, expressing how you feel about them on paper, you download that mental focus onto paper, so that you can devote your cognitive energy toward your daily life, making you more efficient.

A study conducted at the University of Arizona looked at expressive writing and heart rate variability (HRV), which is a sign of the body's ability to ramp up and ramp down its stress responses based on outside events. Participants who spent 20 minutes doing expressive writing every day for just a three-day period experienced improved HRV, which signaled that their autonomic nervous systems were producing a relaxation response.[8] Essentially, they noticed that those who wrote every day were physically less stressed.

Anthony is a client of mine whose story I shared in chapter 11. I highly encourage you to read his story. He is a great example of writing for emotional wellness. As a part of his evening routine, he writes. He explains, "I use writing as an outlet to get my thoughts from the day out and it helps me stay mindful of my actions and choices." Writing is one of many intentional actions Anthony takes on a daily basis to stay connected, grounded, rested, and emotionally well.

You should be aware that writing about a traumatic event immediately after it occurs could make you angry as you recall the details of the event. But in my experience, after an initial 24-hour waiting period and one good sleep cycle, anything you can do to get your thoughts out is going to help you move forward. A high percentage of first responders are introverts and need time to think about things on their own. That is totally fine. Many introverts will sleep more and process a trauma internally. One way to begin to think through an event without having to process with another person immediately is to write your experience down.

If you are up for it, grab a piece of paper and begin to write down everything that happened. Ask yourself what you were feeling, why you were feeling that way, what senses you were using, what was stressing you out, what you were proud of, what you wish had not happened, and most importantly, what you want to learn from this experience. Once you have done some writing, set the paper down, and plan to write again later.

When you are ready, it is also incredibly important to process a traumatic experience with a peer supporter or a counselor. So reach out and get help because processing takes time, and the sooner you can begin to heal after a trauma, the sooner you will begin to see the sunlight again.

Prolonged Exposure/Imaginal Exposure

Another treatment for trauma that I use in my practice is known as *prolonged exposure* or *imaginal exposure* because it requires the individual to think about the traumatic memory for an extended period of time for the sake of removing the fear and anxiety associated with the memory.[9] Often with prolonged exposure, individuals delve into the traumatic memory repeatedly and explore what happened to gain control over the memory, rather than avoiding the memory, as is common with PTSD.

Think of the brain as a filing cabinet. Typically, memories are filed in sequence and in context.[10] Traumatic memories are filed differently, so feelings such as grief or loss sit right beneath the surface and can come back immediately and powerfully. Imaginal exposure helps put these memories in the right "file" so that they are in context and no longer intrusive. This gives the individual control over the memory instead of allowing the memory to have control of the individual.

The natural response of many with traumatic memories is to try to "get away" or "not think about it." This is *avoidance*. According to Matthew Tull, professor of psychology at the University of Toledo, when a person avoids the traumatic memories, thoughts, and emotions, he shuts off his ability to process those experiences. So the goal of prolonged exposure to the event is to reduce a person's anxiety and make the person okay facing the event rather than avoiding it.[11]

Because traditional masculine expectations can force men to avoid showing emotions, men may avoid confronting traumatic memories out of fear of feeling the emotional pain associated with them. This avoidance is one of the symptoms of PTSD, which I believe is reinforced by male stereotypes. Military studies found that processing trauma may be harder for men for this reason. Psychologist Elizabeth Nielson, Morehead University, and others noted that "the emphasis on mental fortitude within both military culture and traditional masculine ideology leads to an avoidance of disclosure and speaking about traumatic experiences, which may have implications for veterans' trauma processing within treatment."[12]

For many individuals with traumatic memories, the memories will come back when they are vulnerable, often when they are sleeping. Not only does this disrupt their sleep, but it makes many people feel that they need assistance from alcohol or other substances to get around the memories that are interrupting their sleep. This can lead down a dangerous path of substance abuse.

To confront the memory, rather than let it control a person's life, prolonged imaginal exposure can help. Prolonged exposure calls to mind the events of the trauma, so that the individual can think through them while he is awake, so that the fear and anxiety connected with those memories do not disrupt his daily life or sleep anymore.

It is important to note here that prolonged/imaginal exposure is effective only when the individual is in a relaxed state. This relaxation facilitates the processing of the memory, so it can become less disruptive.

So the goal of prolonged exposure is to enable the individual to have thoughts about the trauma, conversations about the trauma, and experience triggers associated with the trauma without experiencing the intense anxiety that disrupts life.

How Does Prolonged/Imaginal Exposure Work?

In a session, the counselor will encourage the individual to remember the trauma as vividly as possible. The individual's eyes are closed, and all his energy is focused on remembering the details of the trauma.

The individual then begins to narrate what happened in the present tense, as if he is walking through the event right now all over again. The counselor will record the conversation as the first responder narrates the event. The individual will then replay the recording every day for 30 to 90 minutes. This repeated exposure to the event will open up new layers of memory about the event.

As the first responder listens, he will take a deep breath. Then he will answer questions such as the following:

1. Did you remember anything not previously recalled? What was it?
2. Was it easier or more difficult than you anticipated to hear this again?
3. Would anything else have helped in that situation?
4. What are you feeling right now after listening to it?
5. Do you have any other thoughts?

What Happens During Prolonged/Imaginal Exposure?

During prolonged exposure/imaginal exposure, the individual experiences *emotional processing*—organizing that memory and learning that thinking about the trauma is not dangerous. According to E. B. Foa, K. R. Chrestman, and E. Gilboa-Schechtman, authors of

Prolonged Exposure Therapy for Adolescents with PTSD, prolonged exposure also creates habituation, and this treatment is provided for 9 to 12 sessions that are 90 to 120 minutes each. By repeatedly remembering and reliving the details of the experience, the individual's brain will become accustomed to it, and anxiety will fade. Foa and coauthors also note that the Center for the Treatment and Study of Anxiety (CTSA) is highly supportive of prolonged exposure since they have seen its benefits for reducing anxiety in many patients.[13] The exposure to the traumatic event created in prolonged exposure is similar to the saying, "Time is a great healer." If you think about something enough times, your brain becomes familiar with it, and the emotional responses associated with it diminish.

According to the American Psychological Association, imaginal exposure also helps the brain discriminate between remembering and being traumatized, thus reducing the fear associated with the traumatic memory. The association explains that sufficient repetitions of the same experience will help someone realize that remembering a trauma is not the same as experiencing the trauma. That person's desire to avoid the trauma and everything that serves as a reminder of it will subside.[14] It will also give the individual a sense of increased mastery over the trauma, enhancing a sense of self-control and personal competence as he stops avoiding the event and its triggers and begins facing it head on.

Often when a person experiences a traumatic event, he will generalize fear from that specific trauma, fearing that this event could occur in other situations and other areas of his life. One benefit of prolonged/imaginal exposure is *differentiation*. This is when the brain decreases generalization of fear from that specific trauma to similar but safe situations. In first responder careers, the ability to differentiate "danger" from "nondanger" is a key intuition. But often trauma can increase fear and anxiety, making first responders jumpy and fearful. So being able to differentiate between a specific traumatic event that was fearful and unsafe and an event that is safe, where the first responder is able to relax, is a key part of healing.

Reflection Questions

1. Did any of these treatments stand out as potentially useful to you?
2. If so, who do you need to contact to begin utilizing these treatments in your life?
3. Grab a piece of paper or a journal and write yourself a letter describing how you want to be feeling emotionally three years from now.

Notes

1. "USF College of Nursing Study Reveals Brief Therapy Treats PTSD Symptoms," *USF Nursing News*, July 2, 2014, https://hscweb3.hsc.usf.edu/nursingnews/usf-college-of-nursing-study-reveals-brief-therapy-treats-ptsd-symptoms/.

2. E. C. Nielson et al., "Traditional Masculinity Ideology, Posttraumatic Stress Disorder (PTSD) Symptom Severity, and Treatment in Service Members and Veterans: A Systematic Review," *Psychology of Men & Masculinities* 21, no. 4 (January 27, 2020): 578–592, https://doi.org/10.1037/men0000257.
3. D. Hernandez (licensed clinical psychologist), in discussion with the author about accelerated resolution therapy, 2020.
4. K. E. Kip et al., "Brief Treatment of Co-Occurring Post-Traumatic Stress and Depressive Symptoms by Use of Accelerated Resolution Therapy," *Frontiers in Psychiatry* 4 (March 8, 2013), https://doi.org/10.3389/fpsyt.2013.00011.
5. K. E. Kip et al., "Brief Treatment of Co-Occurring Post-Traumatic Stress and Depressive Symptoms by Use of Accelerated Resolution Therapy."
6. K. E. Kip et al., "Randomized Controlled Trial of Accelerated Resolution Therapy (ART) for Symptoms of Combat-Related Post-Traumatic Stress Disorder (PTSD)," *Military Medicine* 178, no. 12 (December 2013): 1,298–1,309, https://doi.org/10.7205/milmed-d-13-00298.
7. J. W. Pennebaker, "Writing About Emotional Experiences as a Therapeutic Process," *Psychological Science* 8, no. 3 (May 1, 1997): 162–166, https://doi.org/10.1111/j.1467-9280.1997.tb00403.x.
8. K. J. Bourassa et al., "Impact of Narrative Expressive Writing on Heart Rate, Heart Rate Variability, and Blood Pressure After Marital Separation," *Psychosomatic Medicine* 79, no. 6 (July 8, 2017): 697–705, https://doi.org/10.1097/psy.0000000000000475.
9. M. Tull, "Exposure Therapy for Treating Post-Traumatic Stress Disorder Symptoms," VeryWellMind.com, February 25, 2020, https://www.verywellmind.com/exposure-therapy-for-ptsd-2797654.
10. D. Lisak, "The Neurobiology of Trauma," Arkansas Coalition Against Sexual Assault, YouTube video, February 5, 2013, 34:30, https://youtu.be/pyomVt2Z7nc.
11. Tull, "Exposure Therapy for Treating Post-Traumatic Stress Disorder Symptoms."
12. Nielson et al., "Traditional Masculinity Ideology, Posttraumatic Stress Disorder (PTSD) Symptom Severity, and Treatment in Service Members and Veterans."
13. E. B. Foa, K. R. Chrestman, and E. Gilboa-Schechtman, *Prolonged Exposure Therapy for Adolescents with PTSD: Emotional Processing of Traumatic Experiences: Therapist Guide* (Oxford: Oxford University Press, 2009).
14. American Psychological Association, "Prolonged Exposure," May 2017, https://www.apa.org/ptsd-guideline/treatments/prolonged-exposure.

Chapter 13

YOUR PLACE IN THE BROTHERHOOD

A few years ago, I had the opportunity to shadow Aurora Fire Department's Extrication Training, in which Private Stawikowski, Captain Garner, Private Koerbrel, and Lieutenant Hasenheyer prepared and observed recruits as they tackled three staged obstacles with three specific tools. As I watched, I observed the brotherhood at its finest. At the first station, recruits used an axe to cut through a roof. At the second station, recruits used a 26-pound saw with various blades to cut through rebar. At the third station, recruits worked with a hammer and wedge to open a 1,000-pound metal door. Each station required precise skill and tireless physical exertion.

First, I watched closely as a team worked to saw through rebar. Captain Garner instructed them, "Don't bounce the saw. Hold it steady to the rebar. Keep your elbows in." The consistently applied pressure with the right tool—a saw made for rebar, as opposed to one made for wood—would create a breakthrough and make an entry point for them to begin their rescue. Meanwhile, they had to be aware of smoke so it did not choke up their saw.

Next, I focused on the group at the metal door. The two men working to open the door labored for what seemed like forever to get the door to budge only inches. Covered in sweat, they counted together as one placed the wedge between the door and the doorframe and the other pounded it with his hammer. It took perfect timing and teamwork. They were told, "Do as little as you need to get the job done. Conserve your energy." Ultimately, the door budged, and they were able to enter. Completely exhausted, the recruits took a break. But if this had been a real fire, this would only have been the beginning of their work. The real rescue work would require them to forget their exhaustion and move forward with complete intensity of effort.

During these exercises, their primary goal was to protect their brothers. Their secondary goal was to locate and secure civilians. They multitasked—listening to commands from the chief, hearing their radios go off, looking for people, and putting out fires. Hasenheyer exhorted them, "Learn how to talk and work at the same time." Through the chaos, their attention and care for one another impressed me. Truly, this was the beginning of a brotherhood for these recruits, and each one had a part to play in this precise art of rescuing others.

As I observed these fire recruits making breakthroughs, it dawned on me that it is often because of the brotherhood that firefighters make breakthroughs in their personal lives.

Making Personal Breakthroughs Using Firefighting Tactics

As I watched these firefighters making breakthroughs while fighting fires, I could not help but notice that their tactics can be applied to breaking through personal struggles. For example, when it comes to addiction, the rebar saw tactics apply perfectly: Hold steady, and once you gain some momentum, do not let up. So often people bounce in and out of addictions because they think their progress permits them to have a little bit of freedom. The alcoholic says, "I haven't had a beer in three weeks, so I can have this one right now." No! Holding the saw to the rebar and not bouncing it was the only way to cut through it. Similarly, holding clean and not tinkering with the addictive substance is the only way to become completely free and achieve a breakthrough.

The level of protection these firefighters offered one another has something to say to those struggling with exhaustion or difficult memories. It surprised me that the primary objective of the recruits was to keep their peers safe. And this should be our approach when it comes to breaking through exhaustion or even a traumatic memory. If you are not okay, you cannot help your family. If you are exhausted or wrestling with a trauma, your first priority needs to be self-care. Just as these recruits had to keep their personal team safe and strong, so you need to keep yourself strong by getting the rest and help you need first. Then you will be strong to help others, which is another breakthrough.

Third, all recruits were told to step back and do a personal 360° size-up. This advice applies to individuals working for breakthroughs in any area of life. If there is an obstacle you are struggling to overcome, look for another entry point. If you are struggling at home and communication seems to be strained, try a gentler approach. Maybe you could open up about your personal situation and stresses and possibly thank your spouse for her support as you take on your challenging career. Maybe you could find something your spouse enjoys and decide you like it too. Then bond over that. What are other ways you could approach a stuck situation? You may find you have a wooden door waiting, when you thought you had to cut through rebar, leading to another breakthrough.

What the Recruits Walked Away With

In their debrief, Captain Garner urged recruits, "Use what information you know to make plans A, B, and C on the way to the call." He also encouraged them to "take a step back and do a personal 360° to assess the whole situation." For example, those who had been using the saw to cut through the rebar may have found a wooden door, which would have made an easier entry point and required a slightly different tool.

Lieutenant Hasenheyer reminded them as they used the axe to penetrate a hole in the roof that their real-life goal was to "get a hole in the roof." He challenged them to think beyond what they learned in their training manuals. The recruits were encouraged to "do a cost versus benefit analysis." What were the risks involved in each task, and was the benefit worthwhile? For example, if it were a commercial fire, would it be worth risking your life to save an empty building?

When it comes to breaking through roofs, walls, doors, or personal struggles, keep fighting, get a team, and take care of yourself. You will find you always feel most exhausted right before your barrier gives way.

Using Your Unique Strengths in the Brotherhood

In your department, there is only one you. And by being you, you will find that you are happier and that your team is stronger. According to researchers Ed Diener and Martin Seligman, people can find greater happiness in life by using their unique signature strengths. Seligman and his colleagues discovered that when people work within their signature strengths, this staying true to oneself helps them find authentic happiness. Diener and Seligman identified 24 strengths present in individuals throughout the world that, when used, made them happier:

1. Curiosity/interest in the world
2. Love of learning
3. Judgment/critical thinking/open-mindedness
4. Ingenuity/originality/practical intelligence/street smarts
5. Social, personal, and emotional intelligence
6. Perspective
7. Valor and bravery
8. Perseverance/industry/diligence
9. Integrity/genuineness/honesty
10. Kindness and generosity
11. Loving and allowing yourself to be loved
12. Citizenship/duty/teamwork/loyalty
13. Fairness and equity
14. Leadership
15. Self-control
16. Prudence/discretion/caution
17. Humility and modesty
18. Appreciation of beauty and excellence

19. Gratitude
20. Hope/optimism/future-mindedness
21. Spirituality/sense of purpose/faith
22. Forgiveness and mercy
23. Playfulness and humor
24. Zest/passion/enthusiasm[1]

Do you recognize any of these strengths in your personal life? If so, take note. They will make you simultaneously happier and more fulfilled if you use them.

Maybe you are great at making people laugh. You might have the strength, "Playfulness and humor." Maybe you are great at pointing people toward a higher power when they are in distress. You might have the strength, "Spirituality/sense of purpose/faith." Maybe you are great at keeping perspective when hard things happen. You might have the strength, "Perspective." And you are probably the only one with your unique combination of strengths.

When you focus on your strengths, you bring something to the table that only you can bring. On a call, your strengths will help the people around you. In your firehouse, your strengths will help the people around you. In your family, your strengths will help the people around you. Your strengths help them, and using your strengths helps you become happier.

Your job is not to copy a leader or colleague you respect. Your job is to be you. Just think of how great the combination of your strengths and your peers' strengths will be when you each fulfill your unique role.

When a peer needs insight, no one can help him like you can. Your unique combination of experiences, relationships, and strengths will enable you to listen, encourage, and relate to your peer in a way that only you can. Combine your strengths with your memories, your years of service, and your relationships at home, and you are a fingerprint. Unlike anyone else, you can contribute to your team. So bring your strengths. Bring you. You will be happy you did.

Introverted Leadership

One strength many first responder leaders possess is introversion. It may seem odd to call introversion a strength, but it truly can be. When picturing a great leader, many people envision someone who is full of charisma speaking at the front of a large crowd.[2] But introverted leaders bring something unique to the table that I want to highlight here.

Leadership style is a product of the individual personality. Many first responders are introverts, and the leadership skills that come from introverts can be quietly powerful. According to Margarita Tartakovsky, associate editor, PsychCentral, there are four ways that introverts excel in leadership:

1. *Introverts prepare well.* They consider all the angles, questions that need to be asked, and people who need to be addressed before an event. By thinking things through thoroughly, they can lead calmly and precisely. This provides a safe, stable environment in the department.
2. *Introverts are present.* Because they are prepared and focused on being with you, they typically do not multitask. This builds trust with coworkers/employees because the leader listens to them. Having a singleness of focus empowers leaders to see things they might miss and to build relationships that go beyond task-based interactions so they can build staff members who are effectively trained and emotionally supported.
3. *Introverts push themselves.* They are always looking for ways to grow and sharpen their skills. Their commitment to excellence will show up in their departments and will earn the respect of their subordinates.
4. *Introverts practice.* Introverts are great at rehearsing possibilities to get better and improve outcomes. In a job where countless scenarios could arise on a daily basis, having a leader who has looked at things and rehearsed them from multiple possible angles provides departments with methodical, not impulsive, responses to any situation.[3]

Leaders who are introverts can leverage the things that they are naturally good at when leading subordinates where they need to go. Introverted leaders can listen and empathize. In an article for *Forbes*, Chris Myers notes that introverts can look beneath the surface at staff performance problems and work to find a real solution rather than a flippant "suck it up and do it" response. They also think deeply but act with purpose. Introverts think before they act, so they can strategize the best plan of action and be intentional to speak up when they have thought it through. Introvert leaders are also strong at remembering that sometimes less is more. Myers also notes that they bring a subtler approach to problem solving, which can fix issues without conflict or resistance.[4]

Balance is key. Remember that while personality types direct our actions, they do not have to define us. Introversion and extroversion are a spectrum. Introverts possess traits that some of the best leaders have. Introverts may just need the encouragement to speak up and drive change.[5] You bring such an amazing uniqueness to your role. I hope you can see it because your department needs you to be uniquely you, empowered to lead each of them and appreciate them for being uniquely them.

Peer Support

In the brotherhood, there are places for you to help brothers that are not actually work-related. When I served as the clinical director of program development at Illinois Firefighter Peer Support, we discovered that there are so many unique ways that firefighters can help other firefighters.

First responders often join this field because they like to help people. They want to make a difference in their community and world. But the longer they work in a first responder career, the more they realize the world is a messed-up place. They get knocked down. They see things they wish they had never seen. They get cynical. They may even feel depressed or tempted to turn to alcohol or other substances or behaviors to numb the pain or wake them up from the numbness they constantly feel.

If you have been here before or find yourself in this place at this very moment, so have your peers. Most of the seasoned firefighters I have spoken with started their careers with excitement and ambition before becoming jaded and cynical, finally turning to someone like a therapist or a peer supporter for help.

Want to know a secret? You cannot make it alone as a first responder. Your peers are your greatest resource. There are areas where their strengths shine, and there are areas where your strengths shine. Likewise, you have a breaking point, and they have a different breaking point. Differences can either drive you apart or they can bond you as a team. When you are struggling, it might be a peer's strongest moment. When they are hitting their breaking point, you might be maximally motivated. Help each other.

Researchers Diener and Seligman have shown that the happiest people are the most relationally connected people. They noted that those with more close relationships, more time spent with friends and family, more regular habits of opening up about their personal feelings, and more conversations where they listen and respond in positive ways reported having the highest levels of happiness.[6] Other researchers have observed that those who did not self-disclose in close relationships and kept their topics of conversation impersonal reported more loneliness.[7]

The glue that bonds relationships together is vulnerability. The two factors that play the greatest role in your ability to connect with others to form solid, protective relationships are 1) *self-disclosure*, which is sharing your story and being open about your emotions,[8] and 2) *active-constructive responding*, which is listening carefully to others and responding in positive ways.[9] This is why I think peer support is so powerful for firefighters.

In order to help your fellow firefighters to survive and thrive in this field, they need to talk about what they are going through, and they need to know someone knows how they feel. More than anyone else, a fire brother understands what you are going through. The brotherhood can help you break out of the prisons in your mind by showing you that your struggle is valid, you are not alone, and there are steps you can take to become emotionally well.

Stuffing emotional pain in the hidden places of the heart prevents healing. When you are vulnerable with trusted people, you bring light to dark, unhealthy thoughts. When you are vulnerable, it creates safe space for others to be vulnerable, and trust is built on your team. Knowing each other's stories is how unity is built.

Teamwork is the only way to survive as a whole, well individual in a career as trying as this one. Ask yourself who might benefit from hearing your story.

Brotherhood in the First Responder Careers

As with sports, the first responder jobs are filled with high-pressure situations. But one misstep in a high-pressure situation can mean the difference between life and death.

And sometimes the wrong combination of information, equipment, people, and timing can make a mistake inevitable.

Think back to a mistake you made. Remember how you felt and how you wished you could make it right. Your peers can relate. They have been there. Experience puts first responders in a unique position to help their peers recover from mistakes. That is why they call it "the brotherhood."

Getting back to the brotherhood starts with your word choice. When a peer makes a mistake, how you respond can make the difference between him feeling safe to talk about it and him feeling ashamed enough to bury his feelings. Use words to encourage and build him up rather than tear him down. This is your team, and you want him to win.

The brotherhood is broader than just one department. It spans the entire fire service. Supporting your peers is about bringing back the brotherhood. Even if your department does not have your back, your peers across all first responder fields do.

In relating to a peer, we should be asking, "Do I have his back?" Rather than throwing your peer under the bus with a sideways joke or comment, consider how you can encourage him. Try to point out where he *did* succeed. Share your personal experience with something similar. Remind him that everyone has bad days. Keep him focused on the long-term game. Speak positively about his mistake to others. The job of a first responder is hard enough without peers tearing you down. Do you have his back?

Core Values Assessment

If you would like to benefit from leveraging your personal values in your life, here is an exercise I use with my counseling clients:

1. Look at the core values assessment and identify three of the top values in your personal life (fig. 13–1).
2. Write a mission statement for yourself based on those three values.
3. Then write out specifically how you are going to use those values through action and behavior in your daily life. Try not to use the word in your definition.
4. If possible, write on paper, in a planner, or in a journal how you used those values in your current week, or if you did not use them, how you plan to use them in the coming week.

My clients have found this exercise to be helpful in choosing what is most important in their daily lives. When you find that you have been operating outside of your key values, you will be able to recognize feelings of being torn, tired, and not productive. But when you operate within your values, you will feel productive, effective, and passionate. I believe that shifting your efforts in the right direction to be who you uniquely are, and to contribute your unique attributes to the brotherhood, will help you be more passionate and happier about life in general.

Hudson Clinical Counseling — Jada Hudson, M.S., LCPC, CADC

Core Values Assessment
Common Personal Values

Accomplishment	Discipline	Individuality	Pleasure	Self-love
Abundance	Discovery	Initiative	Power	Self-mastery
Accountability	Diversity	Inner peace	Practicality	Self-reliance
Accuracy	Education	Innovation	Preservation	Self-trust
Achievement	Efficiency	Integrity	Privacy	Sensuality
Adventure	Environment	Intelligence	Problem solving	Service
Approval	Equality	Intensity	Professionalism	Simplicity
Autonomy	Excellence	Intimacy	Progress	Sincerity
Balance	Exploration	Intuition	Prosperity	Skill
Beauty	Fairness	Joy	Punctuality	Solitude
Challenge	Faith	Justice	Purpose	Speed
Change	Faithfulness	Knowledge	Quality over quantity	Spirituality
Clarity	Family	Leadership	Quantity over quality	Stability
Cleanliness, orderliness	Flair	Learning	Reciprocity	Standardization
Collaboration	Flexibility	Love	Recognition	Status
Commitment	Forgiveness	Loyalty	Regularity	Straightforwardness
Communication	Freedom	Meaning	Relaxation	Strength
Community	Friendship	Merit	Reliability	Success
Compassion	Frugality	Moderation	Resourcefulness	Systemization
Competence	Fulfillment	Modesty	Respect for others	Teamwork
Competition	Fun	Money	Responsibility	Timeliness
Concern for others	Generosity	Nature	Responsiveness	Tolerance
Confidence	Genuineness	Nurturing	Results	Tradition
Connection	Good will	Obedience	Romance	Tranquility
Conservation	Goodness	Open-mindedness	Rule of Law	Trust
Content over form	Gratitude	Openness	Sacrifice	Trustworthiness
Cooperation	Hard work	Optimism	Safety	Truth
Coordination	Harmony	Patriotism	Satisfying others	Unity
Creativity	Healing	Peace, Non-violence	Security	Variety
Credibility	Holistic Living	Perfection	Self-awareness	Vitality
Decisiveness	Honesty	Perseverance	Self-confidence	Wealth
Democracy	Honor	Persistence	Self-esteem	Wisdom
Determination	Improvement	Personal Growth	Self-expression	
	Independence	Personal health	Self-improvement	

Jada Hudson (M.S., LCPC, CADC) is the owner of Hudson Clinical Counseling, serving women, children, and first responders. She specializes in depression, substance abuse, suicide, pediatric death, and PTSD. Contact Jada at 630.815.3735 or visit www.hudsonclinicalcounseling.com

Figure 13–1. Identify three top values using this core values assessment list.
Source: Adapted from Brené Brown, "List of Values," Dare to Lead (New York: Random House, 2018), https://daretolead.brenebrown.com/wp-content/uploads/2019/02/Values.pdf.

Reflection Questions

1. What are some of the strengths listed earlier in this chapter that you can see in yourself?
2. How can you put those strengths to work in your job?
3. What are some things you uniquely bring to your department?
4. Who is someone in your department you need to speak up for or make room for in the brotherhood?
5. What are three small ways you can do that this week?

Notes

1. E. Diener and M. E. P. Seligman, "Very Happy People," *Psychological Science* 13, no. 1 (January 1, 2002): 81–84, https://doi.org/10.1111/1467-9280.00415.
2. Chris Myers, "An Introvert's Guide to Leadership," *Forbes*, August 14, 2016, https://www.forbes.com/sites/chrismyers/2016/08/14/an-introverts-guide-to-leadership/#117fc70e6d8b.
3. M. Tartakovsky, "4 Things Introverts Do That Makes Them Effective Leaders," *PsychCentral.com*, July 8, 2018, https://psychcentral.com/blog/archives/2013/09/28/4-things-introverts-do-that-makes-them-effective-leaders/.
4. Myers, "An Introvert's Guide to Leadership."
5. Myers, "An Introvert's Guide to Leadership."
6. Diener and Seligman, "Very Happy People."
7. S. R. Asher and V. A. Wheeler, "Loneliness and Social Dissatisfaction Questionnaire—Modified," *APA PsycTests Dataset* (1985), https://doi.org/10.1037/t04785-000.
8. Diener and Seligman, "Very Happy People."
9. T. Niederkrotenthaler et al., "Predictors of Psychological Improvement on Non-Professional Suicide Message Boards: Content Analysis," *Psychological Medicine* 46, no. 16 (September 22, 2016): 3,429–3,442, https://doi.org/10.1017/s003329171600221x.

CHAPTER 14

PATERNALISTIC/MATERNALISTIC LEADERSHIP, TRAINING, AND THE IMPORTANCE OF ALLIANCES

Brad's Story

Brad was involved in a training accident in which he had a near-death experience. The trauma that he carried around after the incident, however, was actually the result of his realization that his fire chief and his department did not have his back.

After the incident, Brad had many post-traumatic stress disorder symptoms, including frightening thoughts about his own death and the death of his children. To avoid thinking about the accident, he wore his headphones constantly. He was always on edge. He was having problems sleeping and was exhibiting angry outbursts, along with depression and a loss of interest in things he used to love to do.

When we got to the heart of the matter, Brad was most traumatized because he felt his department had abandoned him. He wanted to have the incident investigated because he felt that they were negligent in this training. According to Brad, the incident was pushed under the rug and not investigated properly. This left him with the feeling that his department did not care about him in a situation in which he could have died.

The influence of leaders on their subordinates is magnified in the fire service and other first responder careers because of the intensity of the situations faced on a daily basis. Leaders have the opportunity to prepare their subordinates, lead them through intensity and adversity, equip their bodies to respond with physiological toughness, and model what healthy coping and resiliency looks like. Amid innumerable potentially traumatic events (PTEs), leaders can direct the steps of their subordinates in a way that either points toward or away from PTSD.

For firefighters, PTEs can arise any day and at any time. Such events expose firefighters to danger and potential loss of life within ambiguous circumstances.[1] Without preparation, individuals naturally assess PTEs as a threat to their physical or psychological health. But leaders can train their subordinates to assess their circumstances through a lens of hope and strength, which impacts their emotional wellness after the event has taken place.

Trajectories for People Facing a PTE

Individuals facing a PTE will either turn toward resilience or toward PTSD based on their challenge/threat appraisal and their coping flexibility/coping inflexibility.[2] Here are the possible trajectories for someone facing a PTE:

- Perceive PTE as a Challenge + High Coping Flexibility = Resilience
- Perceive PTE as a Threat + High Coping Flexibility = Recovery, then Resilience
- Perceive as a Challenge + Coping Inflexibility = Delayed Recovery or PTSD
- Perceive as a Threat + Coping Inflexibility = PTSD

Challenge Appraisal Versus Threat Appraisal

What determines the appraisal an individual makes when he faces a potentially traumatic event? According to research conducted by Joseph Geraci, clinical psychologist, US Military Academy, and others, in a *challenge appraisal*, the individual assesses the event or task and sees that "my resources (internal and external) exceed the demands of the PTE about to be experienced." In challenge appraisal, the individual's belief system remains intact. Geraci and others note that in a threat appraisal, the individual assesses that "my resources (internal and external) do not exceed the demands of the PTE about to be experienced," leaving the individual's belief system shattered.[3]

When an individual assesses a PTE as a challenge, his mental state is regulated, he is emotionally satisfied and interested, and he shows improved performance. Physiologically, his blood pressure does not rise, his heart pumps blood efficiently to enable action (increased output), and adrenaline is released.[4] He is ready mentally and physically to fight.

When an individual assesses a PTE as a threat, his mind triggers an emergency response and raises cortisol (stress hormone) levels. He may feel anger, fear, guilt, shame, and sadness; he can either attack or withdraw and will show diminished performance. Physiologically, his blood pressure rises as his arteries restrict flow and do not increase output.[5] Sadly, he ramps up without an ability to perform at his best because of the perceived threat to his life or well-being.

Other Factors Influencing Challenge Appraisal Versus Threat Appraisal

In appraising a challenge versus. threat situation, an individual typically considers three factors:[6]

Chapter 14 • Paternalistic/Maternalistic Leadership, Training, and the Importance of Alliances

- How much effort do I need to mobilize?
- How certain/uncertain is the context?
- How dangerous is this situation?

If the individual sees that he has the energy to put forth the effort, a reasonable expectation of the outcome, and an ability to survive the dangers, he will perceive it as a challenge, leading to improved performance and probable resilience after the event. But if the individual sees that the PTE will require more effort, have an uncertain outcome, and be harmful, he will probably drift toward threat appraisal, reducing performance and increasing the likelihood of developing PTSD after the event.[7]

Leaders can train their subordinates to assess PTEs as challenges by providing them with hardy, realistic training and talking them through the training one step at a time. This will help the subordinates understand why they are being asked to undertake specific training tasks and will also help them feel ready if a similar real-life situation occurs.

Setting also plays a role in how the individual assesses the PTE. Consider how the event may shape an individual's reputation. An individual will see an event as a challenge if he feels adequately prepared and if the event is public or in view of his peers and leaders. How others will perceive him is important to him. He will see the PTE as a chance to prove himself and will be more likely to see it as a challenge. But in order for him to see the event as a challenge, his skills must moderately or at least marginally exceed the requirements of the PTE. If the event is too easy, he will see it as boring. If the PTE vastly exceeds his abilities, it will be a threat to him.[8] So leaders should stage training exercises accordingly.

Self-talk also plays a role in the individual's perception of the event. As individuals talk to themselves, the use of the second person, i.e., "You can do this," has proven to be more effective than the first person, i.e., "I can do this," because it creates some distance between the self and the PTE.[9] He takes the role of "impartial observer," shifting his focus to be able to see external insights. Leaders should model the use of the second person and help subordinates find adequate self-talk to make it through a PTE.

Self-talk is one of many ways a firefighter can protect himself mentally when facing a potentially traumatic event. Saying positive things to oneself, such as, "You can do it," or "Easy does it" shifts the individual from the weight of perfectionism, protecting him mentally from threat appraisal.[10]

Another factor influencing challenge appraisal or threat appraisal is the individual's level of autonomy. If he has enough space to problem solve on his own, he will step up to the challenge, but if he is not given enough space to innovate, he may see that he does not have adequate support and may view job complexity as a threat rather than a challenge.[11] If he is given space to innovate, he may come up with solutions that overcome the PTE, and his overcoming will help him strengthen his assessment of his own internal resources. His next PTE will then be more likely to be seen as a challenge rather than a threat.

Within this desire for autonomy is a desire for at least a small measure of control over his circumstances. The individual needs to see that he has some power over the outcome. And when the individual is given a position of responsibility, it changes him psychologically, and he steps up to behave like a leader. This position of responsibility and the need to help others is often associated with the feelings of power that he needs.[12]

Leadership That Prevents PTSD

For Brad, I wonder if leadership could have influenced his psychological outcomes after his training incident. Leaders can reduce the likelihood of their subordinates developing PTSD in two ways. First, subordinates must feel that their leader is their resource and knows more than they do. Second, they also must feel that their leader really does care about them. If one of those ingredients is missing, a subordinate is at risk for developing PTSD in a traumatic incident.

Even though the individual is a subordinate, his sense of collaboration with his superiors influences his perception of challenge versus. threat. Individuals who are invited to collaborate with superiors feel they have access to more capabilities, skills, and provisions. This increased sense of the resources available helps a person assess a PTE as a challenge rather than a threat. But if individuals experience a sense that their skills are superior to the people they are working with, they are more likely to experience threat assessment. If you are a leader, train your subordinates, stretch them, and let them see your strengths as well.

In Brad's case, both aspects of a healthy response—seeing his leader as a resource and feeling like his leader really cared about him—were missing. He had a feeling going into the training that things were not right, but he went along with it because the training officer told him to. In the end, he felt like his life was at stake and the leaders were negligent. Brad's overall feeling was that the department did not care about him. He felt that his department covered up the problem, with the chief turning his back when Brad tried to formally complain. He experienced violent nightmares about arguing with his chief and generally felt abandoned.

Training Subordinates for Physiological Toughness

Leaders influence the cognitive preparation of their subordinates and their responses to traumatic events. They can also empower their subordinates to train their *bodies* to be ready for the events.

Studies have shown that individuals can actually train their bodies to respond with all of the physiological responses—for exmaple, the right chemicals and blood flow—to empower effective action. Psychologist Simon Moss notes that stressful experiences *plus* a sense of control *plus* adequate time to recover result in a body trained to be physiologically tough. For example, individuals who willingly undertake stressful situations in which they have control, such as swimming in cold water, show rapid and intense spikes in adrenaline in response to stressful tasks. These adrenaline spikes tend to enhance performance. In addition, Moss notes that people who endured stressful experiences earlier in life, coinciding with a sense of control followed by sufficient recovery, establish a pattern of physiological response to subsequent challenging events. This can serve to enhance emotional stability and performance.[13]

What happens to the body when physiological toughness is achieved? During challenging or stressful events, individuals who have developed physiological toughness exhibit a sharp, rapid rise in response hormones called *catecholamines* (i.e., adrenaline and noradrenaline) but a limited increase in cortisol. Their bodies experience the following:

- Rise in heart rate
- Augmented mental activity
- Relatively constant blood pressure

Catecholamines, particularly noradrenaline, also facilitate the conversion of fats to energy, which increases muscular activity.

When an individual has developed physiological toughness, his mind releases chemicals and his nervous system responds to enable him to adapt to challenging events. He shows less fear or avoidance in stressful contexts and instead performs optimally. He tends to conceptualize these events as challenges rather than threats. The elevated levels of adrenaline somehow prevent fearful or avoidant responses. In a nutshell, a physiologically tough person releases more adrenaline, which equates to improved performance, and less cortisol. Elevated cortisol equates to increased stress and decreased performance.[14]

Understanding physiological toughness has implications for training subordinates. If individuals have exposure to stressful situations with control of the stimulus (e.g., being allowed to jump at their own timing), combined with appropriate recovery period, their bodies will adapt to these controlled stressors, and their performance will improve in the next stressful event.

Other factors can also influence physiological toughness: (1) early experiences, (2) passive toughening, (3) active toughening, and (4) aging.[15]

Early Experiences

According to research conducted by Richard Dienstbier, University of Nebraska–Lincoln, stressful experiences early in life, if coupled with opportunities to respond actively and to recover sufficiently, tend to promote physiological toughness. In particular, in a series of studies, young rats or mice were exposed to stressful conditions or electrical stimulations. Those exposed to early stressful experiences actually developed larger adrenal glands.[16] Based on the results of this study, individuals with adverse childhood experiences and sufficient recovery may make great first responders.

Passive Toughening

Passive toughening can develop physiological toughness as well. This occurs when an external challenge happens and the individual is given some control.[17] When individuals undertake stressful tasks but are granted a sense of control, they develop physiological toughness. If the task is dull, boring, or too easy, however, stress hormones are released, especially in type A personalities. This may be because they experience a sense of impending punishment.

Beware of training exercises that are not sufficiently challenging. These training exercises will not build the confidence of the trainees and may come across as punishment. When you train your subordinates, make sure the training is challenging and realistic. Explain the significance of the training to everyday life.[18]

Active Toughening

Active toughening can also develop physiological toughness. These deliberate attempts to engage in stressful contexts can promote physiological toughness. Examples include swimming in cold water and aerobic exercise. Individuals who intentionally push themselves to become uncomfortable will train their bodies to step up to the challenge whenever the next uncomfortable or stressful circumstance arises. This can be seen clearly when individuals have the highest levels of catecholamines immediately after their exercise.[19]

Aging

Aging affects physiological arousal. The sensitivity of the receptor sites for neurotransmitters decreases as we age. Consequently, the effects of adrenaline and other neurotransmitters also decrease as we get older.[20]

Training Implications for Leaders

Leaders desiring to improve the performance of their subordinates should remember that challenge appraisal improves performance and well-being of subordinates, while threat appraisal shuts down their physical strength as well as their emotional strength. In training, subordinates need to go through hardy, realistic exercises of what they will face on the job. These training experiences should be equally challenging to situations firefighters will face on the job, but they also need to be safe. Brad's training exercise nearly killed him, but it was the response of his leaders that actually traumatized him.

Firefighters need realistic training to discover their own personal internal resources, gain confidence in their leaders, and to build trust with their team. Researchers call this building their "psychological body armor."[21] This armor is made up of their resources, their social support (including family members), and their trust in their leader. Training may be the strongest influence on how individuals assess a PTE, and therefore on their likelihood of developing PTSD or choosing resiliency.

Brad's training accident left him feeling abandoned by his leaders, which harkened back to some of his childhood pain and abuse. For him, this experience with his department reinforced the idea that people abandon you. He felt layers of resentment, anger, and deep sadness. Being let down by his department was not the first time he had felt this way, but it certainly deepened his pain.

Therapy was quite helpful for Brad. He was willing to talk about many different areas of his life, including family of origin issues. We even talked about his father, who was former military, and the fact that he felt very disconnected growing up in his family, which was not emotionally expressive. This lack of family connection was an important piece of the puzzle. When first responders have unresolved childhood trauma and they become parents themselves, they carry their own deficits and are often unable to connect with their children. Next we returned to his chief, who Brad felt had minimized the dangers of the incident and never acknowledged the serious implications of hanging him out to dry. Brad felt that the chief had zero empathy or concern for him, and it caused him long-term problems such as anxiety, depression, and feelings of uncertainty about his own skills and abilities.

Obviously, Brad took charge of his uncertainty by coming to talk to me. He took control of what he was responsible for—himself. But if Brad's leaders had seen the situation from the viewpoint of supporting Brad, his emotional wellness would have been different after the event. Leaders are privileged to have one of the most powerful places in the lives of their subordinates to build them into resilient, courageous men and women. Alternately, the actions of the leaders can teach their subordinates that they do not care, and that their subordinates need to face PTEs alone. This is traumatizing for the subordinates. Leaders get to choose how they train and respond to their subordinates.

Of equal importance to training and responding is the effectiveness of leaders as role models for their subordinates. As a leader, you set the tone for the well-being of your subordinates. Your words matter. The individual needs to see that he matters to you as a person. So interpersonal conversation should happen frequently in the downtime before a PTE. Likewise, he needs to see that as a leader, you are giving him opportunities to develop skills, as opposed to telling him, "Just go do it." This will develop a culture of learning orientation instead of a punitive atmosphere. Subordinates will strive to enhance their skills, work toward plausible goals, and view events as challenges.

Brad and I talked about a self-care plan moving forward. Self-care is one of the strongest builders of personal resiliency. For Brad, his self-care plan involved reducing the time that he was working on other projects and increasing time to focus on his family and to connect more deeply with his wife. We planned for him to go fishing and golfing more often because nature is very healing for trauma. Reconnecting in Brad's life meant focusing on real friendships and working less, particularly when it came to extra duties and overtime. He chose to understand that the fire service was just one part of his life, and that there were other things he wanted to find meaning in as well. I am grateful for the opportunity to work with Brad and other open, genuine people in the fire service.

Modeling Vulnerability and Coping Flexibility

Leaders set the tone for the behavior of their subordinates. They can perpetuate the idea that first responders need to "suck it up," or they can model vulnerability, resilience, and healthy self-care. The degree of influence that leaders exert on whether or not their subordinates develop PTSD depends on the way they train their subordinates and how they model healthy responses to PTEs.

Leaders need to pay attention to unity on their teams. Pack mentalities naturally arise in high-intensity jobs, but they need to be shut down at all costs. The team needs each individual's strengths to be maximally effective, and when there is an outsider, if that individual's strengths are not leveraged effectively, he is exposed to more emotional danger and is more likely to become emotionally unwell.

In order to shut down pack mentalities, leaders must open the door to unity through their own vulnerability. The way to model vulnerability is by sharing your personal struggles. This will help your team bond over shared vulnerability, rather than bonding over scapegoating. Are you or your team keeping someone on the outside? It is up to you to bring him in. His performance will improve and so will his well-being.

In addition to modeling vulnerability, leaders should show their subordinates a variety of ways to cope with PTEs.[22] Leaders not only must develop hardy, realistic training and help subordinates build psychological body armor, but they also must model coping flexibility after a traumatic event so that subordinates can see the various ways to process and move forward.

Recalling the possible trajectories for individuals facing PTEs presented previously in this chapter, it becomes clear that coping flexibility can make all the difference in whether or not someone develops PTSD. Individuals should be provided opportunities to talk about the event with their leader. They should be allowed to take time to journal, sleep, be with their families, talk to a peer supporter, take time off, or take advantage of other helpful coping opportunities, depending on the type of event and the needs of the individual.

It is clear that the cognitive assessment of a PTE and the coping flexibility of the individual can influence which way an individual turns—toward resiliency or toward PTSD.[23] But leaders can greatly influence the outcome. Let us look at how leaders can behave toward their subordinates before, during, and after PTEs, so that subordinates head toward resiliency every time.

Preparing Subordinates for PTEs

An individual's "psychological body armor" is made up of three parts: training, social support, and leadership.[24] *Training* means being physically and psychologically prepared to encounter any possible scenario. When an individual is well trained, he is more likely to be resilient. *Social support* means being fortified by positive interpersonal relationships with their peers, family members, and leaders. These relationships strengthen him and provide resources to help him be resilient in the face of trauma. *Leadership* means having trust that your leader is skilled enough to lead successfully through PTE and that your leader cares about each individual's well-being. These three pieces of psychological body armor can make or break the outcomes for a firefighter or other first responder when encountering traumatic circumstances.

In order for a leader to build his subordinates' psychological body armor, he needs to train them well. The goal of training should be to develop hardiness in subordinates. This requires rigorous physical training and realistic simulations of likely situations, including simulations of how to care for wounded peers or dead peers. Training should teach them they can handle it and should help them address any "what if" questions they might have, such as, "What would it be like if I got injured?" or "What if a peer did not make it?" In every phase, leaders should explain how the training connects to a PTE, showing the purpose of the training.[25]

Make sure, however, to provide adequate recovery time after the realistic training so that subordinates can process what they saw and did and grasp how they might handle a similar situation in the real world. Putting them through the wringer, so to speak, will not strengthen them. Giving them hardy training with adequate recovery time is how you strengthen them, just as you would in a workout.

Within very rigorous training, subordinates are pushed to their limits in order to pull together with their departments and develop social support for one another. This adds to a person's list of external resources when facing a PTE. But remember that intense situations can create a pack mentality where one individual becomes the scapegoat for teasing or just an outsider. That individual's performance will diminish if you allow him to remain there. His sense of resources depends on camaraderie. So make sure that every individual is given respect, chances to learn, and mentorship from leaders if his performance falls short. Leaders can seek to develop those who do not handle realistic training simulations well, so they can grow, rather than wither. Remember that vulnerability is what builds trust and cohesiveness in a department.

How to Lead Before a PTE

The role of the leader shifts depending on what stage the subordinate is in with regard to a PTE. Before a PTE, the job of the leader is to prepare the subordinates physically and psychologically for the various scenarios they may face. Equally important, the leader should build a positive interpersonal relationship with each individual subordinate. So in the period before a PTE, the leader should operate as 50% task-oriented and 50% relational-oriented toward his subordinates.[26]

Task-Oriented (50%)
- Set and enforce clear standards.
- Provide rigorous, realistic physical and call-specific training.
- Provide psychological skills-based training.

Relational-Oriented (50%)

- Establish positive interpersonal relationships; learn about subordinates' personal lives (such as families or ambitions).
- Explain the purpose for training and missions with regular one-on-one conversations about how training is preparing them for PTEs.
- Strengthen the bond between the individual and the unit and the individual and his or her family.

What Is Paternalistic/Maternalistic Leadership?

Leaders whose subordinates develop fewer cases of PTSD typically are known for their emotional and social leadership. This is called *paternalistic/maternalistic leadership*. Treat them like they are your children, and you are their parent.

In this style of leadership, leaders display empathy regularly. According to D. Goleman and others in *Empathy: A Primer*, empathy is seeing others' feelings and how they see things, being actively interested in their concerns, listening attentively to understand their point of view, and sensing unspoken emotions. The authors note that "fundamentally, leadership involves relationships. . . . To be effective in those relationships, leaders must understand the perspectives of the people with whom they are working. What leaders need is empathy."[27]

Leaders should strive to pick up on subordinates' feelings through their verbal and nonverbal cues and experience what subordinates are feeling by putting themselves in their shoes. Understanding the perspective of another person and the forces that informed that perspective helps them and their leader work as a team to be resilient. Picking up on the feelings of others and experiencing what they are feeling can serve to humanize both the leader and the subordinate.[28]

Even if empathy does not come naturally for some, it can be developed by practicing curiosity and deep listening. When we are curious about other people, we seek to learn their perspective. We ask questions and carefully listen to their answers in a process called *deep listening*. Try to pay attention to how people answer. Are they excited about your question? Distracted? Irritated? Over time, as you practice using being curious and listening closely, you will be able to more accurately understand how others think and feel.

How to Mitigate Anxiety as a Leader

Subordinates may not openly state that they are anxious, but they will communicate it in negative statements such as, "We're never going to be able to do this." According to Mike Murphy, senior contributor to *Forbes*, this is less a sign of opposition and more an indicator of anxiety. Though this sounds negative, it is actually a sign that the individual is taking ownership of the job or project. Murphy suggests that taking small steps to calm the subordinate can go miles in shifting that individual from freak-out mode to calm trust in their leader.[29]

Murphy suggests some helpful responses that leaders can have to subordinates who are experiencing fears in different areas:[30]

- *Feelings of being overwhelmed.* Do not be tough on them when they are already overwhelmed. Communicate support and listen to their concerns.
- *Fear of making mistakes.* Continue to expose them to hardy, realistic training and assure them that mistakes sometimes happen, but training is helping them become sharper in their skills.
- *Fear of the unknown future.* Assure them that whatever happens, your team will be in this together.
- *Having a list of unfinished tasks at the end of a shift (the "cliff-hanger effect").* Help them hand off responsibilities that can be delegated to the next shift and help them identify tasks that can wait.
- *Having a big project with unclear steps to accomplish it.* Work with them to break the project down into steps: "Let's just take this one step at a time."

To mitigate anxiety as a leader, Murphy suggests taking a diplomatic approach and choosing to mentor subordinates and develop them rather than buckling down on them. He calls it the "Do It Right the Second Time Approach," noting that "eliminating the fear of making mistakes and turning errors into learning opportunities that inspire root-cause problem solving provides all employees with new and valuable learning opportunities."[31]

The cutting-in-half technique, as suggested by Murphy, employs the following steps:[32]

Step 1. Ask employees, "What do you have to accomplish in (time frame) in order to stay on track of (big goal)?

Step 2. Next ask, "What do you have to accomplish in the next 90 days to reach that six-month mark?"

Step 3. Next ask, "What do you have to accomplish in the next 30 days to reach that 90-day mark?

Step 4. Finish by asking, "What do you need to accomplish today to stay on track of it all and make today a successful day?"

When you reduce anxiety for your subordinates, you are making your work environment *psychologically safe*. Laura Delizonna notes in an article for the *Harvard Business Review* that in psychologically safe work environments, employees feel accepted and respected, which leads them to innovation, speaking their minds, creativity, and healthy risk-taking. This often leads to outside-the-box thinking and remarkable solutions to work-related problems.[33]

In contrast, leaders who "throw you under the bus" or allow conflict in the workplace can trigger the fight-or-flight response in the brain, which causes the brain to shut off creative thinking and innovation to focus on survival, according to Delizonna. The author notes that when leaders build trust, confidence, and psychological safety, the brains of their subordinates will release more oxytocin, which promotes bonding and will send their minds into "broaden and build mode." In this mode teams are motivated, resilient, and persistent. Delizonna also notes that signs of a psychologically safe work environment include humor, divergent thinking, and solution-finding.[34] The environment you create as a leader not only affects your subordinates in terms of their level of anxiety, but it also affects their innovation. This in turn directly impacts the productivity and performance of your department.

How do you create a psychologically safe environment in your department?

1. *Communicate with your team.* Connection begins with communication. Healthy communication takes place when both individuals get to contribute to the conversation.[35]
2. *Give them time.* Face-to-face conversations affirm your care for them and communicate that they are valued.[36]
3. *Approach conflict as a collaborator,* not an adversary. Work together with members of your department to find solutions, allowing them the respect and autonomy to help you find solutions.[37]
4. *Replace blame with curiosity.* Instead of jumping on them for a mistake or a choice you did not agree with, ask questions to gain an understanding of your firefighters. They may be more innovative and brilliant than you realized. You can recognize their initiative, even if their decision was not best.[38]

Remember, you get to set the tone for your department, and caring, communicating, and listening go a long way.

How to Lead During a PTE

Communicating with a subordinate during a PTE can be difficult in the intensity of the moment. That is why relationship building should occur *before* a PTE. During a PTE, a leader's job is to make timely decisions and maintain the trust of the subordinates. If there is time to talk, do quick check-ins with subordinates to see how they are doing and to understand the event from their perspective.

The balance of relational and task orientations shifts during a PTE to be 70% task and 30% relational.[39]

Task-Oriented (70%)
- Accomplish the mission.
- Minimize unnecessary trauma.
- Share same risks as subordinates.
- Show physical and moral courage.
- Exhibit clear and rapid decision making.

Relational-Oriented (30%)
- Check the well-being of subordinates.
- Look at the situation from their perspective.

Allowing the subordinates to see your personal strengths during the PTE will allow them to have a greater sense of their external resources. Your primary job during a PTE is to make sure people know where to be, what to do, and who is doing what. Your focus is on execution. If you have time to ask, "Are you doing okay?" during a PTE, do it. If not, demonstrate leadership and plan to check in on them afterward.[40]

How to Lead After a PTE

After a PTE, the demeanor of the unit will shift dramatically, and the leader's job is to help subordinates process what happened. Your job as a leader is to help subordinates learn from the PTE, to model for them the ways they can process what happened, and to show them that you care about them.[41] Try to talk to each person about what went right and what went wrong. Allow them to give you feedback on how you did as their leader as well.

After a PTE, the relational- and task-orientation balance shifts again, and leaders should be about 35% task-oriented and about 65% relational-oriented.[42]

Task-Oriented (35%)
- Take lessons learned from PTE.
- Create tactics and training to handle the next PTE.

Relational-Oriented (65%)
- See subordinates as unique with different coping needs.
- Tell subordinates, "Good job."
- Develop self-awareness.
- Model coping flexibility.

After a traumatic event, subordinates will have a renewed sense of social support from a leader who is paternalistic or maternalistic. One encouraging note for leaders who endure tragic, ongoing, or unprecedented circumstances with their departments is that some of the most resilient firefighters come from departments with shared trauma. This is what happened with many firefighters on 9/11. When people see trauma together and are surrounded by social support, they can talk about the shared incident, and the traumatic aftershocks are mitigated. Many can experience post-traumatic growth if the trauma is handled appropriately.

Firefighters can protect themselves mentally from trauma and PTSD by how they think before, during, and after potentially traumatic events. Leaders can model this thinking for them. Because all first responders of every rank, both male and female, experience stress responses, it is imperative for leaders to enable all subordinates to process in healthy ways. When a first responder talks about what he or she is going through with someone who has had similar experiences, it can help legitimize and validate those feelings. Having space to talk about worries and concerns about the behavior of another peer is also important for firefighters.

Leaders should work to build trust, communication, and a reliable flow of information within their departments. Remind subordinates that risk and injury of death is built in to all first responder operations. They have conquered and excelled in many life-and-death situations in the past. They are resilient and intelligent survivors, and they should build on their past experiences.[43]

Are Critical Incident Stress Debriefings Effective?

It may seem intuitive that critical incident stress debriefings (CISDs) should be consistently conducted after PTEs. The effectiveness of this type of debriefing may be limited, however, because it is forced. In addition, it is difficult to evaluate the efficacy of CISDs because control groups and treatment groups cannot easily be randomized and established within the same department and situation.[44] The most important thing to remember about stress debriefings and first responders is that many first responders are introverts. Introverts process thoughts in their own time. They will most likely want to talk to one trusted individual, rather than disclosing their feelings in front of a group. Firefighters may even prefer to talk to friends on their own initiative, so leaders should model talking to friends about what happened, and they should encourage their subordinates to talk to peers or peer supporters. These conversations will help them be resilient, possibly more than a CISD would.

The upside of a CISD, however, is that it opens the door to making it okay to talk about what happened. So if you do not have a CISD, make sure you immediately talk about what happened because subordinates will take their cues from you about what is okay to talk about and what is not okay to talk about. You can make this easier or harder to deal with, based on how you respond.

Modeling coping flexibility and encouraging individuals to find healthy coping mechanisms may be more effective than CISDs for a number of reasons. According to research conducted by David Richards, University of Manchester (UK), CISDs may focus on the wrong things and prevent individuals from talking about anxiety, mood disorders, or even substance use they have begun since the PTE. In addition, mandatory CISDs have been shown to cause greater aggravation and higher levels of stress than authentic conversations between leaders and subordinates or between peers.[45] If a leader was a part of the traumatic event, it can be even more traumatizing for him to lead the CISD. Consider having someone else who is further from the event lead the debriefing. Be aware that some leaders can experience secondary trauma when hearing the details of the traumatic event. On top of CISD, consider referring your subordinates to a counselor or peer supporter, showing them it is okay to seek help.

For individuals who are ready to talk about a traumatic incident, there are some factors that will help the discussion to be productive: receiving validation, gaining the perspective of others, filling in the gaps by getting details to make sense of what happened, and the ability to talk candidly in a safe environment. If you are hosting a group conversation or a personal conversation, try to steer the conversation to validate, fill in gaps for them, and allow them to talk about what happened. But remember, never force them to talk when they are not ready to talk. This will be only counterproductive.

Sadly, some individuals will turn to unhealthy coping mechanisms like substances or behaviors that help them dull the pain or drown their sorrows for a period of time. Individuals striving to be perfect, individuals whose sense of self-worth is based on performance, or those who have not yet identified healthy coping skills will have a harder time coping in healthy ways.[46]

Leaders have the power to guide the lives of their subordinates. They can get them ready, or they can allow them to fail. They can connect and show empathy for them, or they can tell them to "suck it up." Leaders can be friends, and leaders can be like a parent. And leaders can know that their subordinates have the ability to move toward resiliency every time they face a PTE if given the right support. Leaders have a vital role in the well-being of their subordinates by supporting them.

This is beautifully expressed by author S. Kelley Harrell: "Often it isn't the initiating trauma that creates seemingly insurmountable pain, but the lack of support after."[47]

Using Fire and Rescue Alliances

Fighting a fire demands that you go into battle mode. You have a task ahead of you and someone who needs you. Nothing should slow you down. But what if another department arrived before you? Who calls the shots then? Did they do a 360° size-up to your standards? What if they missed something? It can get clunky when multiple cities work together, and one slip up could cost someone his life. Here is how alliances were formed in my neck of the woods—the western Chicago suburbs.

Bob Hoff, chief of the Carol Stream Fire Department and former fire commissioner of Chicago, saw this problem and dreamed of a partnership in which multiple fire teams would work together regularly and know exactly how to act when that time came. He joined with a group of fire chiefs to form the West Suburban Fire/Rescue Alliance. Comprised of the Carol Stream, Wheaton, Winfield, and West Chicago Fire Departments, the West Suburban Fire/Rescue Alliance shares resources, coordinates emergency dispatch, trains together, and assigns specific leadership roles to individuals in each department in order to practice taking orders from leaders in other departments. Others have seen how well it has worked, and now the Bloomingdale and Roselle Fire Departments have joined the alliance.

Battalion Chief Hugh Stott, West Chicago Fire Department, invited me to observe some of the alliance's cooperative trainings. I was impressed by their commitment to setting aside department boundaries for the sake of teamwork and safety.

This formal decision to cooperate has made the fire departments of Wheaton, Winfield, West Chicago, and Carol Stream more efficient, effective, and safe. They share a dispatch system, which sends the closest available fire station to the emergency, even if that station is from another department. And by training together regularly and frequently, they have effectively broken down barriers and built trust among themselves, so that whoever gets to an incident first has the authority and the trust of the other departments to do what needs to be done, ultimately saving lives.

The supervisory level of the alliance is organized into various chief roles. The *incident commander* carries the ultimate responsibility for the operation. The *interior chief* supervises what goes on inside the fire building. The *plans chief* monitors communications, checks on available resources, assists the incident commander, and assigns auxiliary tasks to supporting chiefs. The responsibilities of *safety chief* and *rapid intervention team (RIT) chief* go to one individual until another department arrives and the roles can be separated.

Understanding fire behavior is crucial, and Wheaton Assistant Chief Jeff Benda gave me a lesson on how reading fire can save lives. He noted that "reading the color of the smoke gives firefighters an intuitive sense of what is going on in the fire. Knowing the construction of the building gives insight into fire travel."[48] And knowing the personalities of those in neighboring departments helps when making important action calls. After training together, these departments know how to alter their tactical approaches. The process is clear, smooth, and safe.

The first step in forming an alliance is the decision to cooperate. Leaders must set aside their egos for the sake of unity. As a therapist, I see many conflicts arise from ego. Whether it surfaces in the workplace or in the home, ego interferes with healthy relationships because ego ultimately stems from fear. If someone else is in charge, gets what they want, or decides for the group or the family, you have to let go. And that process of letting go causes anxiety.

How can ego and the tension it causes be replaced with unity? It takes practice. In the home, this may involve foregoing the right to call the shots and trusting that the other person will make an acceptable call. As trust is built, fear subsides, and ego loses its place.

In first responder careers, practice also eliminates ego. What if you knew you could trust the people you were giving up control to? This is the beauty and strength of alliances at work or home. Each department practices letting go of various tasks throughout an emergency response, and each department grows to trust the others, so that ego and anxiety no longer interfere with any emergency situations.

It is true that transitioning to an allied first responder system requires change, which initially may feel awkward. It may cause anxiety. Change gets easier with practice, however, and practice will lessen that anxiety. Groups will soon find unity and trust that is well worth the effort to achieve.

Leaders Gearing Up for Stratospheric Success

In my concluding thoughts about being a good leader, I wanted to share an idea from the book *The Go-Giver* by Bob Burg and John Mann. Leaders motivated by these five laws can greatly influence the success of their subordinates:

1. *The law of value.* Your true worth is determined by how much you give in value rather than what you take in payment.
2. *The law of compensation.* Your income is determined by how many people you serve and how well you serve them.
3. *The law of influence.* Your influence is determined by how abundantly you place other people's interests first.
4. *The law of authenticity.* The most valuable gift you have to offer is yourself.
5. *The law of receptivity.* The key to effective giving is to stay open to receiving. [49]

A leader's style will be shaped by his perception of how true these laws are and how he employs them in his workplace. But whatever his goals and values, and whether he is an introvert or an extrovert, a leader should seek to build value in his department. A leader should seek to serve others and empower them to be effective out on their own, not stuck in their leader's shadow. A leader should influence others by valuing their unique interests and encouraging them to develop or expand on those interests. A leader should offer his or her unique strengths, while caring about each individual's unique strengths as well. A leader should be willing to receive and also willing to give.

By understanding your role as a paternalistic/maternalistic leader, you will shift your perspective into how to develop, build into, and equip your team for success. These laws will empower you and your department to reach its full potential. There is no limit to what you and your subordinates can do if you help them remain emotionally well and resilient throughout their careers.

Reflection Questions

1. How would you describe your relationships with your leaders and peers in your department?
2. Is there anyone with whom you need to cultivate a better relationship?
3. What is one change you want to make in your leadership as a result of this chapter?
4. Have CISDs helped you process trauma? If you are the leader, how would you shift CISDs in your department to make them more effective?
5. What are some of the benefits you can see from working together in an alliance with another department?

Notes

1. J. Geraci et al., "Understanding and Mitigating Post-Traumatic Stress Disorder," *US Army Research* 344 (2011), http://digitalcommons.unl.edu/usarmyresearch/344.
2. Geraci et al., "Understanding and Mitigating Post-Traumatic Stress Disorder."
3. Geraci et al., "Understanding and Mitigating Post-Traumatic Stress Disorder."
4. Geraci et al., "Understanding and Mitigating Post-Traumatic Stress Disorder"; and S. Moss, "Physiological Toughness," SicoTests.com, June 18, 2016, https://www.sicotests.com/psyarticle.asp?id=230.
5. Geraci et al., "Understanding and Mitigating Post-Traumatic Stress Disorder"; and S. Moss, "Physiological Toughness."
6. S. Moss, "The Distinction Between Challenge and Threat Appraisals," SicoTests.com, June 28, 2016, https://www.sicotests.com/psyarticle.asp?id=281.
7. Geraci et al., "Understanding and Mitigating Post-Traumatic Stress Disorder"; and Moss, "The Distinction between Challenge and Threat Appraisals."
8. Moss, "The Distinction between Challenge and Threat Appraisals."
9. Moss, "The Distinction between Challenge and Threat Appraisals."
10. A. P. Doran et al., "Dealing with Combat and Operational Stress," Ceridian Corporation, 2004, https://www.marineparents.com/usmc/downloads/dealingwithcombatandoperationalstress.pdf.
11. Moss, "The Distinction Between Challenge and Threat Appraisals."
12. Moss, "The Distinction Between Challenge and Threat Appraisals."
13. Moss, "Physiological Toughness."
14. R. A. Dienstbier, "Arousal and Physiological Toughness: Implications for Mental and Physical Health," *Psychological Review* 96, no. 1 (1989): 84–100, https://doi.org/10.1037/0033-295X.96.1.84.
15. Dienstbier, "Arousal and Physiological Toughness."
16. Dienstbier, "Arousal and Physiological Toughness."
17. Dienstbier, "Arousal and Physiological Toughness."
18. M. Frankenhaeuser, U. Lundberg, and L. Forsman, "Dissociation Between Sympathetic-Adrenal and Pituitary-Adrenal Responses to an Achievement Situation Characterized by High Controllability: Comparison Between Type A and Type B Males and Females," *Biological Psychology* 10, no. 2 (March 1980): 79–91, https://doi.org/10.1016/0301-0511(80)90029-0.
19. Dienstbier, "Arousal and Physiological Toughness."
20. Dienstbier, "Arousal and Physiological Toughness."
21. Geraci et al., "Understanding and Mitigating Post-Traumatic Stress Disorder"; and Moss, "Physiological Toughness."
22. Geraci et al., "Understanding and Mitigating Post-Traumatic Stress Disorder."
23. Geraci et al., "Understanding and Mitigating Post-Traumatic Stress Disorder."
24. Geraci et al., "Understanding and Mitigating Post-Traumatic Stress Disorder"; and Moss, "Physiological Toughness."

25. Geraci et al., "Understanding and Mitigating Post-Traumatic Stress Disorder"; and Moss, "The Distinction between Challenge and Threat Appraisals."
26. Geraci et al., "Understanding and Mitigating Post-Traumatic Stress Disorder."
27. D. Goleman et al., *Empathy: A Primer*.
28. D. Goleman et al., *Empathy: A Primer*.
29. Murphy, "How to Tweak Your Leadership Style When Your Employees Are Feeling Anxious."
30. Murphy, "How to Tweak Your Leadership Style When Your Employees Are Feeling Anxious."
31. Murphy, "How to Tweak Your Leadership Style When Your Employees Are Feeling Anxious."
32. Murphy, "How to Tweak Your Leadership Style When Your Employees Are Feeling Anxious."
33. L. Delizonna, "High Performing Teams Need Psychological Safety. Here's How to Create It," *Harvard Business Review*, August 24, 2017, https://hbr.org/2017/08/high-performing-teams-need-psychological-safety-heres-how-to-create-it.
34. Delizonna, "High Performing Teams Need Psychological Safety."
35. C. Satterle, "Closing the Communication Gap" (PowerPoint presentation. n.d.).
36. Satterle, "Closing the Communication Gap."
37. Delizonna, "High Performing Teams Need Psychological Safety."
38. Delizonna, "High Performing Teams Need Psychological Safety."
39. Geraci et al., "Understanding and Mitigating Post-Traumatic Stress Disorder."
40. Geraci et al., "Understanding and Mitigating Post-Traumatic Stress Disorder."
41. Geraci et al., "Understanding and Mitigating Post-Traumatic Stress Disorder."
42. Geraci et al., "Understanding and Mitigating Post-Traumatic Stress Disorder."
43. Doran et al., "Dealing with Combat and Operational Stress."
44. K. Barboza, "Critical Incident Stress Debriefing (CISD): Efficacy in Question," *The New School Psychology Bulletin* 3, no. 2 (December 10, 2005), https://pdfs.semanticscholar.org/3384/b3c58069375de1ce081e06f6610289d4cff2.pdf.
45. D. Richards, "A Field Study of Critical Incident Stress Debriefing Versus Critical Incident Stress Management," *Journal of Mental Health* 10, no. 3 (2001): 351–362, https://doi.org/10.1080/09638230124190.
46. Geraci et al., "Understanding and Mitigating Post-Traumatic Stress Disorder."
47. S. K. Harrell, *Gift of the Dreamtime: Awakening to the Divinity of Trauma*, 2nd ed. (Fuquay Varina, NC: Soul Intent Arts, 2012).
48. J. Benda (assistant chief, Wheaton Fire Department), in discussion with the author.
49. B. Burg and J. D. Mann, *The Go-Giver: A Little Story about a Powerful Business Idea* (New York: Portfolio/Penguin, 2015).

CHAPTER 15

HEALTHY RELATIONSHIPS— PREMARITAL AND MARRIAGE

To begin a successful marriage, you need to have a holistic understanding of each other's pasts. This includes things like your childhood family life, your financial situation growing up, your parents' relationship, and your romantic relationship history. We all come from different backgrounds, and our life experiences are what shape us. How a person is raised influences the way he or she will behave in a marriage. Both partners need an understanding of themselves and their own tendencies, as well as those of their partner.

Marriage can be one of the most beautiful and protective experiences in the life of a first responder. I have seen some beautiful partnerships in first responders' lives, but I have also seen some couples struggle because of the strains of this career, different financial expectations, and traumatic experiences. This chapter is designed to equip you to identify the pitfalls in some first responder marriages and work through those in nonconflict conversations, so that you and your partner can have an enduring, beautiful, lifelong union.

Marriage is not supposed to be a relationship of total agreement all of the time. In fact, about 69% of conflict in most marriages will never be fully resolved.[1] Marriage is about learning to manage and respect the differences between partners and live in a peaceful, mutually respectful relationship.

The First Responder Family

First responders often come from families of other first responders, where asking for help is not the norm. They usually are raised in an environment where the first responder instinct is to take care of your own problems and keep them to yourself. These individuals are likely not aware that this is what they are doing all the time because it is what they have grown up with and are used to. They do not know another style of interaction because it is ingrained in their family background as well as their career, encompassing all of their life.

Children

For many couples, starting a family is an important consideration when thinking about marriage. Even if you and your partner do not plan on having children immediately, it is likely to be something that affects the course of your marriage significantly. Deciding whether or not to have children, and then deciding when that may be if you do want to have children, will be two of the more important decisions you will make as a couple. It is a major choice, and starting your marriage knowing how the other person feels about children is critical.

First Responders Need Time to Recover: 24-On/48-Off Shift Patterns

An important thing to consider is the recovery time first responders need between shifts in order to maintain their physical and emotional health. When thinking about having children, you should both agree on your plan for childcare knowing that first responders should not be expected to take on the role of primary caretaker between shifts. This is critical time for them to reset due to the emotionally and physically taxing nature of the job they do. Talk about how you will ensure this time if and when you think about starting a family.

Communication

Communication between a married couple is unlike any other relationship. Couples often find that marriage introduces new types of conversations that they have not previously encountered in life, so they may struggle to communicate effectively with their partners about new topics. Achieving good communication is critical to help couples bond and build on their relationship. Being set up for success from the beginning will be the first step in effectively resolving conflicts or issues throughout the many stages of your marriage.

Paramilitary Communication Style in First Responders

The communication style employed in a first responder's personal life may be heavily based on their paramilitary professional background. Paramilitary communication allows first responders to do their jobs successfully because it is quick, efficient, and based on commands. They are used to using specific commands on the job and are not accustomed to discussing things further in a back-and-forth dialogue. Unfortunately for many couples, this communication style tends to carry over into personal relationships, where it is not a positive way to communicate. This is something first responders will need to be acutely aware of and work to avoid in marriage.

How First Responder Stress Can Impact Intimacy

According to clinical psychologist and marriage counselor Randi Gunther, when stress invades your life, your five senses can become overwhelmed, making it more difficult to differentiate between the welcome sensory inputs and the unwelcome. While senses may sound elementary, they profoundly impact the way you interact with the world and the relationships around you. That is why highly stressed people can find it harder to differentiate between demanding and life-giving sensory experiences in a love relationship.[2] One small rejecting gesture, one day at a time, can suck the love out of your relationship if you let it. And these gestures often come from a person who does want love but is just burned-out.

For many couples, when one partner is stressed, it often shows up as that partner pulling away from one of the five senses—sight, sound, smell, taste, or touch. If the stressed partner stops receiving advances and shuts off from them, the other partner may begin to feel rejected. If this goes on for long, it may become painful or too exhausting for the other partner to put himself or herself out there.[3] So the relationship begins to deteriorate because neither individual is contributing emotionally—one out of exhaustion from stress, and the other out of fear of rejection.

Additionally, the stressed-out person may have a "shorter fuse" when it comes to their partner's idiosyncrasies. An insignificant bad habit may spark an angry outburst from the stressed partner, wounding the other partner and causing the stressed person to feel guilty.[4] All the while the stressed person reassures himself, "It'll be okay. I'll chill out when things get easier at work."

Gunther notes that in stress, the brain goes into overdrive, and thinking can become overwhelming and less clear. Many partners in this situation offer support, but the stressed person may not recognize that the suggestions and supportive offers are helpful because they may only further jumble his thinking.[5] So again, an outburst may come in the form of a rejection of his partner's help. At this point, the partner is starting to shut down.

The stressed-out partner may look for a quick way to de-stress in the form of sex. If the relationship is going okay at that point, a happy partner may be willing to oblige in order to help the stressed partner feel calm for a bit.[6] But if the nonstressed partner or less-stressed partner is emotionally wounded, he or she may be unlikely to provide sex, and the rejection may feel as if it is coming from both sides. Many women want connection before intimacy, and a woman may actually want less sex during times of stress. Instead, healing conversation or even cuddling may help build a bridge and de-stress her.

It is unfortunate how stress can drive people to build patterns that slowly destroy love. It does not have to be that way, however. I always recommend that first responder couples have open conversations about stress and how their homes can be safe, calm places. When couples are comfortable talking about stress and actively working to undertake anti-stress practices in their lives and marriage, they can live in harmony, helping the stressed spouse reconnect with self and others. Love can be the best way for first responders to stay balanced and connected, but it starts by being open about stress and actively fighting its damaging effects on relationships.

When First Responders Have an Overactive SNS

First responders have more fight-or-flight neurotransmitters on a regular basis than the average person. This amount of chemicals will significantly influence your first responder's communication style due to the increased intensity with which they experience the fight-or-flight response when their sympathetic nervous system is activated. When this is happening, the individual cannot access the appropriate communication skills for the home and relationships. Therefore, it may seem like individuals are avoiding conflict or refusing to communicate with their partner. In reality, this biological response is due to the nature of their job and chemical reactions in their bodies.

It is important for you to understand that this is not a personal attack and is not due to anything that has happened in the relationship. When you experience gaps in communication, often a period of cool down is needed for the fight-or-flight reaction to calm down. Only then will you be able to have a successful conversation.

Every relationship will encounter conflict at one point or another. Problems often arise when couples do not understand each other's ways of handling conflict, or when one or both parties struggle to communicate well in stressful situations. Understanding each other's communication styles is an excellent start to managing conflicts that arise, but in the heat of the moment, there are many other factors to consider. The best way to successfully manage conflict in your relationship is to learn how to improve as a team.

The Unique Way First Responders Experience and Respond to Conflict

When it comes to conflict, you will likely find that the first responder in your relationship views things differently than another person would. First responders experience chaos and intense situations on most calls. Therefore, their perception of traumatic things or negative situations is different than the general population. First responders will likely downgrade the significance of any situation due to the way it compares to the very traumatic things they see regularly.

Additionally, many firefighters tend to use sarcasm to deflect conflict. I encourage first responders to adopt an at-home communication style that uses less sarcasm with their families. Sarcasm is a coping skill to help them manage their feelings, but it can be hard for outsiders to understand. It may also make it seem like they do not care about a situation. The most important thing for the spouse to know about these tendencies is that they are not based on personal feelings toward the significant other or spouse but rather are based on a style of communication used at work.

Basics of Listening Well

When you are in the heat of a conflict with your partner, one of the most important factors that will affect the outcome is how well you listen to each other. These are some of the main techniques for communicating with your partner during a conflict or at any other time.

- Your goal is to understand and convey understanding of how the other person is feeling. This is more important than the specific details of the event. Relating to how your partner feels is your top priority.
- Listen more than you talk.
- Do not assume you know what the other person is feeling just because you are feeling a certain way.
- Focus on what the other person is saying, rather than thinking about your personal needs or what you will say next.
- Hear what the person is saying without interrupting or sharing your thoughts.
- After they finish speaking, repeat in your own words what your partner just said, checking for accuracy.

As you briefly recap what they have just said, try to:

- *Reflect.* Verbally restate the content of what the partner has said.
- *Clarify*: Rephrase what the partner has said to improve understanding.
- *Focus*: Help direct the partner's attention to a single topic.
- *Summarize*: Put together key ideas and core feelings into a brief statement.

Finances

According to research, money is the most common source of conflict and argument between couples. There are many aspects relating to money about which couples can easily disagree. Although it can be uncomfortable for couples to talk about finances while dating, a successful marriage will start with two people who have been open with their financial affairs. Discuss the following points with your partner and make sure you can reconcile any possible differences of opinion when it comes to managing your finances as a couple.

Differing "Money Scripts"

I was recently encouraged by Doug Aller,[7] a first responder financial planner, to read the book *Facilitating Financial Health* by Brad Klontz, Rick Kahler, and Ted Klontz.[8] I was impressed by how helpful this type of information is for first responder couples. Most counselors skim over this topic, but looking at it in depth will create the best scenario for an emotionally well individual and marriage. I have seen immediate results with the couples I counsel as we have explored this topic.

The authors of *Facilitating Financial Health* believe that financial paradigms—what they call "scripts"—are taught in childhood. When people look at their family histories, they can see how financial mindsets are passed down and are formative in the financial perspectives of the next generation.[9] Aller pointed out that often financial scripts are different between partners, causing foundational conflicts that can start to be resolved when couples recognize their different scripts.

My first responder father held quite different money scripts than my mother. Aller would say that this is because they grew up in highly different financial situations. My mom was raised in a solid middle-class scenario financially. Her parents were farmers, and I do not believe she ever had a time when her family was financially insecure. Her financial scripts included being conservative in spending, but not at the expense of her family's needs and comfort.

My dad, on the other hand, as I mentioned in chapter 2, grew up in poverty. His lack of food, good shelter, and security directed his financial scripts for the rest of his life. I believe his financial scripts went something like this:

- "If I have money, I won't be bullied, insecure, or depressed."
- "If I have money, I will survive, be respected, and be protected."
- "If I buy used, I can keep more of what I have worked hard for."
- "If I work more hours and save my money, I can give my family a better life."
- "I can put up with a lot because saving money is more important than comfort."

My dad always bought things used whenever possible. We lived in nearly intolerable conditions for the sake of saving money. He saw investments as savings, so he never had a risky portfolio. He avoided taking luxurious vacations and always opted for trips to visit family instead. His spending goal was always zero or as little as humanly possible. In his mind, if you spent money, you lost, and if you saved money, you won. Because of this, he recreated poverty for himself, my mom, and myself. This is common for people with unresolved trauma. They recreate similar experiences for themselves in the future, meanwhile passing some of their dysfunction on to the next generation.

My dad would always say, "You don't get something for nothing." That repeat script echoed in my ears, and even as an adult, I am fairly conservative financially.

But I noticed that my dad had a different standard when it came to education. He believed that education was the ticket out of poverty, so he was quite generous with me when it came to education. He afforded me the opportunity to study abroad in London while I was in college. And he created a fund for my daughter's college education. That was one area he always spent openly.

At the end of his life, my dad was satisfied with the wealth he had created for his family. Though it was not excessive, it was enough to provide a different kind of life for me. I am truly thankful, but I think he could have enjoyed life a little more and had less conflict with my mom and his children if he had written some new financial scripts for himself.

Your Unconscious Beliefs About Money

Maybe you have seen your parents fight about money. Maybe you and your partner experience tension about money. This most likely stems from deep beliefs each partner holds about money.

Authors Klontz, Kahler, and Klontz explain how deep some of these beliefs can be:

> Money scripts are formed in childhood, shaped from both direct and indirect messages we receive about money. From our parents, other significant people in our lives, our circumstances, and society as a whole. We internalize these messages with the immature thinking of childhood as we integrate information in order to help us make sense of the world. Some money scripts are formed at a deep, primal level and become part of our world view. . . . These beliefs, developed for survival and protection in an unpredictable world, are often incredibly resistant to change, especially when there is strong emotion attached to them.[10]

Are there money scripts you hold that are causing tension in your marriage? Aller pointed out what Klontz, Kahler, and Klontz explain so well: "A critical component of financial health is learning to identify and change one's money scripts."

In my practice, I have seen financial stress bring up anxiety, depression, and even marital dysfunction because of how deeply held money scripts can be. Perhaps a critical component of marital health is learning to identify and rewrite your money scripts.

Klontz, Kahler, and Klontz offer some examples of money scripts and an explanation of how they show up in the lives of those who operate by these beliefs.

Money Scripts Examples

1. "More money will make things better." Once essential needs are met—at about $50,000 annual income in the US—people are not necessarily happier with more money. This false belief can often cause people to become workaholics or over-savers, afraid of spending.
2. "Money is bad." This belief usually means people sabotage any financial progress they make.
3. "I don't deserve money." Often this belief is accompanied by low self-esteem. This perpetuates bad financial decisions or even underemployment.
4. "I deserve to spend money." This is usually a belief that people use as rationale to spend beyond their means. These people often live month-to-month or in great debt.
5. "There will never be enough money." Workaholics often take this view, and it can destroy their relationships as they endlessly work to earn money.
6. "There will always be enough money." This belief can come from rich or poor backgrounds, and it often assumes the universe will take care of us, regardless of actions. While these people can be okay financially, this belief lends itself to financial irresponsibility.
7. "Money is unimportant." While it is true that money does not bring happiness, love, or belonging, this belief can contribute to poor financial planning, lack of concern about financial matters, or even laziness.
8. "Money will give me meaning." This script often arises from poverty, and people who believe this often adamantly deny it, though it shows in their financial and work choices.
9. "It's not nice (or necessary) to talk about money." Often people feel shame connected to financial conversations. This can make it difficult to talk about, especially for people who believe this. Some couples have trouble with financial conversations because of these feelings of shame.
10. "If you are good, the universe will supply all your needs." It is the belief that if you do good, you will be provided for. This can be true, but it can also be an excuse for financial laziness.

Source: B. Klontz, R. Kahler, and T. Klontz, *Facilitating Financial Health: Tools for Financial Planners, Coaches, and Therapists* (Erlanger, KY: The National Underwriter Company, 2016).

The list of money scripts could potentially be infinite. Yours may sound something like one of those mentioned, or they could be completely different. Before you and your spouse find yourselves in the middle of financial tension, it may be helpful for you to take some time to reflect. Aller encourages couples to ask themselves: "What are your deep beliefs about money?" "What is your family's financial history, and how do you think that impacted the development of your own money scripts?" Once you have reflected, ask your partner if you can sit down and have an open conversation about where you are coming from.

In this conversation, try to talk safely about how you want to live financially. Aller suggests couples look at their monthly income and set goals together. Klontz, Kahler, and Klontz suggest some items you may want to discuss:

- What are your financial goals, and how do you think you want to go about meeting those goals?
- What would it look like for you and your partner to live within your means?

- Do you want to have agreed upon spending amounts so that if you or your partner are shopping, you call and talk about larger purchases before spending?
- Do you want to contribute to savings and retirement plans?
- What types of investments do you want to choose?
- What is your plan to keep a reasonable and low amount of debt?
- What stresses you out financially, and how can you and your partner work to eliminate financial stress?
- If money is a tool, what would you like it to do for your family?[11]

This should be an ongoing conversation. Maybe you establish an understanding with your partner that you can talk about this openly whenever. Or maybe you set a money planning date once a month specifically to go over your expenses and goals and make sure you are on the same page. Communication will eliminate frustration in the financial aspect of your marriage, and I highly encourage you to start healthy financial conversations right away. I am becoming a firm believer after my conversations with Aller that open communication about finances and money scripts can help first responders overcome the financial tensions that often drive a wedge down the middle of otherwise healthy marriages.

The Emotions of Money

Couples facing financial trouble would benefit from recognizing that the root of money trouble is not usually about money at all.[12] It is actually about the unresolved emotions behind how a person felt about being provided for or not being provided for as a child, according to Aller. When couples come from a place of vulnerability in their money conversations, it can help gain traction and mutual understanding to build on going forward.

If you have had trauma around money in your past, you may have reacted by being overly frugal, not wanting to spend money at all. Or you may have reacted to money trauma by being absolutely passive about paying attention to money. Often individuals who have money trauma in their pasts will marry someone with money trauma and the opposite reaction to it.[13] Thus you have one individual who is tight-fisted and the other who overspends, creating the perfect storm of financial tension in a marriage. But the common ground is money trauma in the past. Talking about where you are coming from will help you and your spouse understand your money values and the history behind them.

My dad had some deep emotions when it came to money, but like many of his emotions, he kept them buried. Because of his unresolved trauma, money was a touchy subject in my household growing up. I have wondered what it would have been like if my dad had sat down and written out his money scripts, gently comparing them with my mom's to find compassion and common ground. Things might have been different if only he had seen how illogical and emotional some of his money scripts were. But he just kept on working to absolve his feelings of financial insecurity.

Illogical Money Beliefs

If you are honest with yourself, you may realize that some of your beliefs are illogical. Or perhaps some of your spending habits are illogical and are based on your deeply-embedded money scripts. Self-reflection can be transformational both for your habits and for your marriage.

The money scripts that people operate by point to beliefs they hold about money, and they often fall into one of four categories: money avoidance, money worship, money status, or money vigilance.[14]

Money Avoidance

According to authors Klontz, Kahler, and Klontz, money avoidance scripts include, "I don't deserve money," "Money is bad," or "Money is unimportant." Usually those who hold these beliefs have lower levels of education, income, and net worth. Often these beliefs contribute to self-destructive money decisions like compulsive buying, hoarding, financial enabling, and workaholism.[15] Typically, wealthy people do not hold money avoidance scripts. Money avoidance scripts are illogical, and the goal should be to rewrite them into scripts about money acceptance and balance.

Money Worship

Klontz, Kahler, and Klontz note that money worship scripts include, "Things would be better if I had more money," "More money will make me happier," or "Money would solve all of my problems." People who hold money worship scripts usually are young, single, less educated, lower income, and higher in debt. They can overspend, overwork, depend on others financially, and even deny financial problems.[16] Money worship scripts are illogical, and the goal should be to write them into scripts about appreciation for money and use of money for appropriate life activities and goals.

Money Status

According to Klontz, Kahler, and Klontz, money status scripts include, "People are only as successful as the amount of money they earn," "If something is not considered the 'best,' it's not worth buying," or "Your self-worth equals your net worth." People who have status money scripts are likely to have grown up in lower socioeconomic classes. These are predictive of overspending, gambling, and possibly financial infidelity. The authors also note that individuals in the top tiers of wealth are higher in this money script and can often derive their self-worth from their net worth, which can contribute to workaholism.[17] Money status scripts are illogical and should be rewritten into scripts about money's limitations, satisfaction, and use of money for others.

Money Vigilance

Klontz, Kahler, and Klontz note that money vigilance scripts include, "You should not tell others how much money you have or make," "Money should be saved, not spent," or "I'd be a nervous wreck if I did not have money saved for an emergency." Individuals who hold these beliefs typically have lower levels of debt and fewer financial disorders.[18] Money vigilance scripts are illogical and should be rewritten into scripts about balance, appropriate spending, and appropriate saving and investing decisions.

Klontz, Kahler, and Klontz state that those who live out these illogical money scripts may develop financial disorders such as excessive dept or bankruptcy, use of money in attempts to control others, lack of savings and investments, anxiety, worry, depression, or resentment about their financial situation, or even ongoing conflict with others about money. The authors also note that illogical money scripts can contribute to compulsive financial behaviors, overspending, preoccupation with material things, lying to others about financial behaviors, underearning, excessive financial risk-taking, gambling, preoccupation with work, money, and spending, and more.[19]

The good news is that money scripts can be changed. You are not stuck in the money scripts you observed and wrote for yourself as a child. You can learn new financial habits and live a financially organized and disciplined life as an adult, at peace with your partner. It takes active-brain decision making, but you can choose to rescript your money scripts.

The Importance of a Budget in Reducing Marital Financial Stress

As mentioned previously in the text, Doug Aller and I have been in an ongoing conversation about first responder finances and how they can help mitigate marital stress. He is a certified financial planner and serves as vice president at New Concept Benefit Group. He specializes in serving first responders and is currently working with 450 active firefighters and more than 80 retirees from various fire departments throughout the state of Illinois. I asked him to share his thoughts about budgets and first responder marriages, so I could include them here in this book. This is what he said:

> In about the year 2000 or so, I met with a firefighter and his wife to gather financial information to put together a financial plan for them. At the time, this firefighter's department had an optional direct deposit program that most of their employees did not use. The process of going to the firehouse to pick up a paper check, then take it to be deposited at a bank seemed very cumbersome and time-consuming to me. But I came to understand why these firefighters were willing to put in the extra work: "Fun Money Checks."
>
> You see, payroll was paid twice per month—usually on the 1st and the 15th. However, this particular department also had a semi-annual payment to each firefighter for a uniform allowance, duty pay, and shift pay. This semi-annual check was usually about $850, so it added about $1,700 to each firefighter's annual income. The firefighters in this department affectionately called these

semi-annual checks "Fun Money Checks" because they were off their spouses' radar.

As the years passed, and I prepared written financial plans for many of these firefighters, they would make me swear not to mention their "Fun Money Checks" in front of their spouses. I would tell them, "Wild horses could not drag it out of me."

I kept the "Fun Money Checks" secret for 17 years. But in 2017, the municipality mandated all fire department personnel switch to direct deposit of paychecks, including all "Fun Money Checks."[20]

Aller continues, noting,

This real-life story illustrates why many Americans—firefighter families included—can face financial stress. Finances, in general, are a common source of anxiety. In fact, money is one of the main issues couples fight about. With baby boomers entering retirement, it has become an even more significant issue. This particular scenario illustrates how financial infidelity is on the rise. According to the National Endowment for Financial Education [in a 2018 Harris Poll], 41% of American adults admit to keeping financial secrets from their partners, including hidden accounts, debt, or spending habits. Millennials are nearly two times more likely to keep financial secrets from their partners than earlier generations.

Over the past 23 years of working with firefighter families, I have seen many different financial circumstances. Unfortunately, there is no "secret sauce" as far as eliminating financial stress in a family's life. However, without a doubt, one of the best methods I have seen to reduce financial stress and take control of your family finances is to have both partners agree upon designing a written monthly budget. Most importantly, they have to agree to adhere to the budget every month.

At first a budget may seem limiting, but a budget properly designed can be liberating because it allocates money for fun and realistic discretionary spending with no guilt. Due to unforeseen circumstances, your budget many change a little bit each month. But a written budget that you both agree to—and stick with—will give you control of your financial lives. As you start to control your financial lives you will feel less stress. Working together to set a budget can help you establish a shared vision for your future.[21]

Money conflict and money secrecy are common, and they can be extremely stressful and painful for couples.[22] Money can be a nonissue in your marriage, or it can become the number one factor that reduces your marital satisfaction.[23] By establishing a budget and healthy financial patterns in your family, you can reduce tension and redirect your children's financial scripts. I truly believe you and your partner can live at peace financially, getting to your goals and enjoying the money you have in a balanced and healthy way.

Setting a Money Agreement in Your Marriage

According to Klontz, Kahler, and Klontz, couples who agree about money have a much lower divorce rate. Couples who argue about money usually have a hard time succeeding financially.[24] That is why I believe every first responder couple needs to make a money agreement. Every couple is different, so your money agreements should be unique to the two of you and your financial situation.

Here is how it could look for you and your partner:

1. *You can set agreements about how much money you want to devote to certain priorities.* This can help you increase spending in your priority areas.[25] For example, if a healthy diet is important to you and your spouse, perhaps you increase your grocery budget a bit to afford space for more organic items.
2. *You establish family spending scripts together.* These can include, "We don't buy big things the day we decide we want them," or "We shop around before we jump into big purchases." This can help you say no to impulsive purchases that one or both of you may regret. These types of unified money patterns can help prevent you and your partner from having tension after a purchase has been made.
3. *You set limits in areas that cause tension.* Doug Aller suggests this to his financial clients. Setting spending limits can help you not spend above certain amounts without the consent of your partner. This way, you agree about bigger purchases. For some couples this limit will be higher than others. For example, a couple could decide that if they are buying something more than $100, the person who is shopping calls and talks about it first. Or the couple could decide that groceries and household goods can be higher than $100 without a conversation, but splurge items over $100 always require a conversation first. Spending limits may work for you as a couple, or they may cause tension, so talk about this as a possibility before deciding it is an absolute for you.
4. *You decide if you want to keep yours, mine, and ours accounts, or if you want everything to be joint accounts.*[26] There is no right or wrong here. Consider what will work best for you as a couple. Some couples keep separate accounts, and each person contributes a percentage, not a dollar amount, to the joint account. For others, they do everything with joint accounts and consider it all "ours." Talk to your partner as soon as possible about what would work best for you both in regard to account setups.
5. *You can set debt limits.*[27] Together, you may decide that you are only comfortable with a certain amount of debt. Klontz, Kahler, and Klontz suggest that you then agree that you will not spend anything that would cause the debt to exceed that limit without having a conversation together or even talking to a financial planner first.

Money scripts can be passed down to the next generation, and they can be harmful or helpful for them. If you decide together to set a budget and establish financial priorities in unison, your children can grow up writing healthy money scripts that will lead to lifelong financial responsibility. The best news is that you and your spouse are not stuck in the patterns of your childhood or in conflict about money scripts. You can rewrite your scripts, and in doing so, redirect the entire financial course of your family.

The Danger of Overworking

First responders have the unique opportunity of having large periods of "time off" between shifts. Because of this, couples are often tempted by the possibility of increasing their income by having first responders take on a second job during their off time. This can be a dangerous situation and should be avoided. It is critical for first responders to maintain time off to engage in self-care. Couples that take the bait for additional income often find themselves overworked and dependent on money that is not sustainable. The best option is to live within your means based on the incomes that are sustainable for your life in the long-term. Value the time between shifts that is necessary for emotional and physical well-being, along with a happy marriage.

Though work can be a beneficial thing, addiction to working and earning can be devastating for relationships. Some see work as an opportunity to earn more to provide for the family. In reality, however, family members of workaholics often see the overworking as a choice to love work instead of loving them, which heightens tension in the family. Klontz, Kahler, and Klontz note that children of workaholics have higher rates of depression and anxiety and a greater sense that things are out of their control. They may see their families as dysfunctional, and they are at a higher propensity for becoming workaholics themselves.[28]

Klontz, Kahler, and Klontz observe that in workaholism, relationships are considered an interruption, and work may be an attempt to feel adequate, strong, able, and appreciated. The authors also note that family members of workaholics feel lonely, rejected, and unnoticed.[29] If you tend toward workaholism, remember that when you constantly pick up more shifts or other jobs, you are investing in your work instead of investing in your family and children, and ultimately it will cause your relationships to suffer. In retirement, workaholics often feel like they are ready to enjoy life with their family members only to realize that their relationships with their family members are distant or too far gone to reconnect after all these years.[30]

While money attainment may be important for you, it cannot be the most important thing. Recognizing where your illogical money scripts are leading you astray and rewriting them can help you redirect the course of your marriage and family before it becomes too late.

Sex and Intimacy

Any relationship between two people involves different ways we show and receive love. Sometimes both people share the same styles of loving, but more often than not couples have to work to understand their partner's needs for sex and intimacy in order to sustain a healthy relationship. Sex is one important part of a marriage, but emotional intimacy is just as important and more so in some ways.

Intimacy in a marriage is a protective force against many sources of conflict, and this includes emotional intimacy as well as physical intimacy. Couples who have established strong emotional bonds that allow them to communicate at any level seem to have diminished symptoms of PTSD and depression.

Why Does Infidelity Happen?

Unfortunately, instances of infidelity can often be traced back to the social norms around being "macho" or "manly" at home. Many men find emotions shameful, so they keep them hidden.

All people need an outlet for emotional expression and interpersonal connection, so if they feel social pressure to eliminate that at home, they will then seek it elsewhere. Additionally, if schedules are opposite or couples have lifestyles where they do not see each other often, this can lead to an increased possibility of infidelity. Couples need to spend quality time together consistently in order to have a happy marriage.

The Connection Between Sex and the Emergency Response

The constant increased adrenaline first responders experience because of their jobs affects their sexual activity. One way to describe it is that first responders can become addicted to the neurotransmitters because of the resulting chemical response in the brain. Sex releases similar chemicals in the brain, which leads to the tendency for first responders to utilize sex in order to achieve that feeling or to forget or distract from uncomfortable feelings. Because of this, there is a higher instance of sex addiction in the first responder community. To manage the constant stress of a first responder job, firefighters need to build a robust menu of self-care options that they employ regularly to de-stress and reconnect.

It is important that both people in the relationship understand the importance of maintaining emotional intimacy throughout their marriage. Because of the differences in displaying emotion and intimacy, first responders need true connection to remain connected and balanced.

Compromising When It Comes to Sex

It is common for sexual activity to decrease after pregnancy, especially on the side of the woman. You each need to be open and honest about your sexual needs and preferences, as well as being open to the needs and preferences of your partner. Sexually inactive couples are more likely to get divorced than sexually active couples, so finding that compromise will be key to a happy and long-lasting marriage.

First Responder Emotional Dangers

Cortisol in the brain is a hormone that is generated as a response to stressful situations, as mentioned previously. Because first responders experience so much ongoing stress in their jobs, they are more susceptible to the long-term effects of large quantities of cortisol. This can cause increasing anxiety and depression and can put immense pressure on a relationship or marriage. Knowing how to de-stress and talk about the stressful experiences they have had on the job can be the difference between relational disaster and relational success for first responders.

Depression is one of the most common effects of long-term stress. The tricky part about depression in first responders is that it can be hard to detect. Because many first responders have instrumental personalities, which tend to avoid emotion and focus on action, it can be difficult to see past the busyness to recognize that someone is struggling. Both men and women with instrumental personalities find it more natural to keep working and pursue achievement and performance, rather than to slow down and face painful feelings.[31] They may not recognize they are running, but they are always on the move, and they are often angry. They are literally wearing themselves out to stay disconnected.

The Importance of Self-Care for First Responders

We have discussed many of the negative side effects that a career as a first responder can have. But the good news is that simple self-care regimens can often be extremely effective in avoiding these effects. Everyone needs regular self-care in order to be a happy, healthy person and also to be a successful and present partner in a relationship. Self-care is what will help prevent the dangerous buildup of cortisol that can lead to many of the negative emotional and physical outcomes of this career. Self-care looks different for everyone, but fundamentally it is any activity that helps you to relax and enjoy your life.

Sometimes partners of first responders have the perception that their significant other spends a lot of time hanging out and relaxing at the firehouse. It can lead to spouses believing that the first responder does not need self-care time at home. This is a common misconception. If you give your partner time for self-care at home, it will benefit his or her emotional wellness, and it will benefit your relationship.

How the Spouse Can Help

Spouses of first responders have the unique opportunity to help their partners manage their emotional health. In times of struggle and following trauma, a married couple who is open and honest can gain increased emotional intimacy. It is also important to note, however, that the approach to emotional openness tends to differ between men and women. Women will often feel glad to engage in openness with their spouse, but men will not generally be drawn to emotional sharing in the same capacity and need more of a gentle push. There is a good chance when a first responder's spouse initially attempts to help, one or both parties in the couple will experience some uncomfortable feelings they need to address. While first responders can deal with *physical* pain better than most, *emotional* pain is much more challenging for them.

Advice for Spouses Who Want to Help

- Remember that your first responder is likely a person of introverted nature and may be averse to intense emotional sharing. It may also take him or her longer to process, and therefore your spouse will need more time before being able to identify and talk about thoughts.
- Help your spouse by acknowledging the importance of rest between shifts. Avoid pressuring your spouse to take on a number of responsibilities as soon as he or she arrives home.
- Express willingness and openness to hear about the issues on the first responder's job. This can be a protective factor for PTSD.
- Keep in mind that regardless of how your spouse responds initially, it is not a personal representation of feelings toward you. A knee-jerk reaction is often associated with higher adrenaline levels and paramilitary communication tendencies. Especially for male first responders, when their spouse is talking, high adrenaline levels make it difficult for him to focus. It may seem like he is not listening or does not care. Use this knowledge to your advantage and accept that you may need to wait a little while in order to have a meaningful conversation.
- Offer to have conversations over a walk. Walking while talking helps to work cortisol out of the body and allows a first responder to focus more on a conversation. This is a technique I use with many of my clients, and it has had great outcomes in their lives.

Increasing Positive Relationships

First responders give so much to their jobs and to the safety of citizens around them, yet sometimes the wholehearted commitment given to the fire service can leave the family feeling shortchanged for attention, time, and emotional presence. How then can a firefighter improve his family life? Two things will help: positive words and positive activities.

When a couple's relationship struggles, one of the first steps they can take to recover is to engage in positive activities together, which build goodwill and increase the commitment to the relationship. Therapist John Gottman studied happy and unhappy couples for more than four decades and identified that couples that had remained happy or had effectively strengthened their relationships all had intentionally engaged in positive communication and actions.[32]

Remember back to the beginning of the relationship when love notes, little gifts, flowers, fun and exciting dates, and romantic dinners were almost daily occurrences? Now fast-forward to the daily grind of going out on calls, juggling social commitments, working out, raising children, keeping a house clean, and paying bills. Who has time for romance? Yet without regular positive activities and communication, both partners can start to feel taken advantage of, ignored, or unappreciated. It is no wonder that The Righteous Brothers wrote a song called "You've Lost that Lovin' Feeling." It can happen unintentionally to any happy couple.

Though this "negative drift," as Gottman identified it, can occur almost accidentally, it does not have to set the trajectory for the entire relationship. Couples seeking to improve their relationship should watch for how many positive comments are said in relation to how many negative comments are said, with the goal of increasing the number of positive comments to outweigh the negative. Gottman believes that in a stable, happy marriage, there is a ratio of at least 5 or more positive interactions for every negative interaction, and this could range upward to a ratio of 20:1 or more.[33] This level of positive interaction may come more naturally for some than others. Even so, I believe that intentionally respectful, kind interactions with a partner not only build your marriage but also teach your first responder that your home is a safe and happy place to be—a place where he or she wants to spend time.

This goes both ways. Both partners should be intentional to be interested, express affection, show the other person that they matter, express appreciation for them, find opportunities for agreement, empathize, apologize, accept the other's perspective, and make jokes.[34] In the process of implementing these things, partners will shift the ratio of negative interactions and positive interactions and build a more emotionally safe relationship. It is these emotionally safe relationships and homes that help protect firefighters from the emotionally tumultuous experiences of a first responder career.

Positive Words Start with You

Though it may seem that your spouse is the one who needs to change, you can improve your relationship by making a list of your own personal pleasing and displeasing actions. The list should include all of the positive things that you could say or do that would be likely to please your partner, as well as all of the things you know displease your partner. Do not show your partner the list; just start implementing the positive actions. See if you can get positive actions to far outweigh the displeasing ones. And notice when your partner does something positive as well.

Positive Actions Are Exciting

Couples who are struggling often stop having fun together. Sadly, if one partner is struggling with an addiction or depression, he or she may spend time recovering rather than having great experiences with a spouse. As a result, both members of the relationship stop engaging in social activities and often begin to live "parallel lives." No longer are they increasing positive words or actions, but they are distant and working just to survive daily life.

Reuniting can begin with making a list of activities you would like to do together and putting a few on the calendar. Brainstorming activities and spending quality time together will build trust, make memories, and increase positive feelings toward the relationship again.

Although the fire service demands so much of its service members, it does not have the power to destroy or repair your relationships. Only you can do that. And the good news is that by being intentional to say positive words and engage in positive activities together, you and your partner can move from bored to excited, from distant to close, and from having "lost that loving feeling" to "letting the good times roll."

Can "Happily Ever After" Really Exist?

In a world of online dating, shows called *Married at First Sight*, and millennials getting married later, it is clear that the institution of marriage means something different today than it did for previous generations.

In an article for *Time*, Belinda Luscombe explains:

> Matrimony used to be an institution people entered out of custom, duty, or a need to procreate. Now that it's become a technology-assisted endeavor that has been delayed until conditions are at their most optimal, it needs to deliver better-quality benefits. Most of us think this one relationship should—and could—provide the full buffet of satisfaction, intimacy, support, stability, happiness, and sexual exhilaration. And, if it's not up to the task, it's quicker and cheaper than ever to unsubscribe.[35]

Surprisingly, however, divorce rates have consistently dropped since the 1980s in every age group *except* older adults.

Luscombe discussed these changes toward increases in divorce in the US, noting, "A report in 2014 found it has doubled among people 50 and older in the past two decades; more men over 65 are divorced than widowed. Only a tenth of the people who divorced in 1990 were over 50. In 2010, it was 25%." Luscombe also notes that some psychologists theorize that shifting technological, cultural, and economic influences have shaped the perception Americans have of marriage and has made the picture of singlehood more attractive.[36]

Others believe that both partners working outside the home has disunified couples by providing each individual with separate social circles and stressors. Still others believe that couples stay together until they reach the "empty nest years" out of desire to provide the best opportunities to their children. Along with social media's ever-connected-to-friends nature, this has made it so that couples can find support and conversation outside the home. Couples then arrive at retirement startled by its shifting social dynamics and in desperate need of marital rediscovery.

So the question remains: Can "happily ever after" exist if a marriage struggles to survive the retirement years?

Psychology professor Eli Finkel of Northwestern University found that Americans view marriage today as both the *most* and the *least* satisfying institution that has ever been.[37] But what if "satisfaction" is the wrong pursuit? What if by asking marriage to satisfy us, we are asking it to do something it can never succeed at?

During working years, couples fill their schedules with meetings, children's sports practices, social outings, and corporate functions, finding satisfaction in work, achievement, and social connection. But upon retirement, couples find themselves staring across the breakfast table wondering what to do with this person they are married to. They spent many years looking forward to "kicking back, relaxing, and traveling," and now they are beginning to realize that they could get a little bored. And they are beginning to realize that they do not know their mate as well as they thought.

How can couples be satisfied and develop marriages that survive the retirement years? Here are four keys to finding marital closeness again:

1. *Commit to stick it out.* As Finkel notes, one study of 700 elderly adults found that 100% of them called their long marriage "the best thing in their life." It also found that 100% of them either stated that marriage was either "hard" or "very, very hard."[38] Yet studies have also shown that couples who are committed to one another for the long run actually find a new kind of closeness to their relationship. If they are willing to practice being good to one another, they begin to rediscover and live at the same level of closeness they had during their courtship.
2. *Develop or discover mutual interests.* Think of activities you both enjoy doing together and do them as often as possible. When you can connect over shared interests, your relationship will be able to endure innumerable changes from the outside.
3. *Keep up your sex life.* When comparing couples who have sex once a week to couples who have sex less than once a month, the happiness level of the frequently intimate was almost three times higher. Sex will make your relationship sweeter as you allow intimacy into your relationship.

4. *Find something meaningful to do with your personal time.* Do not ask your spouse to satisfy your every need. There are many ways to fill your days with satisfying activities. Take the time to mentor someone younger than you, volunteer at your favorite organization, or become a consultant. Pursue gardening or travel. And invite your spouse to join in if he or she wishes.

Though it may have been decades ago, something brought you and your spouse together. Much time may have passed, and many changes may have occurred, but you can rediscover the beauty of an amazing marriage even in the retired years. Start by shifting your focus from getting satisfaction out of your marriage to finding closeness in your marriage. I really do believe that "happily ever after" can exist for firefighter couples and for everyone.

Using Intuition Versus Communication at Home

I mentioned earlier that I have huge respect for fire alliances. When I had the opportunity to attend a cooperative training session with the West Suburban Fire/Rescue Alliance in which they tackled a commercial fire together, I kept my eyes open to see what this taught me about first responders and their communication on the job compared to how they communicate at home.

At the training, Wheaton Fire Chief Bill Schultz invited me to step inside the building to see firsthand what a firefighter experiences in a simulated commercial fire. Since commercial fires are potentially more dangerous and less frequent than residential fires, training sessions like this are invaluable for sharpening a firefighter's ability to navigate large spaces, use the most effective tools and tactics, and communicate, especially with members of other departments.

As I watched these firefighters make their way through the building, I noticed that they used a variety of techniques to orient themselves to their surroundings since visibility was limited. Because it is easy to become disoriented in a large commercial building, the team members were careful to stay together to avoid losing anyone. They carried a rope with evenly spaced knots, indicating how far into the building they were. As they moved forward, these firefighters tapped the floor to make sure it was not hollow. As a standard safety measure, they kept the hose to the right of every room, close to the wall, so they could easily find the exit route. Meanwhile, the RIT team stood by, prepared for rescue if an emergency arose.

After the team cleared the building, they debriefed the training and pinpointed errors in order to eliminate potentially life-threatening mistakes in future fires. Surprisingly, firefighters felt like it took forever to get around the building, but the plans chief in charge of timing the mission pointed out that it took the crew only eight minutes. They were precise and methodical, moving slowly to move smoothly, which ended up being faster than they expected.

This precision is exactly why the West Suburban Fire/Rescue Alliance performs these joint trainings. When fighting a fire, there is little latitude for error. One wrong move can have major consequences, so these firefighters must use their tools, instincts, and familiarity with other firefighters to navigate as safely as possible.

Senses and Equipment

As I observed the firefighters at work within the building, Wheaton Fire Chief Bill Schultz explained what was happening and made sure I did not fall. I noticed how stressful it was being inside. Firefighters were completely unable to see and had to rely entirely on other senses and equipment.

Throughout this exercise, I saw that the firefighters used other senses in collaboration with their equipment. For example, firefighters could tell how big a room was by spraying water and noting how long it took to the water to hit the opposite wall. The seamless integration of senses and equipment made firefighting look like art. Battalion Chief Hugh Stott explained, "I look at our job like a trade. I think there are science and cognitive skills, but there is also a sense and a feeling for what is going on based on experience and prior mistakes."[39]

Firefighters must practice strength, cognitive thought, ability, and innovation during every fire. There is no room for a bad day. Firefighters have to be entirely present, putting aside any personal stress for the sake of the mission. Always bringing their "A game," they behave like professionals—intuitive, intelligent, resilient, and determined. In the fire service, intuition reigns and communication is short, direct, and effective.

Communication in Fires Versus Communication at Home

Fighting fires is not like normal life, however. Transitioning from fighting fires back to home life can be abrupt because of how different the two worlds are. In a fire, stress, adrenaline, setting aside of personal feelings, and command communication drive every movement. In normal life, life and death are not at stake. Normal life is less about intuition and senses. Normal life involves feelings and communication with words, especially at home.

Skills like intuition that are used and valued in the fire service may not necessarily be an asset in a marriage. Two people coming from two perspectives grow closer by talking, so assuming your spouse is on the same page as you may simply create conflict. The intuition of the fire service is admirable and strong, but the communication and openness of healthy relationships is equally admirable and strong. Cultivating both is the goal.

The Secret to Better Relationships Is Not What You Think

The marvelous thing about love is that it is real. This person you dreamed of is in your arms, and she is incredible. She is smart. She is beautiful. She is funny. She makes you feel at ease. You cannot help but fall for her because you are so right together.

> You send flowers to make her feel special.
> You write a note and put it somewhere she will see it.
> You make a list of all the things about her that make you smile, and you read it to her.
> You send thoughtful texts throughout the day to show her she is on your mind.
> You ask her questions because you want to know everything about her.
> She matters, and you want her to know it.

As you two get closer, you get more comfortable. It is nice. You can be yourself even more easily. You do not feel like you are trying as hard. You know her better, so you typically know how she will react to something. It is love, and it is real. Soon you navigate your first disagreement. You talk through something that annoyed you more than you thought it would. You make up and move on.

The "daily-ness" of your relationship can make it easy to get sloppy, however. As you get used to seeing that person, you may not think as often about sending thoughtful texts throughout the day or buying her small gifts as reminders of what she means to you. This is not necessarily bad. The danger, though, is that as you let the positive things slide, you may also be letting some negative feelings or behaviors creep into the relationship as well.

Pulling Weeds and Planting Seeds

It takes daily attention to keep your marriage heading toward "happily ever after." Paul David Tripp, pastor and marriage counselor, uses the analogy of pulling weeds and planting seeds. Every day, couples have to be either pulling weeds or planting seeds in order for their relationship to blossom. Tripp states, "Perhaps one of the fundamental sins that we all commit in our marriages is the sin of inattention."[40]

The Mundanity of Excellence

In 1989, researcher Daniel Chambliss undertook a study of competitive swimmers, distinguishing the Olympic-caliber swimmers from ordinary swimmers. He analyzed their workouts, their diets, and their way of thinking. Coaches would have said that physical characteristics or perhaps time devoted to training set these athletes apart as Olympic-worthy. Instead, something surprising emerged. The athletes who focused on the mundane, small movements and spent their energy mastering these small but significant motions internalized these into habits and performed better. Chambliss concluded that "superlative performance is a confluence of dozens of small skills or activities, each one learned or stumbled upon, which have been carefully drilled into habit and then are fitted together in a synthesized whole."[41]

Every day, these athletes spend their time on perfecting small things, and over time those small things add up to something Olympic-sized. Perhaps it is the same in marriage.

The Mundanity of Relational Excellence

Having a healthy relationship does not necessarily require a major overhaul. You may not need to dive into a big discussion or make major changes. Perhaps all you need is to discover and master the small things that make all the difference.

Success in marriage is not found in the extraordinary. It is found in mastering the mundane—the daily mastering of ordinary love. In the Marine Corps there is a saying, "Brilliance in the basics." Reviving your relationship does not mean you have to read every marriage book and find creative ways to communicate again. It does not mean you have to spend a boatload of money to buy your partner an expensive gift or take them on an expensive trip. It means you master the basics.

Basics of Relational Excellence

1. Maintain eye contact when your partner is talking to you. Put down your phone or look away from the television.
2. Have sex at least once a week.
3. Go on dates regularly, aiming for new experiences and not the same old places.
4. Compliment your partner's appearance.
5. Repeat back to your partner what you heard your partner say to indicate that you were listening and have understood.
6. Be open-minded about your partner's requests when asked to do something.
7. Stop doing the things you know drive your partner nuts.
8. Open up to your partner. Be vulnerable.
9. Keep learning about your partner's interests.
10. Do not assume your partner will always respond in the same way as in the past. People are dynamic.

These small changes in how you relate to your spouse are not rocket science. But when you practice loving and mastering the little things in your relationship, intentionally weeding out the areas that are unhealthy or sloppy, you will find yourself down the road with a set of healthy habits that make your relationship more enjoyable.

Perhaps you not in a place where you feel like you can do some of these basics right now. I would suggest that you choose to "fake it 'til you make it." Even if you do not feel like it, take a small step toward your partner. Give your partner a hug or challenge yourself to write a short note. You might not feel like it, but just like the Olympic swimmers, as you practice small behaviors, they become heartfelt habits. Heartfelt habits become a beautiful "happily ever after."

Reflection Questions

1. What is something you have been reluctant to discuss with your partner until now?
2. What do you think your communication style is? Do you think you and your partner have the same communication style?
3. Have you ever noticed instances of command communication?
4. What motivates you to save money?
5. How does saving money make you feel?
6. Given the choice between saving and spending, which do you usually choose?
7. Would you consider yourself a compulsive spender?

Notes

1. B. Klontz, R. Kahler, and T. Klontz, *Facilitating Financial Health: Tools for Financial Planners, Coaches, and Therapists* (Erlanger, KY: The National Underwriter Company, 2016).
2. R. Gunther, "How Stress Can Bury Love—The Way Back," *Psychology Today*, November 14, 2014, https://www.psychologytoday.com/us/blog/rediscovering-love/201411/how-stress-can-bury-love-the-way-back.
3. Gunther, "How Stress Can Bury Love—The Way Back."
4. Gunther, "How Stress Can Bury Love—The Way Back."
5. Gunther, "How Stress Can Bury Love—The Way Back."
6. Gunther, "How Stress Can Bury Love—The Way Back."
7. D. Aller (financial planner and vice-president, New Concept Benefit Group), email discussion with the author, October 12, 2020.
8. Klontz, Kahler, and Klontz, *Facilitating Financial Health*.
9. Klontz, Kahler, and Klontz, *Facilitating Financial Health*.
10. Klontz, Kahler, and Klontz, *Facilitating Financial Health*.
11. Klontz, Kahler, and Klontz, *Facilitating Financial Health*.

12. Klontz, Kahler, and Klontz, *Facilitating Financial Health*.
13. Klontz, Kahler, and Klontz, *Facilitating Financial Health*.
14. Klontz, Kahler, and Klontz, *Facilitating Financial Health*.
15. Klontz, Kahler, and Klontz, *Facilitating Financial Health*.
16. Klontz, Kahler, and Klontz, *Facilitating Financial Health*.
17. Klontz, Kahler, and Klontz, *Facilitating Financial Health*.
18. Klontz, Kahler, and Klontz, *Facilitating Financial Health*.
19. Klontz, Kahler, and Klontz, *Facilitating Financial Health*.
20. Aller, email discussion with the author.
21. Aller, email discussion with the author.
22. J. Weinberger, "How to Talk to Your Partner About Money (Without a Meltdown)," *Talk Space*, May 14, 2019, https://www.talkspace.com/blog/talk-to-partner-about-money/; and H. Poll, "Celebrate Relationships, but Beware of Financial Infidelity," National Endowment for Financial Education (February 14, 2018), https://www.nefe.org/news/2018/02/celebrate-relationships-but-beware-of-financial-infideltiy.aspx.
23. Klontz, Kahler, and Klontz, *Facilitating Financial Health*.
24. Klontz, Kahler, and Klontz, *Facilitating Financial Health*.
25. Klontz, Kahler, and Klontz, *Facilitating Financial Health*.
26. Klontz, Kahler, and Klontz, *Facilitating Financial Health*.
27. Klontz, Kahler, and Klontz, *Facilitating Financial Health*.
28. Klontz, Kahler, and Klontz, *Facilitating Financial Health*.
29. Klontz, Kahler, and Klontz, *Facilitating Financial Health*.
30. T. E. Joiner, *Lonely at the Top: The High Cost of Men's Success* (New York: Palgrave Macmillan, 2011).
31. Joiner, *Lonely at the Top*.
32. J. Gottman and N. Silver, *Why Marriages Succeed or Fail, and How to Make Yours Last* (Bloomsbury Paperbacks, 2014).
33. Gottman and Silver, *Why Marriages Succeed or Fail, and How to Make Yours Last*.
34. Gottman and Silver, *Why Marriages Succeed or Fail, and How to Make Yours Last*.
35. B. Luscombe, "Staying Married: Marriage Has Changed, but It Might Be Better Than Ever," *Time*, June 2, 2016, https://time.com/4354770/how-to-stay-married/.
36. Luscombe, "Staying Married."
37. E. J. Finkel et al., "The Suffocation of Marriage: Climbing Mount Maslow without Enough Oxygen," *Psychological Inquiry* 25, no. 1 (March 10, 2014): 1–41, https://doi.org/10.1080/1047840x.2014.863723.
38. Finkel et al., "The Suffocation of Marriage."
39. H. Stott (battalion chief, West Chicago Fire Protection District), in discussion with the author.
40. P. D. Tripp, *What Did You Expect?* (Wheaton, IL: Crossway, 2010).
41. D. Chambliss, "The Mundanity of Excellence: An Ethnographic Report on Stratification and Olympic Swimmers," *Sociological Theory* 7, no. 1 (1989): 70–86, https://doi.org/10.2307/202063.

Chapter 16

CANCER

For some, cancer is a distant tragedy that does not directly affect them or their loved ones, while for others it is the enemy that has ruined their family. Sadly, cancer is a reality that is more common among those in the fire service than in other occupations. According to a study authored by R. D. Daniels for the National Institute for Occupational Safety and Health (NIOSH), firefighters have a 9% higher rate of cancer diagnoses compared to the general population, and they have a 14% higher rate of cancer-related deaths than nonfirefighters.[1] In a different study, Daniels and others noted that though these were mainly digestive and respiratory cancers, there were variety of other kinds of cancer such as urinary, prostate, bladder, leukemia, multiple myeloma, breast, and more. There was also found to be a strong correlation between firefighting and cases of malignant mesothelioma.[2]

For many, cancer diagnoses can surface unexpectedly. Although this topic is also included in the later-life section of this book, I am sad to say that cancer can strike in younger years as well. Whatever your age, a cancer diagnosis invokes feelings of desperation, uncertainty, anger, and resentment, mixed with feelings of determination, hope, and courage.[3] For first responders, a cancer diagnosis can shatter their sense of who they are. Responders see themselves as strong, independent, unemotional, and as the rescuers, not the people who need to be rescued. Many cannot bear the thought of being a burden to someone else. Thus cancer disrupts the seemingly impenetrable position of a first responder. The challenge is to recognize that relationships are built on mutuality. First responders are cheating their loved ones if they do not allow them to assist emotionally and practically during cancer treatment.

Emotions of Cancer

First responders who are diagnosed with cancer may feel a variety of difficult emotions: fear, worry, hopelessness, powerlessness, regret, guilt, denial, shame, isolation, loss of self, profound sadness, despair, grief, numbness, emotional paralyzation, anxiety, depression, or belief that they will not be able to cope.[4] When a first responder feels like his body is no longer capable of doing the things it formerly could do, it can make him feel disoriented, like he is losing his identity as a first responder and capable rescuer. Researcher Linda Viney observed that loss or threat of loss to a person's body functioning caused emotional responses of anger, helplessness, hopelessness, and sadness.[5] Experiencing all of these emotions is normal, and it is a part of the grieving process of cancer.[6]

Many firefighters grew up thinking they had to conceal their emotions. This practice only creates internal pain that will eventually explode outwardly. Instead of avoiding emotions, start to pay attention to your body. If your stomach tightens up, you notice you are holding your breath, your fists are clenched, or your muscles are very tense, it is likely a sign that you are feeling anxious, stressed, or afraid.[7] It is very normal to feel strong emotions throughout the cancer treatment process. Just remember that when you have a strong feeling, it is okay to acknowledge it. When you name your feeling, it helps you deal with it appropriately. Then you can move forward, and a new emotion will take its place.

Allowing yourself to feel will allow you to process and move forward. As the news of a diagnosis sets in, the emotions will subside, and life may even return to normal in many ways.[8] When you keep your pain hidden, it only hurts you from the inside out. Choosing to allow yourself to feel fear, pain, stress, and other feelings will help you come to terms with your diagnosis and will help you feel more peace and strength going forward.

Keys to Emotional Wellness in the Face of a Cancer Diagnosis

One key to anchor yourself during cancer treatment is through self-care. You can read in-depth about self-care in chapter 11. Authors Lawrence Robinson, Jeanne Segal, and Melinda Smith suggest that giving yourself a healthy diet, lots of hydration, good rest, and plenty of time spent with close relationships will help bring joy and restore you. The authors note that avoiding sugar is crucial in cancer recovery and recommend that you also try to avoid caffeine, alcohol, and nicotine. All of these things will hinder your recovery and can lead to anxiety and mood crashes.[9]

Nature is also very healing emotionally and physically, and I recommend spending a significant amount of time outdoors while you recover from cancer. Breathing deeply, especially when in nature, is a simple way you can calm yourself down, reduce blood pressure, trigger the physical healing responses in your body, and find peace again.

Self-care is critical for getting through the cancer process. Now more than ever, your body needs you to care for it. I strongly encourage you to walk and think, if possible. Even if you cannot do some of the things you formerly enjoyed, I encourage you to find things that you can do that bring you joy. This could mean you embrace a new hobby or spend more time in a favorite place. Perhaps you take your dog on frequent walks or you take up horseback riding. You could learn something new, create something, or write or blog your memoirs.

In addition to self-care, the strongest pillar of emotional wellness during cancer treatment is relationships. In chapter 14, I discussed the idea of "psychological body armor" as described by Geraci and others. The three keys to remaining emotionally well during turbulent times at work are training, social support, and leadership.[10] Training and leadership create security and strengthen a person's emotional resiliency in the face of traumatic work experiences, but social support strengthens a person's psychological body armor both on and off the job.

According to Robinson and others, having trusted people in your life with whom you can enjoy positive activities not only will help your emotional wellness, but it will also greatly help you keep a positive perspective as you go through the course of your treatment. The authors note that spending time in person with people who care about you will reduce stress.[11]

Breakdown of Relationships During Cancer Treatment

First responders are used to being strong. They may feel exposed and vulnerable during cancer treatment, or they may try to hide their true feelings with a "make the best of it," cheerful façade. But as I mentioned in chapter 1, surrounding yourself with an emotional support web is the way to stay connected and remain emotionally well during hardship. So I encourage people with cancer to allow someone to go with them to each treatment, even if they do not really need assistance. It is important for both physical and emotional health to allow people to show love to you in this way.

Do not worry about feeling like a burden. Reaching out to people for help will only make them feel more connected to you.[12] Just as you feel more connected to people when you help them, others will feel a sense of joy that they can do something to help you as a way to show that they care about you. If you are worried about cancer becoming too much for those who are close to you, consider getting support from a cancer support group or firefighter peer support group as well. You need family, friends, and acquaintances, in addition to your medical providers as you go through cancer. You need others to go through this journey with you so you can be transformed in the best way possible with resilience and personal growth.

Sadly, the strains cancer puts on a marriage are often suffocating and can end in a couple separating or divorcing. But walking through a cancer battle together can be rewarding and strengthening for a relationship if open communication is practiced. Researchers Katarzyna Woźniak and Dariusz Iżycki note that relationships can begin to break down under the shadow of cancer because of poor communication coupled with family members feeling their own grief over the loss of what was, guilt, inability to cope with stress, fear of the unknown, fear of death, uncertainty, shame, feelings of powerlessness, and anger for being put in this position. Cancer is hard on family members as well. Woźniak and Iżycki observe that family members may try to cope by distracting, being busy, turning to their own "comforts," such as alcohol or substance abuse, infidelity, or other types of "acting out" behavior.[13] Again, choosing connection and open communication is very therapeutic for both the individual with cancer and his or her family members.

Cancer and the Family

Whatever the demographics of the individual with cancer, I am a big believer in seeking help from a counselor. According to Darlene Daneker, associate professor at Marshall University, the individual will need not only physical support from an oncology team but also emotional support from friends, family, and counselors. Daneker notes that feelings of anger, grief, guilt, and fear can overwhelm people with cancer, so having open communication is critical.[14]

Family members are overwhelmed with questions, and cancer takes a heavy toll on them emotionally and mentally. Additionally, within the first responder's surrounding community, there are psychological and social dynamics that change when cancer is present. First responders may feel helpless. They may search for meaning. They may have a sense of failure, and they may fear stigmas that come with having a disease. They may feel isolation or even a lack of support from health professionals. Physically, patients may wrestle with the side effects of treatment or the uncertainty that comes with having a disease.

While cancer often wreaks havoc physically, it can also wreak havoc emotionally beneath the surface.[15] So how can first responders and their families overcome cancer's unique emotional wellness challenges?

First responders and their families can move forward by managing fatigue, adjusting to physical changes, and sorting through the financial issues, address changes, and relationship dynamics associated with a cancer diagnosis. But they need not feel alone. By increasing communication with family members, chaplains, and peer supporters, cancer patients can cope with processing the physical pain, thinking about the end of life, and bereavement. In addition, seeking family therapy at regular intervals during recuperation and remission can help families stay unified through such a trying time. Simply understanding that there is a possibility of the patient or members of the family developing situational depression will help when emotions are overwhelming. Gaining a better understanding of the changing physical needs of the first responder and establishing a new "normal" routine will make life more comfortable and less stressful.

Being surrounded by support changes everything for first responders. Cancer can be scary, but it cannot take what you refuse to let it take. So fight for your family and let the diagnosis, treatment, and remission bring you together, not apart.

When Children Are Asking About Cancer

When children are involved, their primary source of emotional pain will come from lack of communication about the disease or avoidance of the topic of death. According to Woźniak and Iżycki, the most important thing for kids whose parents are fighting cancer is communication. Kids do not know whether or not it is acceptable to talk about cancer unless the adults in their lives speak about it openly. Woźniak and Iżycki also note that kids have many questions relating to the cancer diagnosis:[16]

- What causes cancer?
- Is it somehow my fault?
- Will they die? If so, what will life be like after they are gone?
- Is it my fault they are not getting better?

Children can often develop deep and pathological beliefs surrounding cancer if adults do not communicate about it.[17] Their unanswered questions will weigh on them with serious emotional ramifications, if left unaddressed.

Suicidal Ideation in Cancer Patients

In chapter 5, I share some of the factors that can place a person at higher risk of committing suicide. Cancer is one of those factors. Sadly, many people feel afraid and powerless on receiving a cancer diagnosis, so they begin to consider ending their lives. According to Nicholas Zaorsky, Penn State College of Medicine and Penn State Cancer Institute, and others, when compared to the general suicide rate, cancer patients are more than twice as likely to end their lives. The standard population's suicide rate is about 10.5 to 13/100,000, while the cancer patient population's suicide rate is more than 28/100,000.[18] To help first responders with cancer, we need to identify warning signs and provide emotional support, fighting alongside them until the battle is won.

Zaorsky and others have determined that those at greatest risk of committing suicide are those with cancers of the lung, head and neck, testes, bladder, and Hodgkin lymphoma. Of those with cancer who commit suicide, they note that the greatest percentages are male (83%), white (92%), and diagnosed at younger ages than older ages.[19] Everyone is capable of reaching despair sufficient to consider taking his or her own life, however. There are a number of factors involved, so it is important to keep an eye on the emotional wellness of those who have received a cancer diagnosis.

> **Truth:** A cancer diagnosis is not an indication that life is over.

> **Truth:** Many people recover from cancer and go on to live amazing lives.

> **Truth:** Cancer can be absolutely overwhelming, but you can recover with resilience and personal growth.

Helping Those with a Cancer Diagnosis

When someone you know has cancer, it can be difficult to know how to help and how to respond. Getting specific with how you would like to offer support can be comforting for someone with a cancer diagnosis. Offer to pick up dinner for the cancer patient and family. Offer to sit with the person during treatment. Offer to come over and watch a movie with the person or sit in the backyard and just hang out. There are many ways you can be a friend. Offering specific help will indicate that you are serious about wanting to help.

Another way to become sensitive about what someone with a cancer diagnosis is going through is to educate yourself about that type of cancer. Try not to offer advice unless you are asked, though.[20]

A third way to support someone with a cancer diagnosis is to stay connected with him or her, according to Robinson and others. This means providing support throughout the cancer treatment, not just at the beginning when it is top-of-mind. Also try to send the individual a text and indicate your availability to talk. Tell the person when you feel unsure of how to talk about it. If you feel awkward, just say it. Robinson and others suggest trying a hug or a tender touch if words do not come easily. They note that you can send a card if you are not able to be with the patient in person. Finally, they recommend that you try to keep the relationship with the patient as normal as possible. Let them hear your jokes and laugh with you.[21]

There are some things you should avoid when talking to a person who has a cancer diagnosis, including the following:[22]

- Saying you know exactly how they feel
- Comparing their situation to someone else's
- Using empty platitudes like, "Just stay positive"
- Not letting them say how they feel
- Taking things personally when they do not get back to you or when they are angry or upset

Finishing Business

People who receive a diagnosis will feel the need to "finish business" by repairing broken relationships, connecting with long-term friends, or asking for forgiveness.[23] Completing these tasks provides peace of mind to those worried about the end of life. They may also feel a strong search for meaning and spiritually significant answers. They may explore life events, photo albums, journals of history, tape recordings of family members who have gone before, or they may begin looking for answers in spiritual books like the Bible. Without finishing business in these ways, the individual may feel distress over loss of control and loss of his life.

After enduring hardships like cancer, it is normal for people to experience a renewed sense of meaning, a passion for life, and a new sense of personal growth. So take heart if you have received a cancer diagnosis. You are not alone. You can trust others to help you. You can find meaning and purpose from this, and you can go on to live an even richer life after this is over. So fight the good fight, my friend. Your life is worth living.

Reflection Questions

1. Have you or someone you love ever received a cancer diagnosis? How did it feel to hear those words?
2. What feelings came with the diagnosis?
3. Do you find it hard or easy to let people help you?
4. Do you see yourself being comfortable talking about cancer with your family? If not, what is one thing you could say today that would make it more tolerable to talk about cancer with your family?
5. Aside from your family, what sources of support do you have?
6. Do you need to reach out for additional support somewhere? If so, where?
7. Have you felt the need to "finish business" with someone or something as a result of your cancer diagnosis? What is it that you need to finish? Is it possible to do that this week or this month?

Notes

1. R. D. Daniels, "Firefighter Cancer Rates: The Facts from NIOSH Research," *NIOSH Science Blog*, Centers for Disease Control and Prevention, May 10, 2017, https://blogs.cdc.gov/niosh-science-blog/2017/05/10/ff-cancer-facts/.
2. R. D. Daniels et al., "Mortality and Cancer Incidence in a Pooled Cohort of US Firefighters from San Francisco, Chicago, and Philadelphia (1950–2009)," *Occupational and*

Environmental Medicine 71, no. 6 (October 14, 2013): 388–397, https://doi.org/10.1136/oemed-2013-101662.
3. D. Daneker, *Counselors Working with the Terminally Ill*, VISTAS Online, 2006, https://www.counseling.org/Resources/Library/VISTAS/vistas06_online-only/Daneker.pdf.
4. L. Robinson, J. Segal, and M. Smith, "Coping with a Life-Threatening Illness or Serious Health Event," HelpGuide (October 8, 2019), https://www.helpguide.org/articles/grief/coping-with-a-life-threatening-illness.htm.
5. L. L. Viney, "Loss of Life and Loss of Bodily Integrity: Two Different Sources of Threat for People Who Are Ill," *OMEGA—Journal of Death and Dying* 15, no. 3 (November 1, 1985): 207–222, https://doi.org/10.2190/x1b1-3c94-v2d5-amqj.
6. Robinson, Segal, and Smith, "Coping with a Life-Threatening Illness or Serious Health Event."
7. Robinson, Segal, and Smith, "Coping with a Life-Threatening Illness or Serious Health Event."
8. Robinson, Segal, and Smith, "Coping with a Life-Threatening Illness or Serious Health Event."
9. Robinson, Segal, and Smith, "Coping with a Life-Threatening Illness or Serious Health Event."
10. J. Geraci et al., "Understanding and Mitigating Post-Traumatic Stress Disorder," *US Army Research* 344 (2011), http://digitalcommons.unl.edu/usarmyresearch/344.
11. Robinson, Segal, and Smith, "Coping with a Life-Threatening Illness or Serious Health Event."
12. Robinson, Segal, and Smith, "Coping with a Life-Threatening Illness or Serious Health Event."
13. K. Woźniak and D. Iżycki, "Cancer: A Family at Risk," *Menopausal Review* 13, no. 4 (2014): 253–261, https://doi.org/10.5114/pm.2014.45002.
14. Daneker, *Counselors Working with the Terminally Ill*.
15. Daneker, *Counselors Working with the Terminally Ill*.
16. Woźniak and Iżycki, "Cancer."
17. Woźniak and Iżycki, "Cancer."
18. N. G. Zaorsky et al., "Suicide among Cancer Patients," *Nature Communications* 10, no. 1 (2019), https://doi.org/10.1038/s41467-018-08170-1.
19. Zaorsky et al., "Suicide among Cancer Patients."
20. Robinson, Segal, and Smith, "Coping with a Life-Threatening Illness or Serious Health Event."
21. Robinson, Segal, and Smith, "Coping with a Life-Threatening Illness or Serious Health Event."
22. Robinson, Segal, and Smith, "Coping with a Life-Threatening Illness or Serious Health Event."
23. T. A. Rando, *Grief, Dying, and Death: Clinical Interventions for Caregivers* (Braille Jymico, 2008).

Chapter 17

LIVE LIKE YOU WERE DYING

The only way to make sense out of change is to plunge into it, move with it, and join the dance.

—Alan Watts

I recently read an article about having a relationship with death, and it surprised me. It proposed that you cannot truly live the most fulfilling life unless you have a relationship with death. Initially, the idea of having a "relationship with death" sounds unsettling, but I think they might be on to something.[1]

First responders see a lot of deaths and near-death experiences on the job. Many people fear death. But what if fear of death is what is holding so many people back from truly living? What if the thing many people are missing, the thing that makes them live in fear, is a healthy relationship with death?

This does not mean becoming more comfortable taking risks, though taking appropriate risks is a part of growing and pursuing dreams as a human. Having a relationship with death goes deeper than just confronting your fear head on. Having a relationship with death has to do with how you see your relationships, your time, your challenges, those you need to forgive, and even your relationship with a higher power.

The Death of a Situation

Every change in life requires adjusting to a new way of doing things, a new schedule, a new rhythm for daily life, a new set of expectations, or even a new way of relating to the people around you. Marcus Aurelius observed that "loss is nothing else but change, and change is Nature's delight."[2] Change is inevitable, and loss is all-too-regular. Some changes are easy, and other changes will bring up unexpected emotions. This is a natural part of human processing.

Similar to physical death, changes can signal the death of a situation. They usher in the end of the way things were. Whether it is the loss of a job or a shift change, moving across the country or just moving across town, a falling out with a close friend or just a friend moving away, the end of a marriage or just a change of schedules that affects the time you have with your spouse, life's changes often call for us to grieve. It could be loss of what you once thought was your dream job. It could be the loss of someone dear. It could be the loss of a particular level of a particular relationship. Regardless of the loss, it is just another change. Change is natural and necessary, and grief is a natural part of processing change.

As French poet Anatole France noted, "All changes, even the most longed for, have their melancholy; for what we leave behind us is a part of ourselves; we must die to one life before we can enter another."[3]

Rather than living in fear that your situation will change, the same principle applies to the death of a situation. You cannot truly live the most fulfilling life without a relationship with death. You cannot truly live your current situation to the fullest unless you have considered what life would be like if that situation came to an end. As you picture your life without some of the things, people, and rhythms that define your daily life, you may find that you grow in gratitude, pay closer attention to the small things, and demonstrate greater intentionality with how you spend your time.

Escaping the Idea of Death

In first responder fields, it can be easy to become overwhelmed with the prominence of death and life changes. The civilian population has to face the idea of death to a certain extent, but first responders are frequently confronted with the reality of death. Nearly every day, first responders see people pass away, become injured, or lose loved ones. It can be easy to hate the idea of death and grief enough to mentally distance oneself from it. The problem is that when death and change are viewed as an all-too-common enemy, first responders start turning toward escapes instead of talking about them and forming a relationship with them.

Escapes from the reality of death and change may take the form of substance abuse, pornography, keeping busy, goal attainment, affairs, and so on. Omar Khayyam stated, "I drink not from mere joy in wine nor to scoff at faith—no, only to forget myself for a moment, that only do I want of intoxication, that alone."[4] If first responders rely on alcohol or these other distractions to keep their minds off of the pain they have experienced or the deaths they have seen, they usually end up spiraling downward. Instead of growing emotionally, individuals end up with strongholds and addictions in their lives that only keep them enslaved to the pain of their past.

The greatest possible upside to facing death daily is that first responders can actually live more self-actualized lives. When individuals face hardship, they grow. People who are given situations that are beyond their control have only one thing that they can control—their response. Viktor E. Frankl once stated, "When we are no longer able to change a situation, we are challenged to change ourselves."[5]

Pain usually produces the most profound growth in people. So when first responders see pain and process through it, they have the beautiful potential of becoming the most in-the-moment, grateful, and connected people. Looking death in the eye and processing through pain will help first responders move from fear and avoidance to acceptance.

Author William McDonald has chronicled the life and writings of Søren Kierkegaard, who was a Danish philosopher and the youngest of seven children. Kierkegaard lost many family members during his childhood, with only himself and one of his siblings still living by the time he turned 25. This exposure to death caused him to think deeply about life and death, and his thinking became what many people consider to be the foundation of existential thought. After years of pondering the idea of death, he concluded that "the 'healthy' person, the true individual, the self-realized soul, the 'real' man, is the one who has transcended himself."[6]

How does this occur? According to McDonald, Kierkegaard believed that transcending yourself can be done by realizing the truth of one's situation. People who experience hardships and loss "build up strategies and techniques in the face of terror of [the] situation. These techniques can become an armor that can hold the person prisoner. The very defenses that he needs in order to move about with self-confidence and self-esteem become his life-long trap. In order to transcend himself, he must break down that which he needs in order to live."[7]

Kierkegaard's perspective found that people grew up with the idea of avoiding the reality of their own mortality, according to McDonald. This idea that humans grow up, build strengths, live in their identity, and contribute to the world around them, only to die was unsettling and anxiety inducing. But by realizing this truth, humans could transcend themselves. By recognizing that death is inevitable, humans find the meaning in the moments they do have here on earth. They enjoy the social structures and symbols around them, admitting that their successes are built upon "borrowed powers." Individuals live on a finite timeline with the support and structure of those who went before them. So they can live, enjoy relationships, contribute to the world around them, and then face death.[8]

Living Like You Could Lose What You Have

I think that you will find that when you have thought through some of your questions about death, or about any major changes in your life for that matter, it changes the way you live. Thinking about your death does not motivate you to die; it motivates you to live. Thinking about a divorce or loss of a loved one does not motivate you to end a relationship with that person. It motivates you to celebrate what you have.

Often when people have experienced a traumatic event, they find that they have a renewed sense of gratitude for life, warmer and more intimate relationships, and a greater ability to live one day at a time. But you do not need to have your world rocked by trauma to change your thinking.

A Premortem for Life

Once concept used in the world of Fortune 500 company management is the idea of a *premortem*. When someone has an idea, one of the most thorough ways to vet that idea is to write a premortem. When you want to execute something as well as possible, you need to know what it would look like if that idea fell flat and completely failed. What would it cost you and the organization? A premortem is a look at the worst-case scenario—at the failure of an idea or organization—and then working backward from that idea to determine what caused the failure.

Perhaps it is like that with life. Maybe we never quite live until we have a premortem for life. What is the worst-case scenario outcome of our lives? Perhaps our confidence in life comes from the security of knowing what comes at the end. Perhaps our perspective to love others, live in the moment, forgive freely, embrace change, and overcome loss comes from our understanding that our lives are temporary, and death is inevitable. Perhaps how we view death shapes our lives more than we realize.

Here are some categories to consider and some questions to ask yourself:

1. *Spirituality*. Do you have spiritual beliefs? What do those beliefs tell you about death and the afterlife? What do you need to do to settle those beliefs and become okay with the afterlife or your relationship with God? If you do not have spiritual beliefs, what is your thinking about what happens when you die? Are there things you need to settle or questions you need to ask to become okay with the perspective you hold?
2. *Experiences*. Are there some experiences in your life that you would like to have happen before you die? What are they? Do you have control over making them happen? What would it look like if those things never occurred?
3. *Relationships*. Who are some people in your life that you would want to have by your side as you died? What if they were not there? Are there things you need to say or do for them now or in the meantime so that you are settled in your relationship with them?
4. *Places*. Are there places you would like to see or see again before you die? What if you did not see them? What are some things you can do now to satisfy your desire to be there?
5. *Passing*: What is your ideal situation, place, and way that you would like to die? What would be the worst? What would it mean if your life ended the worst way possible?

You can use this similar line of questioning about almost every great thing in life—marriage, jobs, goals, loss of your health, and other relationships:

1. What would you want?
2. What is the worst possible outcome?
3. What would it mean for you if your worst-case scenario came to pass?
4. How do you want to live now in light of that?

The phrase "You do not know what you have until it is gone" is true. When you picture your life and you know in your heart how temporary (and sometimes hard) it can be, you learn to live fully, love deeply, and forgive readily. The fear of losing what you have is no longer a crippling fear but a motivating fear.

This healthy fear motivates you to live more in the moment, appreciate connections with people, and be thankful for what you have. Almost every genre of music has a song about this concept. Songs like "Live Like You Were Dying," by Tim McGraw; "Big Yellow Taxi," by Counting Crows; and "Like I'm Going to Lose You," by Meghan Trainor all echo this same concept. People need to have a relationship with death. We need to understand our finitude and the temporal nature of the things we have.

So ask the questions you need to ask. Have the conversations that need to take place. Do this so that your life does not consist of holding back but consists of living, loving, and giving because you have thought through what it would be like to lose them.

One day at a time, you are spending the moments that make up your life. You get to decide how you treat people. You get to decide where you will invest the best of yourself. You get to decide what you want to make last. Maybe settling your relationship with death will help you define who you want to be and how you want to live. What are you waiting for?

Reflection Questions

1. Take a minute to write a premortem for your life, based on the questions asked above.
2. What are some things you would like people to say about you at your funeral? How do you want them to remember you?
3. What can you do today to shift toward the person you want to be remembered as?
4. Take a minute to think about the most valuable things to you—your spouse, kids, friends, home, and so on. What would life be like without these people or things?
5. What can you do today to express gratitude for the people or things in your life?

Notes

1. J. Hudson, "Your Relationship with Death," FirefighterNation.com, August 24, 2018, https://www.firefighternation.com/firerescue/your-relationship-with-death/#gref.
2. "Marcus Aurelius," InternetPoem.com (accessed June 23, 2021), https://internetpoem.com/marcus-aurelius/quotes/loss-is-nothing-else-but-change-and-change-is-38886/5/.
3. A. France, *The Crime of Sylvestre Bonnard*, trans. Lafcadio Hearn, ed. B. Fishburne and D. Widger (Project Gutenberg, 2016) pt. II, chap. 4, https://www.gutenberg.org/files/2123/2123-h/2123-h.htm#link2HCH0004.
4. O. Khayyam, quoted in E. Becker, *Denial of Death* (United Kingdom: Free Press, 1973), https://www.google.com/books/edition/_/DhgQa6rzEGQC?hl=en&sa=X&ved=2ahUKEwiW6dm12ZfzAhXEEFkFHaGZBeEQ7_IDegQICBAD.
5. V. E. Frankl, *Man's Search for Meaning* (New York: Simon and Schuster, 1985).
6. W. McDonald, "Søren Kierkegaard," *Stanford Encyclopedia of Philosophy*, November 10, 2017, https://plato.stanford.edu/entries/kierkegaard/?source=post_page.
7. McDonald, "Søren Kierkegaard."
8. McDonald, "Søren Kierkegaard."

CHAPTER 18

RETIREMENT AND LATE-ONSET STRESS SYMPTOMATOLOGY (LOSS)

Firefighters often complete their years in the service and then face an onslaught of personal life changes. The combination of events that pile up—the job change, change in finances, change in sleeping habits, changes in relationships, sometimes a change in residence, sometimes a personal injury or illness—can become suffocating. Each of these changes comes with its own set of stressors.

In 1967, psychologists Holmes and Rahe developed the Social Readjustment Rating Scale (SRRS) to quantify the levels of stress precipitated by various major life changes. They assigned a number of life change units (LCUs) to each of the major life changes to help individuals identify which changes were more overwhelming and consumed more emotional energy than others.[1]

Holmes and Rahe designated the SRRS list as follows:[2]

Event	LCUs
Death of a spouse	100
Personal injury or illness	53
Change in financial state	38
Change to a different line of work	36
Outstanding personal achievement	28
Change in living conditions	25
Change in working hours or conditions	20
Change in residence	20
Change in social activities	18
Change in sleeping habits	16
Change in eating habits	15

If you would like to take the test yourself, a simple version of the Holmes-Rahe stress inventory is available online.[3]

Research has shown undeniably that a person who experiences a high number of life change units (300 points or more) within one to two years has a significantly increased likelihood of experiencing a "stress-induced health breakdown"[4] and mental health problems, including depression. So what can a first responder do to weather the storm of life change at the start of retirement?

The answer is to build a new brotherhood. You may not have all the same people in your daily life that you had when you were working, but you do still need a strong circle of friends. While family relationships are important, friendships are vital.[5] In a study of 6,418 retirees from 2008 to 2016, the spousal relationships of the retirees had the strongest impact on their mental health. Interestingly, their friendships had the second strongest impact, stronger even than the individual's other family members.[6]

Friendships alleviate loneliness, provide emotional and social support, give individuals companionship and someone to participate with in mutual interests, open up opportunities to talk about things that are close to your heart, and offer encouragement to continue to take care of yourself.[7] This connectedness brings meaning to older adults, which is crucial for well-being in the second half of life.

Remember, you did not retire to spend the rest of your life doing nothing all by yourself. Instead, you retired to embrace a new life of freedom and flexibility to enjoy the things that interest you and the people around you. So take advantage of the time you have and intentionally surround yourself with a new set of social connections to support, challenge, play, and go forward with you. This does not mean you stop being friends with the brotherhood you have known all these years. It just means finding a new set of relationships to surround yourself with so that you can ground yourself to ride out major changes.

Maybe it is a local softball league. Maybe it is a group at church. Maybe it is a group of other retirees who gather for coffee once a week. Maybe it is an organization you start volunteering with. Maybe it is signing up for first responder peer support so you can talk through your pain with someone who understands. Maybe it is a new job and a new set of coworkers you develop trust with. Whoever it is for you, do what you need to do to find them because your retirement years will be absolutely transformed by these friendships.

Why Retirees Need to Focus on Emotional Wellness

I shared Mark's story in chapter 3. When Mark retired, he shifted to part-time work. This kick-started a life review, and he began thinking about his time in the fire service, the traumatic events he witnessed, and how he wanted to spend the rest of his life. Life reviews are natural and common between ages 38 and the mid-50s, especially when first responders retire.[8]

The most overwhelming thing that surfaces during a life review is trauma. The life changes that occur at retirement can make it hard to keep avoiding the pain of earlier events. The second most overwhelming thing that surfaces during a life review is the need to be needed. Many retired firefighters begin thinking, "Hey, I've been doing all these things, and I've achieved so much, and yet here I am still feeling like I'm not enough." These feelings can become overwhelming, and I often see alcohol addiction and suicidal ideation begin to take root in response.

First responder retirees face new psychological, physiological, and relational dynamics. Because these retired first responders are no longer wrapped up in the daily busyness of first responder life, they often find themselves searching for belonging and discovering that years of commitment to the service have left them feeling somewhat disconnected from family life and longing for a renewed sense of purpose and excitement. The vacancies left in these schedules leave room for them to realize the sadness or pain that was previously masked by busyness. It becomes easy for these retired first responders to fall into depression, substance abuse or addiction, and even suicide if left unchecked. But with a little support, retirees can develop a renewed sense of belonging, perspective, and hope.

Mark is extremely intuitive and a very deep-thinking individual. I believe he came to see me at just the right time for him. At the beginning of his retirement, he wanted to examine all these issues and to process his past experiences. Thinking through his pain, why he felt so disconnected, why he always felt the need to fix everything, and why he always wanted to achieve helped Mark by giving him permission to set boundaries and orient his retirement years in the right direction.

Retirees Facing Depression

The first emotional wellness issue that often emerges in retired first responders is depression. The common Myers-Briggs personality profile for a first responder is ISTP/ESTP. Within these letters is a wealth of knowledge about how these individuals make decisions, spend their time, and react to situations around them. But for the sake of simplicity, note that the letter T stands for "thinking." These instrumental personalities often process information and make decisions based on what they know rather than how they feel about a situation. Thus they are more accustomed to operating without emotion. People often begin exhibiting and developing the opposite traits of their personality type around the ages 40 and 50, however, and "thinkers" may begin to display "feeler" behavior.

As emotions come into play, thinkers often do not know how to deal with it. These emotions can be positive and negative, including new feelings of excitement, freedom, accomplishment, ambivalence, sadness regarding the loss of professional identity, anxiety, pessimism, or several of these at once. First responder retirees not only find themselves with more time to think about their memories and pain, but they also find themselves with more emotion and the discomfort of not knowing how to deal with it.

Thomas Joiner explains the disconnection men feel as they get into the retirement years:

> Men, on average, fail to make the crucial transition from the institutionalized, ready-made friendships of childhood to the earned, worked-for friendships of adulthood. That they fail to do this is reflected in their suicide rates, their divorce rates, and, well, their friendlessness. . . . Like all meaningful relationships, friendships require work to initiate and maintain.[9]

Being disconnected from the department and unearthing new emotions can lead many retired first responders into rescuer's depression (see chapter 2). As previously discussed, firefighters often have instrumental personalities, which means they value self-sufficiency and devalue emotional conversations. So in retirement, not only does the firefighter have much more pain to face, but he also has much fewer people around to talk to. Where will he go with these feelings? This is why retirees need new friends. They may need to talk to a counselor or spend time with a peer supporter. By opening up to someone who can relate to his past and his pain, the first responder can experience healing, process even the most difficult trauma, and be free.

I believe that another reason retirees face depression is because of the emptiness they feel when they stop being needed all the time. When this lifelong pursuit of achievement comes to an end, many firefighters realize that they are lonely. Joiner explains it this way:

> In early adulthood, men, more than women, face a dilemma—a dilemma about which they are not very aware for the next twenty years: namely, they believe that their focus on money and status will buy them a happy future—a belief fueled by a materialistic culture—when in actuality, it is setting them up for the hard fall of lonely years in their forties and beyond. A lucky few can get by on the friendships they made back in the day. But many can't, and over time, men drift away from friendships and simultaneously earn money and status. This leaves them puzzled indeed—they spent years achieving money and status, finally got it, and yet they feel lonely and empty. Lonely at the top. It wasn't supposed to be like that. The trouble for the lonely sex is that it *is* like that.[10]

Retirement can be very lonely for people with instrumental personalities. Firefighters, who often have instrumental personalities, go into this career to serve, not necessarily for money or status. But then something takes over, and they want to become the best at it, and they want to earn every certificate and complete every training. The motivator is not necessarily money and status; it is achievement and productivity. But the outcome is the same. Firefighters can end up very lonely and without friends in retirement because they have been busy chasing their picture of what an ideal firefighter is. Disappointed, they discover that although retirement was supposed to be about leisure and relationships, their relationships are few and far between. This is when they need to build a new brotherhood.

You cannot go through retirement alone. Every phase of life should be lived in connection with trusted relationships, but retirement needs connection even more. Without connection, retirees are prone to depression, loneliness, substance abuse/addictive behaviors, suicidal ideation, and processing traumatic memories in unhealthy ways. Retirement needs a brotherhood, too, in order for firefighters to remain connected and resilient through this new phase.

Retirees Facing Substance Abuse and Addictive Behaviors

Unfortunately, one of the ways retired first responders deal with pent-up emotional pain is through turning to substances such as alcohol and drugs. These addictive substances or behaviors either excite or inhibit the brain's messages, and they often provide a temporary escape from emotional pain. Painful memories or trauma, lonely feelings, negative life circumstances, and more recede into the background when an individual is under the influence. But once the body filters the substance, painful memories, feelings, and circumstances emerge yet again. Continued exposure to addictive substances or behaviors teaches neurotransmitters in the brain to adapt, and addiction sets in. Individuals can no longer function normally without the excitatory or inhibitory response that the substance inspires, so they become dependent. Unfortunately, breaking this addiction requires withdrawal and often depression, but freedom can be achieved with commitment and support.

According to psychologist Karen Jennison, becoming addicted to a substance may be an easy thing to do at the turn of retirement. She acknowledges that drinking alcohol "increase[s] during periods of prolonged exposure to emotionally depleting life change and loss, when supportive needs may exceed the capacities of personal and social support."[11] People who experience change or loss and do not have an appropriate support system to cope with the changes or pain may increase alcohol consumption. Without appropriate support, retired first responders may develop addictions, and the physical ramifications will be worse due to age.

For those who are ages 55 and older, substances have more extreme consequences. Physiologically, retirement-aged adults metabolize alcohol differently than they did when they were younger. Due to the increased alcohol saturation that occurs in an older body, one beer has the same effect as two or three beers would have on a younger person. This makes it easier for them to become addicted.

Alcohol and other substances also worsen residual health problems in older adults, such as heart conditions, blood pressure, or even the effects of obesity. According to "Alcoholism and Aging," alcohol can shorten life expectancy in older adults.[12] It can also destroy families and other relationships, reduce memory capacity, slow brain function, and diminish a person's quality of life.

Even so, substance abuse and addiction can be difficult for doctors to identify because the negative effects of aging can look similar to the negative effects of alcohol. For example, if a person's memory is fading, it may receive a "dementia" diagnosis, when really it is a result of overconsumption of alcohol. Thus relationships, peer support, and accountability are crucial in the retired years, so that struggles are not faced alone, and substance abuse can be halted before it takes root and destroys a retiree's health and relationships.

Retirees Facing Suicide

According to the Surgeon General's National Strategy for Suicide Prevention, older adults are at a higher risk than any other population for committing suicide.[13] Thomas Joiner, one of the world's leading suicidologists, proposed a compelling theory about why people commit suicide, as described in chapter 5. I am particularly moved to speak about suicide to retirees because retirees exhibit many of the conditions Joiner puts forth in his theory of suicide. As discussed previously, Joiner believes that in order for a person to have the desire and ability to commit suicide, they need 1) a sense of being burdensome, 2) thwarted belongingness, and 3) capability to take their own life.[14]

First we will examine the ways retirees may experience "a sense of being burdensome." Retirees may feel that they are a burden if they are not able to do things they used to be able to do. Retirees who feel chronic pain may feel like a burden to those around them because they may not be able to get things done or go on adventures as they used to. Those with chronic illness or cancer may feel like a burden because they need support and care from others in their recovery process. Those who have had a change in employment status, like a shift from regular salary to pension, may feel like they are a burden if they are no longer the primary breadwinner in the family. Their occupational commitments have decreased, which may lead to a sense that they are a burden. If this feeling grows, the individual may feel that his death would be more valuable than his life. This could lead to a desire to commit suicide.

Next we will look at "thwarted belongingness." Retirees may experience thwarted belongingness as they shift from spending their days surrounded by the brotherhood to spending their days with just a few family members around them. As I mentioned in chapter 15, retirees are the population at highest risk for divorce.[15] A divorce is one of the most striking losses of belonging, leaving a person feeling lost and alone. If a person has spent his years pursuing his career over connection with his family, he may feel "lonely in a crowd" and unable to connect with people around him when he retires.[16] These experiences of thwarted belongingness are Joiner's second criterion for desiring to commit suicide.

Concerns about belongingness are why I am a big advocate of counseling and peer support in retirement. Retirement is a time to build a new sense of belonging with other retired firefighters or with a counselor, validating the individual's memories and fire service experiences. These individuals can remind the retiree that he is not a burden but a blessing.

Third, Joiner believes that for a person to commit suicide, he must also have the "ability to commit suicide."[17] The ability to commit suicide evolves over time, as a person sees death, feels personal injuries, and further develops a risk-taking personality. In addition, unprocessed memories combined with hyperarousal and a lack of emotional regulation can lead to this risky behavior or substance abuse. One final factor that enables people to take their own lives is alcohol use. As I mentioned earlier, retirees can easily become drunk, as their bodies process alcohol differently than they did when younger. On average, 29% of people who committed suicide were found with alcohol in their systems.[18]

Retirees who struggle with alcohol and addiction are at a high risk of committing suicide and should be intentionally checked on, included in friend gatherings, and monitored closely.

Emotional Trauma

Unfortunately, a firefighter who retires will be forced by circumstances to examine the emotional trauma of the previous 30 years on the job. First responders can go their whole careers without dealing with much depression, PTSD, suicidal thoughts, or addiction. It is when retirement begins that these issues may arise. Throughout their careers, first responders are filled with roles and responsibilities at the department, in the community, and at home. There are rarely days with nothing to do, and they enjoy the energy of that constant calendar. When this activity stops, they find themselves searching for a sense of self and belongingness. One common occurrence among newly retired first responders is a loss of identity. Thoughts of loneliness may set in, and memories from the past seem to resurface. This can bring up chronic depression or PTSD, and if not addressed, it can lead to failing relationships, substance abuse, or suicide.

When a newly retired firefighter is faced with these mental and emotional issues, it is important that the spouse is aware of what is happening. The retiree will need support and patience while working through this. Counseling is an excellent resource for the retired firefighter to explore. It is important to utilize all resources and also to open up in conversations with your loved ones.

Late-Onset PTSD

Tripp Wilson is a firefighter friend who worked with me as a peer support trainer. He is familiar with the struggles faced by many first responders when they retire. Tripp told me,

> When I retired . . . I would page through my mental life book at night as I fell asleep, and I became unable to turn past the bad pages to see the good ones. Sleepless nights, anger, frustration, and fear became daily emotions for me. I became mean, selfish, self-centered, egotistical, and difficult to be around.[19]

This experience is not uncommon. A few years ago, I received a call from a retired firefighter from an area that did not have many counselors. He had received my name from a friend and called me because he had recently retired and was being crippled by horrific memories of calls he had experienced 30 years ago. In tears, he told me about his nightmares and the ruminating thoughts he had about burned bodies. He called them "the mannequins" because they had become a regular part of his thought life over the previous few months. He had been a highly respected firefighter, and he had not thought about these calls at all until retirement. He told me that the more he thought about these traumatic calls, the more he just wanted to return to work. He felt that being busy was the only way to escape the haunting memories that played out in his mind every day.

I told him that he is not alone. Time and time again I have seen retired first responders come into my office telling me that since they retired, they have been having intrusive thoughts and nightmares. They are experiencing things that they never experienced while they were working. They tell me they try to stay busy to keep their mind off of it. They often feel like they need a drink or substance to clear their heads or help them relax. They try not to talk about it because they think that no one would understand. They avoid places, things, or people that make them remember. They startle easily. And they possibly think about harming themselves.

Sometimes when firefighters come into retirement, if they have not processed the traumatic memories they experienced during the working years in healthy ways, their minds begin using their spare time to organize thoughts, occasionally bringing up truly traumatic things. My advice to my clients is always this: The sooner you process traumatic events, the sooner you can find meaning and healing from them. If you put aside processing them in earlier years, you may experience later-life processing, which can include late-onset PTSD.

Much research has been conducted by Horesh and others to analyze late-onset PTSD in veterans, but this phenomenon is extremely common in first responders as well. Standard PTSD symptoms often occur in the first three years after a traumatic experience. But PTSD symptoms can also occur 20 years or more after the event occurred. In fact, research by Horesh and others indicates that nearly 17% of veterans have PTSD that emerges later in life. One theory as to why this happens is that the memories have an "incubation period" in which the memories are present, but the symptoms do not show up until later, according to Horesh and others.[20] This is called late-onset PTSD.

Another group studied Dutch veterans of military action in World War II, the former Dutch East Indies, and Korea. They sampled 774 Dutch military veterans—576 with a military disability pension and 198 community sample veterans. Researchers quantified those who met the criteria for a PTSD diagnosis in 1992 and again in 1998. They discovered that in 1992, 27% of the 576 disabled veterans met the criteria for a PTSD diagnosis, while in 1998, 29% met the criteria. The community sample also showed an increase in PTSD diagnoses from 8% in 1992 to 9% in 1998. Clearly PTSD was not diminishing over the years, but rather, the prevalence of PTSD in the aging population was rising as the veterans retired.[21]

Late-onset PTSD has the symptoms of PTSD—difficulty sleeping; re-experiencing traumatic memories so that they are intrusive to daily life; avoidance of places, people, or situations that remind the first responder of the trauma; and feeling disconnected or emotionally numb. If you experience any of these symptoms, I strongly encourage you to see a counselor who specializes in first responder care. These memories can be processed, and you can heal and live an amazing retirement.

Late-onset PTSD is nothing to be afraid of. It is something you can act on right now. Instead of fearing a potential traumatic backlash in later life, I always recommend processing trauma in healthy ways in your current season of life. As I have mentioned all throughout this book, first responders tend to use busyness to push aside pain. This is a part of your training, and it is extremely helpful on a call when you need to focus on the task at hand. When you consistently push aside painful memories throughout your life, however, you deny yourself the opportunity to process them in healthy ways, which your brain desperately needs to do. Your brain wants to file them and make meaning out of them, and denying yourself the opportunity to do so does not mean they go away for good. Your brain still needs a chance to heal from them.

Some individuals have an easier time recognizing how certain memories fit into the bigger picture of life and process them willingly. Others have more difficulty with the process of sorting through these traumatic memories and may develop symptoms of late-onset PTSD.[22]

Processing painful memories does not require taking significant time out of the things you love just for thinking. It does not always mean you have to spend months and months thinking, although some people need more time than others. Processing traumatic memories can be as simple as letting yourself feel the feelings whenever they come. It means you write down and sort out what happened that night. It means you talk to someone—a counselor or peer supporter—when you need to talk about it. It means you choose to be present and connect with yourself, even if it hurts.

There are many evidence-based treatments for PTSD that can be helpful if you experience trauma, PTSD, or even late-onset PTSD. I detail these treatments in chapter 12. Processing trauma is one of the most powerful things you can do for your emotional wellness in retirement. You will find that not only is your retirement less filled with anxiety, but it is also more meaningful because you have identified how each trauma has made you better, more resilient, more thankful, closer to your family members, and even more hopeful in the world.

> **Truth:** When you process trauma, you grow.

> **Truth:** The sooner and more thoroughly you process your traumatic memories, the sooner you find meaning from them.

> **Truth:** Ignoring traumatic memories does not make them go away.

Late Onset Stress Symptomatology (LOSS)

Traumatic memory review is quite common at retirement, but not all traumatic memory reviews turn into late-onset PTSD. Some experience a much gentler version of reminiscing, called late onset stress symptomatology (LOSS). It is such a new concept in the field of psychology that it has not been researched with specific focus on first responders yet, but it has been researched in the military. Stan McCracken specializes in working with military service members, and we had a lengthy discussion about how I recognized LOSS in my first responder clients. He affirmed that the first responder clients who have come to me with their late-life stressful memories are likely experiencing the same late-onset stress symptomatology that veterans often experience.[23]

Late-onset stress symptomatology is a term given by researchers from the Department of Veterans Affairs.[24] Many retiring veterans experience the same thing. Often the progression begins with a traumatic event early in your career. You function successfully in your career without symptoms of stress, trauma, or PTSD. Then as you face the challenges and changes of aging, you begin to experience thoughts, feelings, reminisces, and symptoms of stress as you retire.[25]

These sound quite similar to late-onset PTSD, but McCracken points out that LOSS is *subclinical*, meaning that the symptoms of LOSS (intrusive memories, thoughts, feelings, and reminisces) are not as disruptive, overwhelming, or full of anxiety as the symptoms of PTSD or late-onset PTSD. Instead, LOSS is a part of the process of growing from trauma and finding post-traumatic growth, as discussed in chapter 10. McCracken notes that LOSS is milder than PTSD because the reactions to the traumatic event are significantly reduced. He further explains that these reminisces, or even dreams, in later life are a natural part of thinking about what happened, and though they can be upsetting, retirees do not necessarily re-experience the events with fear, avoidance, sleep disturbances, anxiety, or disruption to daily life.[26]

According to research conducted by Rajdip Barman and Mark Detweiler, it is possible for LOSS to develop into PTSD, late-onset PTSD, or other anxiety disorders like depression or paranoia, however. Barman and Detweiler note that LOSS is a processing of events, but if it begins to interfere with daily life, sleep, cognitive function, or other aspects of a retiree's mental and physical health, it may begin to move toward late-onset PTSD and may require clinical support.[27] It is very important, however, that clinicians do not misdiagnose LOSS as PTSD, late-onset PTSD, or even dementia.[28]

Why Do These Memories Show Up During Retirement?

Davison and others refer to late-onset stress symptomology as "late-life emergence of early-life trauma." These researchers set up a focus group of 47 Korean Conflict and

Vietnam War veterans. They interviewed these veterans to identify what, exactly, LOSS is, and which late-life stressors may bring LOSS about.[29]

One theory is that current stressful incidents in the life of the retiree—for example, transitioning to retirement, the loss of a spouse, kids moving away, or declining physical ability—can invoke those same feelings of helplessness or powerlessness, reminding the individual of how they felt during the traumatic event.[30] Another theory is that retirees do a lot of reflecting. Reflecting brings back both the good and the painful memories.[31] A third theory is that as the brain ages, it tends to bring up traumatic memories.[32]

Davison and others conducted focus groups and heard the following things from their veterans:

- "Maybe I have more time on my hands or maybe as we get older—when we're young and we're 40 or 50 you're not thinking of older years or death or illness."
- "As we now reach a certain point in our lives, certain things come to mind. . . . We seem to reflect more."
- "At our age, there should be some type of therapy for us. We remember more about the war now than ever. We are older and have more time to reflect."

The consensus from the 47 veterans they interviewed was that retirement brought time, reflection, and unique stressors that made traumatic memories surface again.[33]

I believe the strongest cases of LOSS (and late-onset PTSD for that matter) show up in people who have put off processing traumatic events. As Davison and his colleagues explain, "Veterans who tended to 'block out' thoughts or memories over the course of their lives might be more vulnerable to particularly intense, unwanted, or intrusive memories of combat or wartime experience later in life."[34]

As I mentioned earlier, the more readily you allow yourself the time to process memories as they come up throughout your life, the sooner you will find meaning out of the experiences. If you deny yourself the opportunity to process by staying busy, you will eventually process the memories when you are not as busy—that is, in retirement—because your brain needs to make sense of them.

So, yes, spare time in retirement is one of the biggest reasons why LOSS comes up in retirement. But perhaps the biggest contributor to LOSS is not the amount of time retirees have on their hands, but rather the processing that happens in the brain around age 65.

Erikson's Cognitive Stages

Psychologist Erik Erikson believed the process of aging is a cognitive developmental process, in which the brain actually encounters a psychosocial crisis at the turn of each developmental age, grows from it, and carries forward a new virtue and a stronger sense of the world. In Erikson's framework, the brain analyzes the world around it and makes a decision about life and how the world works. Each stage asks a fundamental question, which the individual answers based on personal experiences.[35]

Erikson's Cognitive Stages

Stage	Age	Phase	Basic Virtue
1	0–1.5	Trust vs. Mistrust	Hope
2	1.5–3	Authority vs. Shame	Will
3	3–5	Initiative vs. Guilt	Purpose
4	5–12	Industry vs. Inferiority	Competency
5	12–18	Identity vs. Role Confusion	Fidelity
6	18–40	Intimacy vs. Isolation	Love
7	40–65	Generativity vs. Stagnation	Care
8	65 and older	Integrity vs. Despair	Wisdom

Source: E. H. Erikson, Identity: Youth and Crisis, Austen Riggs Monograph No. 7 (New York: W.W. Norton & Co., 1968).

In Erikson's stage 8, which is 65 and older, the mind performs a life review. Events that occurred long ago will inevitably come up as a normal part of processing and making meaning out of life's events. According to Erikson, around retirement age, the human brain goes through a crisis he called "integrity vs. despair," in which the person evaluates his life to make a judgment if it was a life well lived. Was his life "meaningful and satisfying"?[36]

Often this question is triggered by the major life changes I listed at the beginning of the chapter—for example, retirement, death of a spouse, changes in friendships, an injury or illness, starting a new job, moving, or changes in employment. In fact, one 72-year-old veteran shared that when he lost his spouse, he began to be "plagued" by fellow soldiers' deaths from decades ago.[37] So life reviewing is a natural cognitive process that occurs around age 65, but it is also a reaction to life events that trigger self-analysis.

In Erikson's integrity vs. despair phase, if the person feels his life was morally upright and meaningful, he will feel a great sense of fulfillment from a life well lived and a powerful sense of integrity. If, however, he sees that he has made some morally questionable decisions, if he feels his life has been wasted, or if he has regrets, he may feel a strong sense of despair.[38]

LOSS can be a natural process of integrating memories into one's sense of how his life has gone, as a part of the integrity vs. despair life review. Should you personally experience LOSS, you may find that the most healing thing may be to find a group of other retired first responders to talk to about what you have seen and be validated in how you are feeling. Together you can build a sense of integrity.

Ron's Story

A friend of mine, Ron, is a retired firefighter from the Chicago Fire Department. Ron experienced LOSS when he retired, and I asked him to share his thoughts about retirement for other firefighters to read. Here is what he said:

When I retired, I moved out of state. I moved away from old friends and family, and I have found that I miss the old brotherhood more than I would have thought. I have found that it is difficult to build a new brotherhood because it seems unnatural or forced, but I still feel the need for support, so I'm being intentional to pursue new friendships in our new community.

Friends are so important and helpful. I have found them better than family because there are no expectations on what should be—only a genuine fondness and appreciation for a fellow person. Last week, my wife and I hosted our good friends Jim and Donna for a couple of days at our new home. It was glorious! There is nothing like old friends—good times, laughter, and love. Retirees need good friends, and if they don't live near those friends, they need to get together with those friends whenever they can. Friendship is really important in retirement.

Next, I have been surprised to find that I am way more emotional than when I was on the job. With the macho personality I have lived with for so long, it kind of freaks me out. It is amazing how many times I have what people would call "flashbacks" of some of the more horrendous incidents I experienced on the job. . . . I would have thought the spectacularly vile scenes would be more up-front, but I find them more subtle and not as painful as I expected. There is just no explaining it, but it doesn't hurt as much as I thought it would. I have dealt with these memories by talking to old friends and people who care about mental health. For a good retirement, mental health really is important.

Notice how Ron said his flashbacks were not as up-front as he expected? That is because Ron experienced LOSS, which is a part of the natural development of the human brain over time. I have huge respect for Ron and for how he worked to build new friendships and get together with old friends in his retirement. I believe that his community will be one of the most powerful fortresses for his emotional wellness into the future.

Another fortress against being overwhelmed by LOSS in retirement is rituals or ceremonies. McCracken believes that rituals, as he has seen in the military, are a key part of emotional wellness in the face of LOSS. Rituals, and I would include first responder ceremonies and funerals, annual celebrations, and the like, provide space for grieving, remembering, and processing. McCracken notes that rituals give permission to feel emotions connected with past events, which is why they are significant, and it is critical for first responders to take time to reconnect.[39]

What Can I Do to Overcome LOSS?

If you experience LOSS, first recognize that your feelings are normal. LOSS happens to many retired individuals. You are in good company. To begin processing these memories, I recommend writing your thoughts with pen and paper, which helps you organize your memories and begin healing them. And I recommend walking meditation. This can bring balanced thinking as you activate both hemispheres of your brain. Third, I recommend talking to a counselor who can help you with evidence-based treatments like accelerated resolution therapy or prolonged exposure to put the memories into their proper places in your mind and process them fully. Finally, I recommend getting the support of friends and family. Opening up to safe people can be truly healing.

You are not alone. Retirement can be a beautiful time in your life. You just need to sort some memories first.

Reflection Questions

1. Who are five people you want to be intentional to keep friendships with during your retirement years?
2. What are some things you can invite them to do with you to keep your friendship?
3. Have you found yourself reaching for alcohol or other addictive substances or behaviors to deal with your pain lately? Who do you need to talk to about this?
4. Have you noticed yourself doing a life review in this stage?
5. What are some of the memories that keep recurring to you?
6. Why do you think these memories keep coming up?
7. What can you do to process these memories more fully so you can find meaning from them?
8. What is one truth from this chapter that is helpful for you to remember?

Notes

1. T. H. Holmes and R. H. Rahe, "The Social Readjustment Rating Scale," *Journal of Psychosomatic Research* 11, no. 2 (August 1967): 213–218, https://doi.org/10.1016/0022-3999(67)90010-4.
2. Holmes and Rahe, "The SRRS."
3. American Institute of Stress, "The Holmes-Rahe Stress Inventory," n.d., https://www.stress.org/holmes-rahe-stress-inventory-pdf.
4. American Institute of Stress, "The Holmes-Rahe Stress Inventory."
5. Y. Chen and T. H. Feeley, "Social Support, Social Strain, Loneliness, and Well-Being among Older Adults: An Analysis of the Health and Retirement Study," *Journal of Social and Personal Relationships* 31, no. 2 (2014): 141–161, https://doi.org/10.1177/0265407513488728.
6. H. J. Lee and M. E. Szinovacz, "Positive, Negative, and Ambivalent Interactions with Family and Friends: Associations with Well-Being," *Journal of Marriage and Family* 78, no. 3 (2016): 660–679, https://doi.org/10.1111/jomf.12302.
7. R. Blieszner, A. M. Ogletree, and R. G. Adams, "Friendship in Later Life: A Research Agenda," *Innovation in Aging* 3, no. 1 (January 2019), https://doi.org/10.1093/geroni/igz005.
8. E. H. Erikson, *Identity and the Life Cycle* (W. W. Norton & Co., 1980).
9. T. E. Joiner, *Lonely at the Top: The High Cost of Men's Success* (New York: Palgrave Macmillan, 2011).

10. Joiner, *Lonely at the Top*.
11. K. M. Jennison, "The Impact of Stressful Life Events and Social Support on Drinking among Older Adults: A General Population Survey," *International Journal of Aging and Human Development* 35, no. 2 (September 1, 1992): 99–123, https://doi.org/10.2190/f6g4-xlv3-5kw6-vmba.
12. M. Altpeter and J. Campbell, "Alcoholism and Aging: A User-Friendly Curriculum," *Substance Abuse* 14, no. 3 (1993): 129–136.
13. US Department of Health and Human Services (HHS), *2012 National Strategy for Suicide Prevention: Goals and Objectives for Action*, Office of the Surgeon General and National Action Alliance for Suicide Prevention, HHS, September 2012, https://www.ncbi.nlm.nih.gov/books/NBK109917/.
14. T. E. Joiner, *Why People Die by Suicide* (Cambridge, MA: Harvard University Press, 2007).
15. B. Luscombe, "Staying Married: Marriage Has Changed, but It Might Be Better Than Ever," *Time*, June 2, 2016, https://time.com/4354770/how-to-stay-married/.
16. Joiner, *Lonely at the Top*.
17. Joiner, *Why People Die by Suicide*.
18. C. Smith, "The Link Between Alcohol Use and Suicide," May 19, 2020, https://www.alcoholrehabguide.org/resources/dual-diagnosis/alcohol-and-suicide/.
19. F. T. Wilson III (Illinois Firefighter Peer Support member), in discussion with the author.
20. D. Horesh et al., "The Clinical Picture of Late-Onset PTSD: A 20-Year Longitudinal Study of Israeli War Veterans," *Psychiatry Research* 208, no. 3 (August 15, 2013): 265–273, https://doi.org/10.1016/j.psychres.2012.12.004.
21. A. J. Dirkzwager, I. Bramsen, and H. M. van der Ploeg, "The Longitudinal Course of Posttraumatic Stress Disorder Symptoms among Aging Military Veterans," *Journal of Nervous and Mental Disease* 189, no. 12 (2001): 846–853, https://doi.org/10.1097/00005053-200112000-00006.
22. C. M. Potter et al., "Distinguishing Late-Onset Stress Symptomatology from Posttraumatic Stress Disorder in Older Combat Veterans," *Aging & Mental Health* 17, no. 2 (2013): 173–179, https://doi.org/10.1080/13607863.2012.717259.
23. S. McCracken (Crown Family School of Social Work, Policy, and Practice, University of Chicago), in discussion with the author, October 22, 2019.
24. National Center for PTSD, "Late-Onset Stress Symptomatology (LOSS) Scale," US Department of Veterans Affairs, January 28, 2020, https://www.ptsd.va.gov/professional/assessment/adult-sr/loss_scale.asp.
25. E. H. Davison et al., "Late-Life Emergence of Early-Life Trauma: The Phenomenon of Late-Onset Stress Symptomatology Among Aging Combat Veterans," *Research on Aging* 28, no. 1 (2006): 84–114, https://doi.org/10.1177/0164027505281560.
26. McCracken, in discussion with the author.
27. R. Barman and M. B. Detweiler, "Late Onset Stress Symptomatology, Subclinical PTSD or Mixed Etiologies in Previously Symptom-Free Aging Combat Veterans," *Journal of Traumatic Stress Disorders & Treatment* 3, no. 4 (July 28, 2014), https://doi.org/10.4172/2324-8947.1000132.
28. Barman and Detweiler, "Late Onset Stress Symptomatology, Subclinical PTSD or Mixed Etiologies in Previously Symptom-Free Aging Combat Veterans"; and Davison et al., "Late-Life Emergence of Early-Life Trauma."

29. Davison et al., "Late-Life Emergence of Early-Life Trauma."
30. M. D. Buffum and N. S. Wolfe, "Posttraumatic Stress Disorder and the World War II Veteran: Elderly Patients Who Were in Combat or Were Prisoners of War May Have Special Health Care Needs That May Not Be Obvious," *Geriatric Nursing* 16, no. 6 (1995): 264–270.
31. E. H. Erikson, *Identity: Youth and Crisis*, Austen Riggs Monograph No. 7 (New York: W. W. Norton & Co., 1968).
32. C. T. Allers, K. J. Benjack, and N. T. Allers, "Unresolved Childhood Sexual Abuse: Are Older Adults Affected?" *Journal of Counseling & Development* 71, no. 1 (September/October 1992): 14–17, https://doi.org/10.1002/j.1556-6676.1992.tb02163.x.
33. Davison et al., "Late-Life Emergence of Early-Life Trauma."
34. Davison et al., "Late-Life Emergence of Early-Life Trauma."
35. Erikson, *Identity: Youth and Crisis*.
36. Erikson, *Identity and the Life Cycle* (New York: W. W. Norton & Co., 1980).
37. L. Hyer et al., "Posttraumatic Stress Disorder: Silent Problem Among Older Combat Veterans," *Psychotherapy: Theory, Research, Practice, Training* 32, no. 2 (1995): 348–364, https://doi.org/10.1037/0033-3204.32.2.348.
38. E. H. Erikson, *Identity and the Life Cycle*.
39. McCracken, in discussion with the author.

Chapter 19

PLANNING FOR YOUR IDEAL RETIREMENT

In chapter 4 I introduced Brett, a firefighter who was transitioning into retirement. Brett felt unsettled and disconnected from his wife, and he was unmotivated by the idea of retiring and just sitting on the beach. In my experience, many first responders feel this way. People are supposed to feel productive and passionate, moving things forward and creating meaning their entire lives. The idea of a passive retirement is a modern construct.

So when a first responder comes to me around the time he is planning to retire, we immediately work on having a plan. In my experience, people do not generally hate work. They hate the politics. They hate the inflexible structure of shift work. They hate long work hours and the missed family experiences, but they generally like work itself. So what can first responders do to find meaningful productivity in the later years of life?

One option might be to find a different job with fewer hours per week. Another option might be volunteering somewhere with flexible work hours. In my work with some of the younger guys, I always try to emphasize male friendships. Once firefighters retire, it is harder for them to make the same depth of friendships.

In Brett's case, the first thing he had to do before he could build his dream retirement was to face the trauma and emotions he had been avoiding for years. It is extremely hard to build emotional wellness when someone struggling with alcohol addiction keeps drinking. In therapy, we have difficulty confronting pain and discomfort when alcohol is still a crutch. But when Brett removed alcohol from his life, he immediately was confronted with his pain. As Brett and I have worked together, he has begun to realize a new vision for his future, his connection with his wife and kids, and his years of retirement. The best truly is ahead for him.

Making the Most of the Retired Years

According to Mitch Anthony, author of *New Retirementality*, after overcoming the emotions of retirement and painful memories experienced in the service, many retirees find excitement and purpose by pursuing second careers, meaningful leisure planning, or volunteerism. Retired first responders now have the beautiful opportunity to create a new vision for the future. By identifying values—strengthening long-term relationships, establishing new relationships, building fitness, mastering finances, giving back to others—retirees can take steps to frame their retirements around the pursuit of those values, notes Anthony.[1]

After years of work and experience accumulation, retirees are ready to pass on information, coaching, and support to the younger generations. Giving back to younger generations provides new purpose, self-identity, and fulfillment.[2] Retired first responders have so much to offer to younger first responders, and it truly is a win-win for both. Retirees feel purpose, and active first responders feel empowered and equipped.

A New Definition of Retirement

Retirement is a relatively new concept, and it can almost seem unnatural when it finally happens. It brings in unexpected transitions, especially for first responders. As with everything else in life, it is best to be prepared before entering a new situation. Armed with the proper tools, first responders and their spouses can make the most out of their new lives together and live the happy retirement dream that they envisioned.

Anthony notes that retirement has been evolving since its conceptualization back in the early 20th century. The millions of Americans who have retirement on the horizon likely have grown up with the belief that retirement is a life that is filled with bickering couples, long lunch dates, and weekly card games. If they were lucky, it also included a condo somewhere warm.[3] Not only is this idea outdated, but it is also not healthy or realistic.

People are now retiring with more passion and energy. Retirement is a second life—another chance to do something you love and spend your time doing things that make you feel fulfilled. It could be a second career, further education, being more creative, or volunteering. The options are endless. This precious time should not be wasted. It should be invested in your happiness and in the world around you.

Retirement Planning

The road to get to this happy and fulfilling retirement needs to be taken seriously. It requires planning and necessary conversations with those important to you, especially your spouse. When people hear the term *retirement planning*, they often think of the financial aspects involved. Finances are only one piece to the retirement puzzle, however. Being prepared for retirement is also about being prepared in your marriage, being prepared emotionally, and being prepared to spend your time wisely. These necessary conversations should take place as early as five years before you retire.

Anthony notes that it is important to take advantage of this time in retirement.[4] You and your spouse will have a healthier retirement if you are able to create a plan for how you will use your time and energy in new ways. First responders are accustomed to solving problems and helping people. A healthy retirement for first responders may involve using their time to leave a positive impact on future generations.

As explained in the previous chapter, psychologist Erik Erikson's theory of psychosocial development is made up of eight stages of human development from birth to death.[5] The seventh stage he explores is generativity versus stagnation. He explains that adults begin to search for ways to leave their mark on the next generation. If adults successfully tackle this stage, they will feel a strong sense of worth, which is quite important for retired first responders. This could include taking care of grandchildren, volunteering at schools, or mentoring younger first responders. Unfortunately, if retired first responders are not able to find their own way to contribute, they may feel stagnant and unimportant. This could lead in turn to more unhealthy emotional wellness issues. By taking the appropriate time to plan for retirement, first responders and their spouses can take care to avoid these common problems.

Changing Roles

You will be creating a new life together, and many roles will be changing. Typically, first responders retire earlier than the national standard age of 65. Because of this, your spouse may still be in the workforce. The roles in the house are sure to change, and the goal of retirement planning is to make it an easier transition. Who will do the laundry? Who is in charge of dinner? Should we make a new budget? Is the retiree going to take up a new hobby or a second career?

You have most likely spent the majority of your marriage with specific roles. You have organized your work schedules and social calendars to align with each other, you each have specific household duties, and you may only get a few times a week to spend quality time with just the two of you. As human beings, we crave routine, normalcy, and balance. Firefighter couples have been forced to make the 24-hour shift a normal part of life, and it works. So now how do you deal with the wealth of time? More importantly, how do you deal with the loss of the firehouse life?

As with all relationship issues, communication is the best way to start. The more we can talk and listen to one another, the more chance we have at successfully overcoming the hurdle. What will life look like beyond the world of 24-hour shifts? Both people in the relationship may have an idea of what it will look like, but do the two visions align?

The Five Cs of Successful Aging

According to Anthony, successful aging means living with vitality, continuing to challenge oneself mentally and physically, continuing to move forward, trying new things, and treating oneself as if there are many more years ahead.[6]

Anthony has observed that there are five keys to navigating the retirement years successfully. He calls them the "Five Cs of Successful Aging," which are as follows:

- *Curiosity.* Learning new things and growing your awareness.
- *Challenge.* Keeping your mind sharp by trying something new.
- *Connectivity.* Living and working in close proximity to friends and family.
- *Creativity.* Finding ways to solve problems, make something new, or fix things.
- *Charity.* Giving ourselves or our resources to help others.[7]

What is something you are curious about that you could learn more about? What is something new or challenging you could begin to pursue? Who do you want to connect with in the coming months and years? What is one problem or project you want to fix or finish? What is one cause that matters to you that you could become more involved with in the coming months and years?

One of the ways retirees can begin to envision successfully aging into their future is to picture someone who has aged well. Anthony refers to these individuals as "retire-mentors." They set the example for how to approach this new, final season of life. They go and do something with their retirements. They use their time and energy for their good and the good of others.

Financing a Healthy Retirement

Having a healthy retirement means building a sustainable financial foundation. By defining retirement goals, considering your possible pension, and analyzing personal savings and investments, a retiree can determine what level of risk can be tolerated and what future plans will look like.

A financially sustainable retirement should start with making a retirement budget and taking it for a test-drive to see how it fits the family's needs. Paying off debt and setting aside an emergency fund will bring peace of mind. Simplifying financial commitments will relieve pressures on the budget. Then the retiree can assess what other streams of income will provide for the family—pension, fixed annuities, spouse's social security, and other personal investments. These streams of income should aim to cover fixed expenses and provide additional budget space to pursue personal interests.

If moving out of state is a possibility, the individual should consider tax laws in prospective states. In some states, pension and retirement income is not taxed. Elsewhere, it may be. If the spouse will be taking social security, the individual may want to consider delaying as long as possible so that the social security payments increase. Beginning social security at age 62 will reduce benefits, while delaying social security until after age 67 will increase payments. Those who seek early retirement may want to consider seeking employment or even starting a business to allow continued investments for future years.

Retirement presents so many changes emotionally, physiologically, physically, and financially. By approaching it with financial and emotional preparation, however, you can find meaning, purpose, and excitement in this new phase of life.

Bridging the Gap from Full-Time Work to Retirement

Maybe you love your job, and the idea of retirement does not truly interest you. But retirement is coming, and you are not that excited about it. Or maybe you are disenchanted with your job these days, and work is not where you want to be anymore. For you, retirement sounds like a well-deserved break. Whatever you think of retirement, you are not alone. According to a 2017 Gallup study, 15% of people are "engaged" or passionate about their work, their companies, or both. The other 85% find themselves either "not engaged" or "actively disengaged" in their jobs. For these 85% of disengaged workers, retirement is exciting. It is the end of an era and the beginning of something much more freeing.[8]

Anthony asks the question, "Do you really want to quit working?"[9] You should ask yourself if retirement is what you are truly looking for. What if freedom is really what you are looking for? If so, this could reframe your entire perspective on your retired years.

Your pension as a firefighter may afford you the freedom to spend your retired years playing with family, traveling the world, or even taking up golf. But Anthony points out that many people spend their careers looking forward to getting out of "the race" only to discover that they are bored with not being in it. They feel purposeless and irrelevant. According to Anthony, "Retirement is an unnatural condition! Even if you can afford to retire, the worst thing you can do is withdraw completely from the track of relevance!"[10]

For the majority of human history, retirement was not even a concept. Yet in our day and age, we are expected to hit age 62 or maybe 65 and dive headfirst into a nonemployed life of golf, travel, grandchildren, and newspaper crossword puzzles. But what if instead of viewing retirement as the end of a miserable career, you viewed retirement as the start of an occupation that excites you, makes you happy to get out of bed in the morning, and still contributes to your financial goals? Get ready—this different kind of work might be exciting and incredibly fulfilling for you.

Crossing the Bridge

Past generations retired abruptly. They hit retirement age and simply quit working. Retirement was a huge transition from a life of deadlines, coworkers, and productivity to a life of leisure, family, and nonproductive pursuits. Anthony encourages those coming close to retirement, noting, "We need to decide when we are done. It should be a matter of personal policy, not corporate or government policy. The first order of business, then, is to eradicate artificially imposed finish lines in our life and begin designing our own track to run on."[11]

Rather than "jumping off a vocational cliff," we have the choice of just slowing down. Maybe you want to work as long as you can. Maybe you want to work part-time, so you have more time for your family or personal interests. Maybe you want to start a different, more fulfilling job. Maybe you want to volunteer some of your time. You do not *have* to retire, at least not in the traditional sense of the word.

Working for Your Well-Being

Many people are deciding to work longer than the designated "retirement age" because of the psychological and relational benefits that come from working. Anthony suggests, "Maybe the real truth is that you want to quit what you are currently doing to be able to do something else."[12]

It used to be that retirement was a badge of honor. If someone worked to the finish line, he could boast about it. On the flip side, people who had to work past 62 were looked down on, as if they had not been able to prepare financially. Today, the retirement game is completely different. In fact, "those who have to work will not be the losers, because they are still in the game—they will find that work keeps them vital, involved, and healthier," according to Anthony.[13]

This concept is illustrated in the movie *The Intern*. Robert Di Niro stars as a retired widower looking for meaning and relationships, and he finds himself working in an internship in a budding online retail company. His professionalism and insight lead him to become the personal intern for the CEO, played by Anne Hathaway. As he shows up to work every day, he transitions into a much more meaningful life of new relationships, interesting work, and fulfillment as he passes on wisdom to the younger generation. You may be surprised how good for your well-being the right work can be.

Motivated by Autonomy

If you take a good look in the mirror, my guess is that retirement is not entirely about finishing your work; it is about doing what you really want to spend your time doing. It is actually about having the autonomy to call your own shots. Getting organized financially and eliminating debt while working as a firefighter will give you the autonomy you want to work at another job that you feel is fulfilling after you retire.

Successful Retirement

According to Anthony, there are four pillars for identifying the characteristics of successful retirements:[14]

1. *Vision.* Success = Retiring *to* something. Failure = Retiring *from* something.
2. *Balance.* Success = Finding the balance between vocation and vacation. Failure = Going from bingeing on work to bingeing on leisure.
3. *Work.* Success = Keeping yourself plugged into meaningful pursuits. Failure = Devolving to boredom and aimlessness.
4. *Aging.* Success = Focusing on growing and well-being. Failure = Just taking what comes.

Spend some time envisioning what you want to retire to. Think about what vocation interests you, so that you can enjoy that occupation as a balanced part of your life with your family and friends, downtime, sleep, health/fitness, and personal growth.

It is true that aging is a part of retirement, but it does not have to be the defining factor. In fact, viewing your life as moving in a positive direction can help you mentally and physically to combat any deterioration in health. Recalling Anthony's "Five Cs of Successful Aging," you need to stay connected to others; challenge yourself; be curious and willing to learn; create and enjoy hobbies like fixing cars, rearranging your house, or even picking up a guitar; and be charitable toward others. These steps will help you live a fulfilled retirement of balance and emotional wellness.

Reflection Questions

1. What are some things you envision when you think about retirement?
2. What are some things you do not like about your work life, schedule, responsibilities, and so on?
3. How would you like your life, schedule, responsibilities, and so on to look different in retirement?
4. What are some things you do like about your work life, schedule, responsibilities, and so on?
5. How can you continue doing these things you like during retirement?
6. If you are about to retire, identify five people you would like to be intentional to form friendships with or to continue relationships with after retirement. What first step do you need to take with each of them?
7. If you have already retired, what has been your biggest change or struggle?
8. What is one thing you can do to help overcome the difficulty of change you are facing?
9. Who is one person you could see being a good "retirementor" for you?

Notes

1. M. Anthony, *New Retirementality: Planning Your Life and Living Your Dreams...at Any Age You Want* (Hoboken, NJ: Wiley & Sons, 2014).
2. Anthony, *New Retirementality*.
3. Anthony, *New Retirementality*.
4. Anthony, *New Retirementality*.
5. E. H. Erikson, *Identity and the Life Cycle* (New York: W. W. Norton & Co., 1980).
6. Anthony, *New Retirementality*.
7. Anthony, *New Retirementality*.
8. Gallup, *State of the Global Workplace* (New York: Gallup Press, 2017), https://www.gallup.com/workplace/257552/state-global-workplace-2017.aspx.
9. Anthony, *New Retirementality*.
10. Anthony, *New Retirementality*.
11. Anthony, *New Retirementality*.
12. Anthony, *New Retirementality*.
13. Anthony, *New Retirementality*.
14. Anthony, *New Retirementality*.

Chapter 20

Epilogue: What Now?

As I said at the beginning, I love first responders, and my passion is helping them reconnect with their emotional wellness.

I have immense respect for first responders. They are deep-thinking and intelligent individuals. First responders are not perfect, but they work hard for their families and for civilians, and they try to be better human beings. Their desire to be the best they can be and to learn and grow from past experiences is remarkable. Often I find they just need a little direction, and they will soar.

So I hope the truths in the pages of this book will help you soar. Here is what I want you to know going forward:

1. Your service is noticed. You have made a difference. The people you have touched on the job are so glad you chose to become a first responder. I know this job is not always easy on your body or your mind, but I see how you give of yourself day in and day out to help those in need. I respect you and thank you for it.
2. You deserve self-compassion. Identifying what you need and taking care of yourself is the way you will be able to weather the storms of traumatic calls, department drama, and countless sleepless nights as a first responder. This job is emotionally exhausting, and self-care is the avenue to long-term emotional wellness and resiliency. Make sure you take time to care for yourself.
3. You have a good life ahead of you. Your job is the best job in the world, but now is the time for you to be planting seeds for a healthy retirement. Spend time connecting with your family now. Invest in your friendships this week. Slow down and think carefully about how you are doing. Talk to a counselor if you need to. When you build your support web now, you set yourself up for a good life ahead. When you neglect yourself now, you set yourself up for doing the work of processing and healing later. You have a good life ahead of you. Take some time now to make it great.

Bibliography

Abbot, C., E. Barber, B. Burke, J. Harvey, C. Newland, M. Rose, and A. Young. *What's Killing Our Medics? Reviving Responders.* Ambulance Service Manager Program. Conifer, CO: Reviving Responders, 2015. http://www.revivingresponders.com/originalpaper.

Achor, S. and M. Gielan. "Resilience Is About How You Recharge, Not How You Endure." *Harvard Business Review*, June 24, 2016. https://hbr.org/2016/06/resilience-is-about-how-you-recharge-not-how-you-endure.

Adelson, R. "Stimulating the Vagus Nerve: Memories Are Made of This." *PsycEXTRA Dataset* 35, no. 4 (2004). https://doi.org/10.1037/e362742004-023.

Alexander, B. K. "Addiction: The View from Rat Park," accessed August 9, 2021. http://www.brucekalexander.com/articlesspeeches/rat-park/148-addiction-the-view-from-rat-park.

Allers, C. T., K. J. Benjack, and N. T. Allers. "Unresolved Childhood Sexual Abuse: Are Older Adults Affected?" *Journal of Counseling & Development* 71, no. 1 (September/October 1992): 14–17. https://doi.org/10.1002/j.1556-6676.1992.tb02163.x.

Altpeter, M., and J. Campbell. "Alcoholism and Aging: A User-Friendly Curriculum." *Substance Abuse* 14, no. 3 (1993): 129–136.

American Addiction Centers. "Depression Among First Responders." *American Addiction Centers Blog*. August 8, 2014. https://americanaddictioncenters.org/blog/depression-among-first-responders.

American Counseling Education. *Stressors & Stress Response.* American Counseling Education, 1994.

American Institute of Stress. "The Holmes-Rahe Stress Inventory." n.d. https://www.stress.org/holmes-rahe-stress-inventory-pdf.

American Pain Society. "Yoga and Chronic Pain Have Opposite Effects on Brain Gray Matter." ScienceDaily.com, May 15, 2015. http://www.sciencedaily.com/releases/2015/05/150515083223.htm.

American Psychiatric Association. *Diagnostic and Statistical Manual of Mental Disorders.* 5th ed. Washington, DC: American Psychiatric Publishing, 2013.

American Psychological Association, "Prolonged Exposure." May 2017. https://www.apa.org/ptsd-guideline/treatments/prolonged-exposure.

Amodeo, J. "Are You Codependent or Just a Caring Person?" *Psychology Today*, November 4, 2017. https://www.psychologytoday.com/us/blog/intimacy-path-toward-spirituality/201711/are-you-codependent-or-just-caring-person.

Anthony, M. *New Retirementality: Planning Your Life and Living Your Dreams...at Any Age You Want.* Hoboken, NJ: Wiley and Sons, 2014.

Arpin, S. N., and C. D. Mohr. "Transient Loneliness and the Perceived Provision and Receipt of Capitalization Support Within Event-Disclosure Interactions." *Personality and Social Psychology Bulletin* 45, no. 2 (2018): 240–253. https://doi.org/10.1177/0146167218783193.

Asher, S. R., and V. A. Wheeler. "Loneliness and Social Dissatisfaction Questionnaire—Modified." *APA PsycTests Dataset*, 1985. https://doi.org/10.1037/t04785-000.

Ashton, C. "Yoga to Transform Trauma: Leadership Training and Intensive with Catherine Ashton." Illumine Chicago 2014. https://illuminechicago.com/events/yoga-to-transform-trauma-leadership-training-intensive-with-catherine-ashton/.

Asprey, D. "Stimulate Your Vagus Nerve to Improve Memory, Says New Study." *Dave Asprey* (blog). June 26, 2018. https://blog.daveasprey.com/vagus-nerve-affects-memory/.

Babyak, M., J. A. Blumenthal, and S. Herman. "Exercise Was More Effective in the Long Term than Sertraline or Exercise Plus Sertraline for Major Depression in Older Adults. (Therapeutics)." *Evidence-Based Mental Health* 4, no. 4 (November 2001): 105–106.

Bacon, I., E. McKay, F. Reynolds, and A. McIntyre. "The Lived Experience of Codependency: An Interpretative Phenomenological Analysis." *International Journal of Mental Health and Addiction* 18, no. 3 (June 26, 2018): 754–771. http://doi.org/10.1007/s11469-018-9983-8.

Barboza, K. "Critical Incident Stress Debriefing (CISD): Efficacy in Question." *The New School Psychology Bulletin* 3, no. 2 (December 10, 2005). https://pdfs.semanticscholar.org/3384/b3c58069375de1ce081e06f6610289d4cff2.pdf.

Barman, R., and M. B. Detweiler. "Late Onset Stress Symptomatology, Subclinical PTSD or Mixed Etiologies in Previously Symptom-Free Aging Combat Veterans." *Journal of Traumatic Stress Disorders & Treatment* 3, no. 4 (July 28, 2014). https://doi.org/10.4172/2324-8947.1000132.

Baumeister, R. F., K. D. Vohs, J. L. Aaker, and E. N. Garbinsky. "Some Key Differences Between a Happy Life and a Meaningful Life." *Journal of Positive Psychology* 8, no. 6 (August 20, 2013): 505–516. https://doi.org/10.1080/17439760.2013.830764.

Beattie, M. *Codependent No More: How to Stop Controlling Others and Start Caring for Yourself.* Center City, MN: Hazelden Publishing, 1992.

Benedek, D. M., C. Fullerton, and R. J. Ursano. "First Responders: Mental Health Consequences of Natural and Human-Made Disasters for Public Health and Public Safety Workers." *Annual Review of Public Health* 28, no. 1 (2007): 55–68. https://doi:10.1146/annurev.publhealth.28.021406.144037.

Bentley, M. A., J. M. Crawford, J. R. Wilkins, A. R. Fernandez, and J. R. Studnek. "An Assessment of Depression, Anxiety, and Stress Among Nationally Certified EMS Professionals." *Prehospital Emergency Care* 17, no. 3 (February 15, 2013): 330–338. https://doi.org/10.3109/10903127.2012.761307.

Bergland, C. "Longer Exhalations Are an Easy Way to Hack Your Vagus Nerve." *Psychology Today*, May 9, 2019. https://www.psychologytoday.com/us/blog/the-athletes-way/201905/longer-exhalations-are-easy-way-hack-your-vagus-nerve#:~:text=A%20myriad%20of%20breathing%20patterns,and%20calm%20 one's%20nervous%20system.

Bernert, R. A., J. S. Kim, N. G. Iwata, and M. L. Perlis. "Sleep Disturbances as an Evidence-Based Suicide Risk Factor." *Current Psychiatry Reports* 17, no. 3 (2015): 15.

Bisek, A. "When a Child Dies: Understanding Emergency Responder's Reactions." 2012. http://www.whenachilddies.com/.

Blascovich, J., and J. Tomaka. "The Biopsychosocial Model of Arousal Regulation." *Advances in Experimental Social Psychology* 28 (1996): 1–51. https://doi.org/10.1016/s0065-2601(08)60235-x.

Blieszner, R., A. M. Ogletree, and R. G. Adams. "Friendship in Later Life: A Research Agenda." *Innovation in Aging* 3, no. 1 (January 2019). https://doi.org/10.1093/geroni/igz005.

Bodnar, L. M., and K. L. Wisner. "Nutrition and Depression: Implications for Improving Mental Health Among Childbearing-Aged Women." *Biological Psychiatry* 58, no. 9 (July 27, 2005): 679–685. https://doi.org/10.1016/j.biopsych.2005.05.009.

Bourassa, K. J., J. J. Allen, M. R. Mehl, and D. A. Sbarra. "Impact of Narrative Expressive Writing on Heart Rate, Heart Rate Variability, and Blood Pressure After Marital Separation." *Psychosomatic Medicine* 79, no. 6 (July 8, 2017): 697–705. https://doi.org/10.1097/psy.0000000000000475.

Bozionelos, N., and G. Bozionelos. "Instrumental and Expressive Traits: Their Relationship and Their Association with Biological Sex." *Social Behavior and Personality: An International Journal* 31, no. 4 (2003): 423–429. https://doi.org/10.2224/sbp.2003.31.4.423.

Bremner, J. D., S. Mishra, C. Campanella, M. Shah, N. Kasher, S. Evans, N. Fani, A. J. Shah, C. Reiff, L. L. Davis, V. Vaccarino, and J. Carmody. "A Pilot Study of the Effects of Mindfulness-Based Stress Reduction on Post-traumatic Stress Disorder Symptoms and Brain Response to Traumatic Reminders of Combat in Operation Enduring Freedom/Operation Iraqi Freedom Combat Veterans with Post-Traumatic Stress Disorder." *Frontiers in Psychiatry* 8 (August 25, 2017). https://doi.org/10.3389/fpsyt.2017.00157.

Brown, B. "List of Values," in *Dare to Lead* (New York: Random House, 2018), https://daretolead.brenebrown.com/wp-content/uploads/2019/02/Values.pdf.

Bryant, R. A., M. L. Moulds, and R. M. Guthrie. "Acute Stress Disorder Scale: A Self-Report Measure of Acute Stress Disorder." *Psychological Assessment* 12, no. 1 (March 2000): 61–68. https://doi-org.du.idm.oclc.org/10.1037/1040-3590.12.1.61.

Buckley, T., A. Stannard, R. Bartrop, S. McKinley, C. Ward, A. S. Mihailidou, M.-C. Morel-Kopp, M. Spinaze, and G. Tofler. "Effect of Early Bereavement on Heart Rate and Heart Rate Variability." *American Journal of Cardiology* 110, no. 9 (August 2, 2012): 1,378–1,383. https://doi.org/10.1016/j.amjcard.2012.06.045.

Buffum, M. D., and N. S. Wolfe. "Posttraumatic Stress Disorder and the World War II Veteran: Elderly Patients Who Were in Combat or Were Prisoners of War May Have Special Health Care Needs That May Not Be Obvious." *Geriatric Nursing* 16, no. 6 (1995): 264–270.

Burg, B., and J. D. Mann. *The Go-Giver: A Little Story About a Powerful Business Idea.* New York: Portfolio/Penguin, 2015.

Calhoun, L. G., and R. G. Tedeschi, eds. *Handbook of Posttraumatic Growth: Research and Practice.* Routledge, 2014.

Campos, M. "Heart Rate Variability: A New Way to Track Well-Being." *Harvard Health Blog.* Harvard Health Publishing, October 24, 2019. https://www.health.harvard.edu/blog/heart-rate-variability-new-way-track-well-2017112212789.

Capaldi, C. A., H. Passmore, E. K. Nisbet, J. M. Zelenski, and R. L. Dopko. "Flourishing in Nature: A Review of the Benefits of Connecting with Nature and Its Application as a Wellbeing Intervention." *International Journal of Wellbeing* 5, no. 4 (2015): 1–16. https://doi.org/10.5502/ijw.v5i4.1.

Caparrotta, M. "Dr. Gabor Maté on Childhood Trauma, The Real Cause of Anxiety and Our 'Insane' Culture." *Human Window,* July 1, 2020. https://humanwindow.com/dr-gabor-mate-interview-childhood-trauma-anxiety-culture/.

Carter, R., S. Aldridge, M. Page, and S. Parker. *The Human Brain Book: An Illustrated Guide to Its Structure, Functions, and Disorders.* London: Dorling Kindersley Ltd., 2019.

Center for Health and Wellbeing. "The Seven Dimensions of Wellbeing." Accessed August 10, 2021. https://yourhealthandwellbeing.org/about/sevendimensions/.

Chambliss, D. "The Mundanity of Excellence: An Ethnographic Report on Stratification and Olympic Swimmers." *Sociological Theory* 7, no. 1 (1989): 70–86, https://doi.org/10.2307/202063.

Chen, Y., and T. H. Feeley. "Social Support, Social Strain, Loneliness, and Well-Being Among Older Adults: An Analysis of the Health and Retirement Study." *Journal of Social and Personal Relationships* 31, no. 2 (March 2014): 141–161. https://doi.org/10.1177/0265407513488728.

Cloitre, M., D. W. Garvert, B. Weiss, E. B. Carlson, and R. A. Bryant. "Distinguishing PTSD, Complex PTSD, and Borderline Personality Disorder: A Latent Class Analysis." *European Journal of Psychotraumatology* 5, no. 1 (September 15, 2014): 25,097. https://doi.org/10.3402/ejpt.v5.25097.

Cloud, H. *The Law of Happiness: How Ancient Wisdom and Modern Science Can Change Your Life.* New York: Simon and Schuster, 2011.

Coan, J. A., H. S. Schaefer, and R. J. Davidson. "Lending a Hand: Social Recognition of the Neural Response to Threat." *Psychological Science* 17, no. 12 (December 1, 2006): 1,032–1,039. https://doi.org/10.1111/j.1467-9280.2006.01832.x.

Conner, D. R. *Managing at the Speed of Change.* New York: Villard Books, 1992.

Cooney, G. M., K. Dwan, C. A. Greig, D. A. Lawlor, J. Rimer, F. R. Waugh, M. McMurdo, and G. E. Mead. "Exercise for Depression." *Cochrane Database of Systematic Reviews* 9 (September 12, 2013). https://doi.org/10.1002/14651858.CD004366.pub6.

Cross, C. L., and L. Ashley. "Trauma and Addiction: Implications for Helping Professionals." *Journal of Psychosocial Nursing and Mental Health Services* 45, no. 1 (January 2007): 24–31. https://doi.org/10.3928/02793965-20070101-07.

Csikszentmihalyi, M. *Finding Flow: The Psychology of Engagement with Everyday Life.* New York: Basic Books, 1997.

Daneker, D. *Counselors Working with the Terminally Ill*. VISTAS Online, 2006. https://www.counseling.org/Resources/Library/VISTAS/vistas06_online-only/Daneker.pdf.

Daniels, R. D. "Firefighter Cancer Rates: The Facts from NIOSH Research." *NIOSH Science Blog*. Centers for Disease Control and Prevention, May 10, 2017. https://blogs.cdc.gov/niosh-science-blog/2017/05/10/ff-cancer-facts/.

Daniels, R. D., T. L. Kubale, J. H. Yiin, M. M. Dahm, T. R. Hales, D. Baris, S. H. Zahm, J. J. Beaumont, K. M. Waters, and L. E. Pinkerton. "Mortality and Cancer Incidence in a Pooled Cohort of US Firefighters from San Francisco, Chicago, and Philadelphia (1950–2009)." *Occupational and Environmental Medicine* 71, no. 6 (October 14, 2013): 388–397. https://doi.org/10.1136/oemed-2013-101662.

Davidson, R. J., J. Kabat-Zinn, J. Schumacher, M. Rosenkranz, D. Muller, S. F. Santorelli, F. Urbanowski, A. Harrington, K. Bonus, and J. F. Sheridan. "Alterations in Brain and Immune Function Produced by Mindfulness Meditation." *Psychosomatic Medicine* 65, no. 4 (July 2003): 564–570. https://doi.org/10.1097/01.psy.0000077505.67574.e3.

Davison, E. H., A. P. Pless, M. R. Gugliucci, L. A. King, D. W. King, D. M. Salgado, A. Spiro III, and P. Bachrach. "Late-Life Emergence of Early-Life Trauma: The Phenomenon of Late-Onset Stress Symptomatology Among Aging Combat Veterans." *Research on Aging* 28, no. 1 (January 1, 2006): 84–114. https://doi.org/10.1177/0164027505281560.

Decker, K. P., S. P. Deaver, V. Abbey, M. Campbell, and C. Turpin. "Quantitatively Improved Treatment Outcomes for Combat-Associated PTSD with Adjunctive Art Therapy: Randomized Controlled Trial." *Art Therapy* 35, no. 4 (2018): 184–194. https://doi.org/10.1080/07421656.2018.1540822.

Delizonna, L. "High Performing Teams Need Psychological Safety. Here's How to Create It." *Harvard Business Review*, August 24, 2017. https://hbr.org/2017/08/high-performing-teams-need-psychological-safety-heres-how-to-create-it.

Diener, E., and M. E. P. Seligman. "Very Happy People." *Psychological Science* 13, no. 1 (January 1, 2002): 81–84. https://doi.org/10.1111/1467-9280.00415.

Dienstbier, R. A. "Arousal and Physiological Toughness: Implications for Mental and Physical Health." *Psychological Review* 96, no. 1 (1989): 84–100. https://doi.org/10.1037/0033-295X.96.1.84.

Dirkzwager, A. J., I. Bramsen, and H. M. van der Ploeg. "The Longitudinal Course of Posttraumatic Stress Disorder Symptoms Among Aging Military Veterans." *Journal of Nervous and Mental Disease* 189, no. 12 (2001): 846–853. https://doi.org/10.1097/00005053-200112000-00006.

Dockett, L., and R. Simon. "The Addict in All of Us." *Psychotherapy Networker*, July/August 2017. https://www.psychotherapynetworker.org/magazine/article/1102/the-addict-in-all-of-us.

Doran, A. P., T. A. Gaskin, W. R. Schumm, and J. E. Smith. "Dealing with Combat and Operational Stress." Ceridian Corporation, 2004. https://www.marineparents.com/usmc/downloads/dealingwithcombatandoperationalstress.pdf.

Douillard, J. *The 3-Season Diet: Solving the Mysteries of Food Cravings, Weight-Loss, and Exercise*. New York: Harmony Books, 2000.

Earl, N. "Chill 101: How to Activate the PNS" HealthVibed.com, 2017. http://healthvibed.com/relaxation-101-how-to-activate-the-pns/.

Envisage Technologies. "Gender in First Response: How Biology and Physiology Shape Male and Female Responders." Envisagenow.com, May 16, 2018. www.envisagenow.com/gender-differences.

Erikson, E. H. *Identity and the Life Cycle.* New York: W. W. Norton & Co., 1980.

Erikson, E. H. *Identity: Youth and Crisis.* Austen Riggs Monograph No. 7. New York: W. W. Norton & Co., 1968.

Evans, J. H. "Race, Religion, and the Pursuit of Happiness." In *Theological Perspectives for Life, Liberty, and the Pursuit of Happiness: Public Intellectuals for the Twenty-First Century.* A. M. Isasi-Díaz, M. McClintock Fulkerson, and R. Carbine, eds. London: Palgrave Macmillan, 2013. 13–22.

Fay, J. "Resiliency for First Responders." First Responder Support Network, 2018. https://www.frsn.org.

Fay, J. "With Help, Life Gets Better. PTSD." *The Squad Room*, episode 7. Audio podcast, June 29, 2015. http://www.thesquadroom.net/episode7/.

Fields, R. D. *Why We Snap: Understanding the Rage Circuit in Your Brain.* New York: Dutton, 2016.

Figueroa-Fankhanel, F. "Measurement of Stress." *Psychiatric Clinics of North America* 37, no. 4 (2014): 455–487. https://doi.org/10.1016/j.psc.2014.08.001.

Finkel, E. J., C. M. Hui, K. L. Carswell, and G. M. Larson. "The Suffocation of Marriage: Climbing Mount Maslow Without Enough Oxygen." *Psychological Inquiry* 25, no. 1 (2014): 1–41. https://doi.org/10.1080/1047840x.2014.863723.

Flores, P. J., and L. Mahon. "The Treatment of Addiction in Group Psychotherapy." *International Journal of Group Psychotherapy* 43, no. 2 (2014): 143–156. https://doi:10.1080/00207284.1994.11491213.

Foa, E. B., K. R. Chrestman, and E. Gilboa-Schechtman. *Prolonged Exposure Therapy for Adolescents with PTSD: Emotional Processing of Traumatic Experiences: Therapist Guide.* Oxford: Oxford University Press, 2009.

Forman, T. "Self-Care Is Not an Indulgence. It's a Discipline." *Forbes*, December 13, 2017. https://www.forbes.com/sites/tamiforman/2017/12/13/self-care-is-not-an-indulgence-its-a-discipline/#30e901c8fee0.

France, A. *The Crime of Sylvestre Bonnard.* Translated by Lafcadio Hearn, and edited by Brett Fishburne and David Widger. Project Gutenberg, 2016. https://www.gutenberg.org/files/2123/2123-h/2123-h.htm#link2HCH0004.

Frankenhaeuser, M., U. Lundberg, and L. Forsman. "Dissociation Between Sympathetic-Adrenal and Pituitary-Adrenal Responses to an Achievement Situation Characterized by High Controllability: Comparison Between Type A and Type B Males and Females." *Biological Psychology* 10, no. 2 (March 1980): 79–91. https://doi.org/10.1016/0301-0511(80)90029-0.

Frankenhuis, W. E., and C. de Weerth. "Does Early-Life Exposure to Stress Shape or Impair Cognition?" *Current Directions in Psychological Science* 22, no. 5 (September 25, 2013): 407–412. https://doi.org/10.1177/0963721413484324.

Frankl, V. E. *Man's Search for Meaning.* New York: Simon and Schuster, 1985.

Froeliger, B., E. L. Garland, and F. J. McClernon. "Yoga Meditation Practitioners Exhibit Greater Gray Matter Volume and Fewer Reported Cognitive Failures: Results of a Preliminary Voxel-Based Morphometric Analysis." *Evidence-Based Complementary Alternative Medicine* (December 5, 2012): 1–8. https://doi.org/10.1155/2012/821307.

Gallegos, A. M., H. F. Crean, W. R. Pigeon, and K. L. Heffner. "Meditation and Yoga for Posttraumatic Stress Disorder: A Meta-Analytic Review of Randomized Controlled Trials." *Clinical Psychology Review* 58 (December 2017): 115–124. https://doi.org/10.1016/j.cpr.2017.10.004.

Gallegos, A. M., W. Cross, and W. R. Pigeon. "Mindfulness-Based Stress Reduction for Veterans Exposed to Military Sexual Trauma: Rationale and Implementation Considerations." *Military Medicine* 180, no. 6 (June 2015): 684–689. https://doi.org/10.7205/MILMED-D-14-00448.

Gallup. *State of the Global Workplace*. New York: Gallup Press, 2017. https://www.gallup.com/workplace/257552/state-global-workplace-2017.aspx.

Garbern, S. C., L. G. Ebbeling, and S. A. Bartels. "A Systematic Review of Health Outcomes Among Disaster and Humanitarian Responders." *Prehospital and Disaster Medicine* 31, no. 6 (2016): 635–642. https://doi.org/10.1017/s1049023x16000832.

Geraci, J., M. Baker, G. Bonanno, B. Tussenbroek, and L. Sutton. "Understanding and Mitigating Post-Traumatic Stress Disorder." *US Army Research* 344 (2011). http://digitalcommons.unl.edu/usarmyresearch/344.

Gladwell, M. *The Tipping Point: How Little Things Can Make a Difference*. New York: Little, Brown, 2002.

Goff, S., D. Thomas, and M. Trevathan. *Are My Kids on Track?: The 12 Emotional, Social, and Spiritual Milestones Your Child Needs to Reach*. Grand Rapids, MI: Baker Publishing Group, 2017.

Goleman, D., R. Boyatzis, R. J. Davidson, V. Druskat, and G. Kohlrieser. *Empathy: A Primer*. Book 6 in series, *Building Blocks of Emotional Intelligence*. Florence, MA: More Than Sound, 2017.

Gorski, T. T., and M. Miller. *Staying Sober: A Guide for Relapse Prevention*. Independence Press, 1986.

Gottman, J., and N. Silver. *Why Marriages Succeed or Fail*. Bloomsbury Paperbacks, 2014.

Grand Rapids Community College. "Seven Dimensions of Wellness." n.d. https://www.grcc.edu/humanresources/wellness/sevendimensionsofwellness.

Greenberg, D. M., S. Baron-Cohen, N. Rosenberg, P. Fonagy, and P. J. Rentfrow. "Elevated Empathy in Adults following Childhood Trauma." *PLoS ONE* 13, no. 10 (October 3, 2018). https://doi.org/10.1371/journal.pone.0203886.

Grieve, S. M., M. S. Korgaonkar, S. H. Koslow, E. Gordon, and L. M. Williams. "Widespread Reductions in Gray Matter Volume in Depression." *NeuroImage: Clinical* 3 (2013): 332–339. https://doi.org/10.1016/j.nicl.2013.08.016.

Griffin, B. J., N. Purcell, K. Burkman, B. T. Litz, C. J. Bryan, M. Schmitz, C. Villierme, J. Walsh, and S. Maguen. "Moral Injury: An Integrative Review." *Journal of Traumatic Stress* 32, no. 3 (2019): 350–362. https://doi.org/10.1002/jts.22362.

Gunther, R. "How Stress Can Bury Love—The Way Back." *Psychology Today*, November 14, 2014. https://www.psychologytoday.com/us/blog/rediscovering-love/201411/how-stress-can-bury-love-the-way-back.

Hanks, T. "Critical Incident/Traumatic Events Information." Northern Illinois Critical Incident Stress Management Team, January 2007. http://www.thecrisisdoctor.com/assets/informationandsymptoms2011-11-24.pdf.

Hari, J. "Everything You Think You Know About Addiction Is Wrong." Filmed June 2015 in London. TED Global video, 14:33. https://www.ted.com/talks/johann_hari_everything_you_think_you_know_about_addiction_is_wrong?language=en.

Harrell, S. K. *Gift of the Dreamtime: Awakening to the Divinity of Trauma*. 2nd ed. Fuquay Varina, NC: Soul Intent Arts, 2012.

Hawkins, K. A., J. L. Hames, J. D. Ribeiro, C. Silva, T. E. Joiner, and J. R. Cougle. "An Examination of the Relationship Between Anger and Suicide Risk Through the Lens of the Interpersonal Theory of Suicide." *Journal of Psychiatric Research* 50 (March 2014): 59–65. https://doi.org/10.1016/j.jpsychires.2013.12.005.

Hawkley, L. C., and J. T. Cacioppo, "Loneliness Matters: A Theoretical and Empirical Review of Consequences and Mechanisms." *Annals of Behavioral Medicine* 40, no. 2 (2010): 218–227. https://doi.org/10.1007/s12160-010-9210-8.

Helm, B. "Friendship." *Stanford Encyclopedia of Philosophy*, August 7, 2017. https://stanford.library.sydney.edu.au/entries/friendship/.

Hernandez, D. In discussion with the author about accelerated resolution therapy. 2020.

Hernandez, D. Email communication with the author about yoga and the brain. October 30, 2020.

Hettler, B. "Six Dimensions of Wellness." *National Wellness Institute*, 1976. https://nationalwellness.org/resources/six-dimensions-of-wellness/.

Heyman, M., J. Dill, and R. Douglas. "The Ruderman White Paper on Mental Health and Suicide of First Responders." White paper, Ruderman Foundation, 2018. https://rudermanfoundation.org/white_papers/police-officers-and-firefighters-are-more-likely-to-die-by-suicide-than-in-line-of-duty/.

Holmes, T. H., and R. H. Rahe. "The Social Readjustment Rating Scale." *Journal of Psychosomatic Research* 11, no. 2 (1967): 213–218. https://doi.org/10.1016/0022-3999(67)90010-4.

Holt-Lunstad, J., T. B. Smith, M. Baker, T. Harris, and D. Stephenson. "Loneliness and Social Isolation as Risk Factors for Mortality: A Meta-Analytic Review." *Perspectives on Psychological Science* 10, no. 2 (2015): 227–237. https://doi.org/10.1177/1745691614568352.

Hölzel, B. K., S. W. Lazar, T. Gard, Z. Schuman-Olivier, D. R. Vago, and U. Ott. "How Does Mindfulness Meditation Work? Proposing Mechanisms of Action from a Conceptual and Neural Perspective." *Perspectives on Psychological Science* 6, no. 6 (October 14, 2011): 537–559. https://doi.org/10.1177/1745691611419671.

Hom, M. A., I. H. Stanley, M. L. Rogers, M. Tzoneva, R. A. Bernert, and T. E. Joiner. "The Association Between Sleep Disturbances and Depression Among Firefighters: Emotion Dysregulation as an Explanatory Factor." *Journal of Clinical Sleep Medicine* 12, no. 2 (2016): 235–245. https://doi.org/10.5664/jcsm.5492.

Honneth, A., and J. Farrell. "Recognition and Moral Obligation." *Social Research* 64, no. 1 (Spring 1997): 16–35. https://www.jstor.org/stable/40971157?seq=1#metadata_info_tab_contents.

Horesh, D., Z. Solomon, G. Keinan, and T. Ein-Dor. "The Clinical Picture of Late-Onset PTSD: A 20-Year Longitudinal Study of Israeli War Veterans." *Psychiatry Research* 208, no. 3 (August 15, 2013): 265–273. https://doi.org/10.1016/j.psychres.2012.12.004.

Hudson, J. "Exercise Does More Than You Think." Wellness Supplement, *Fire Engineering*. September 22, 2018. https://digital.fireengineering.com/fireengineering/201809/MobilePagedArticle.action?articleId=1423595#articleId1423595.

Hudson, J. "Your Relationship with Death." FirefighterNation.com. August 24, 2018. https://www.firefighternation.com/firerescue/your-relationship-with-death/#gref.

Hustad, M. "Surprising Benefits for Those Who Had Tough Childhoods." *Psychology Today*, March 7, 2017. https://www.psychologytoday.com/intl/articles/201703/surprising-benefits-those-who-had-tough-childhoods?amp.

Hyer, L., M. N. Summers, L. Braswell, and S. Boyd. "Posttraumatic Stress Disorder: Silent Problem Among Older Combat Veterans." *Psychotherapy: Theory, Research, Practice, Training* 32, no. 2 (1995): 348–364. https://doi.org/10.1037/0033-3204.32.2.348.

InternetPoem.com. "Marcus Aurelius." Accessed June 23, 2021. https://internetpoem.com/marcus-aurelius/quotes/loss-is-nothing-else-but-change-and-change-is-38886/5/.

Isasi-Diaz, A. M., and J. H. Evans. "Race, Religion, and the Pursuit of Happiness." Essay in *Theological Perspectives for Life, Liberty, and the Pursuit of Happiness: Public Intellectuals for the Twenty-First Century*. London: Palgrave Macmillan, 2013.

Jancin, B. "AAS: Acute Suicidal Affective Disturbance Proposed as New Diagnosis." *Clinical Psychiatry News* MDedge.com, May 28, 2015. https://www.mdedge.com/psychiatry/article/100017/depression/aas-acute-suicidal-affective-disturbance-proposed-new-diagnosis/.

Jennison, K. M. "The Impact of Stressful Life Events and Social Support on Drinking Among Older Adults: A General Population Survey." *International Journal of Aging and Human Development* 35, no. 2 (September 1, 1992): 99–123, https://doi.org/10.2190/f6g4-xlv3-5kw6-vmba.

Joiner, T. E. "The Clustering and Contagion of Suicide." *Current Directions in Psychological Science* 8, no. 3 (1999): 89–92. https://doi.org/10.1111/1467-8721.00021.

Joiner, T. E. *Lonely at the Top: The High Cost of Men's Success*. New York: Palgrave Macmillan, 2011.

Joiner, T. E. *Why People Die by Suicide*. Cambridge, MA: Harvard University Press, 2007.

Jones, H. M. "The Philosophical Basis for Caring, Compassion, and Interdependence." *In the Pursuit of Happiness*. Cambridge, MA: Harvard University Press, 1953.

Kamkar, K. "First Responders Suffering from 'Moral Injury.'" SafetyMag.com, January 7, 2019. https://www.thesafetymag.com/ca/news/opinion/first-responders-suffering-from-moral-injury/187489/.

Killeen, C. "Loneliness: An Epidemic in Modern Society." *Journal of Advanced Nursing* 28, no. 4 (1998): 762–770. https://doi:10.1046/j.1365-2648.1998.00703.x.

Khayyam, O. Quoted in E. Becker, *Denial of Death*. United Kingdom: Free Press, 1973. https://www.google.com/books/edition/_/DhgQa6rzEGQC?hl=en&saX&ved=2ahUKEwiW6dm12ZfzAhXEEFkFHaGZBeEQ7_IDegQICBAD.

Kim, S. H., S. M. Schneider, M. Bevans, L. Kravitz, C. Mermier, C. Qualls, and M. R. Burge. "PTSD Symptom Reduction with Mindfulness-Based Stretching and Deep Breathing Exercise: Randomized Controlled Clinical Trial of Efficacy." *Journal of Clinical Endocrinology & Metabolism* 98, no. 7 (2013): 2,984–2,992. https://doi.org/10.1210/jc.2012-3742.

Kim, S. H., S. M. Schneider, L. Kravitz, C. Mermier, and M. R. Burge. "Mind-Body Practices for Posttraumatic Stress Disorder." *Journal of Investigative Medicine* 61, no. 5 (2013): 827–834. https://doi.org/10.2310/jim.0b013e3182906862.

King, M. L., Jr. *Strength to Love*. United States: Beacon Press, 2019. https://www.google.com/books/edition/Strength_to_Love/FHKEDwAAQBAJ?hl=en&gbpv=0.

Kip, K. E., L. Rosenzweig, D. F. Hernandez, A. Shuman, K. L. Sullivan, C. J. Long, J. Taylor, S. McGhee, S. A. Girling, T. Wittenberg, F. M. Sahebzamani, C. A. Lengacher, R. Kadel, and D. M. Diamond. "Randomized Controlled Trial of Accelerated Resolution Therapy (ART) for Symptoms of Combat-Related Post-Traumatic Stress Disorder (PTSD)." *Military Medicine* 178, no. 12 (December 2013): 1,298–1,309. https://doi.org/10.7205/milmed-d-13-00298.

Kip, K. E., K. L. Sullivan, C. A. Lengacher, L. Rosenzweig, D. F. Hernandez, R. Kadel, F. A. Kozel, A. Shuman, S. A. Girling, M. J. Hardwick, and D. M. Diamond. "Brief Treatment of Co-Occurring Post-Traumatic Stress and Depressive Symptoms by Use of Accelerated Resolution Therapy." *Frontiers in Psychiatry* 4 (March 8, 2013). https://doi.org/10.3389/fpsyt.2013.00011.

Kjaer, T. W., C. Bertelsen, P. Piccini, D. Brooks, J. Alving, and H. C. Lou. "Increased Dopamine Tone During Meditation-Induced Change of Consciousness." *Cognitive Brain Research* 13, no. 2 (April 2002): 255–259. https://doi.org/10.1016/s0926-6410(01)00106-9.

Klontz, B., R. Kahler, and T. Klontz. *Facilitating Financial Health: Tools for Financial Planners, Coaches, and Therapists*. Erlanger, KY: The National Underwriter Company, 2016.

Knutson, S. "How the Human Stress Response Explains Away 'Bipolar Disorder.'" Mad in America, April 1, 2018. https://www.madinamerica.com/2018/04/how-the-human-stress-response-explains-away-bipolar-disorder/#fn-154832-2/.

Lee, R. "Codependency: The Helping Problem." Psych Central, 2018. https://psychcentral.com/lib/codependency-the-helping-problem/.

Lee, H. J., and M. E. Szinovacz. "Positive, Negative, and Ambivalent Interactions with Family and Friends: Associations with Well-Being." *Journal of Marriage and Family* 78, no. 3 (2016): 660–679. https://doi.org/10.1111/jomf.12302.

Leenaars, A. A. "Review: Edwin S. Shneidman on Suicide." *Suicidology Online* 1 (March 8, 2010): 5–18. http://www.suicidology-online.com/pdf/SOL-2010-1-5-18.pdf.

Levine, P. "The Absence of an Empathetic Witness." Unitarian Universalist Association, August 4, 2020. https://www.uua.org/worship/words/quote/absence-empathetic-witness.

Ley, D. J. "Surviving Childhood Adversity Builds Empathy in Adults." *Psychology Today*, September 18, 2020. https://www.psychologytoday.com/us/blog/women-who-stray/202009/surviving-childhood-adversity-builds-empathy-in-adults.

Linehan, M. M., H. E. Armstrong, A. Suarez, D. Allmon, H. L. Heard. "Cognitive-Behavioral Treatment of Chronically Parasuicidal Borderline Patients." *Archives of General Psychiatry* 48, no. 12 (1991): 1,060. https://doi.org/10.1001/archpsyc.1991.01810360024003.

Lisak, D. "The Neurobiology of Trauma." Arkansas Coalition Against Sexual Assault. YouTube video, 34:30. February 5, 2013. https://youtu.be/pyomVt2Z7nc.

Luscombe, B. "Staying Married: Marriage Has Changed, but It Might Be Better Than Ever." *Time*, June 2, 2016. https://time.com/4354770/how-to-stay-married/.

Macy, R. J., E. Jones, L. M. Graham, and L. Roach. "Yoga for Trauma and Related Mental Health Problems: A Meta-Review with Clinical and Service Recommendations." *Trauma, Violence, and Abuse* 19, no. 1 (December 9, 2015): 35–57. https://doi.org/10.1177/1524838015620834.

Marx, B. P., J. P. Forsyth, G. G. Gallup, T. Fusé, and J. M. Lexington. "Tonic Immobility as an Evolved Predator Defense: Implications for Sexual Assault Survivors." *Clinical Psychology: Science and Practice* 15, no. 1 (2008): 74–90. https://doi.org/10.1111/j.1468-2850.2008.00112.x.

Mayanil, S. "Yoga and PTSD." In discussion with the author, October 10, 2020.

McCracken, S. In discussion with the author, October 22, 2019.

McDonald, W. "Søren Kierkegaard." *Stanford Encyclopedia of Philosophy*, November 10, 2017. https://plato.stanford.edu/entries/kierkegaard/?source=post_page.

McGonigal, K. "The Willpower Instinct." YouTube video, 54:02. February 1, 2012. https://www.youtube.com/watch?v=V5BXuZL1HAg&t=1s.

Melemis, S. M. "Relapse Prevention Plan and Early Warning Signs." Addictions and Recovery, March 15, 2017. https://www.addictionsandrecovery.org/relapse-prevention.htm.

Miller, J. J., K. Fletcher, and J. Kabat-Zinn. "Three-Year Follow-Up and Clinical Implications of a Mindfulness Meditation-Based Stress Reduction Intervention in the Treatment of Anxiety Disorders." *General Hospital Psychiatry* 17, no. 3 (May 1995): 192–200. https://doi.org/10.1016/0163-8343(95)00025-m.

Miller, R. R. "About iRest." iRest Institute, accessed June 25, 2021. https://www.irest.org/about-irest-institute. https://www.irest.org/about-irest-institute.

Mind Tools. "What Is Stress?" Mindtools.com. Accessed August 9, 2021. https://www.mindtools.com/pages/article/newTCS_00.htm.

Mittal, C., V. Griskevicius, J. A. Simpson, S. Sung, and E. S. Young. "Cognitive Adaptations to Stressful Environments: When Childhood Adversity Enhances Adult Executive Function. *Journal of Personality and Social Psychology* 109, no. 4 (2015): 604–621. https://doi.org/10.1037/pspi0000028.

Moss, S. "The Distinction Between Challenge and Threat Appraisals." SicoTests.com, June 28, 2016. https://www.sicotests.com/psyarticle.asp?id=281.

Moss, S. "Physiological Toughness." SicoTests.com, June 18, 2016. https://www.sicotests.com/psyarticle.asp?id=230.

Murphy, M. "How to Tweak Your Leadership Style When Your Employees Are Feeling Anxious." *Forbes*, June 2, 2019. https://www.forbes.com/sites/markmurphy/2019/06/02/how-to-tweak-your-leadership-style-when-your-employees-are-feeling-anxious/#8009d33b19de.

Myers, C. "An Introvert's Guide to Leadership." *Forbes*, August 14, 2016. https://www.forbes.com/sites/chrismyers/2016/08/14/an-introverts-guide-to-leadership/#117fc70e6d8b.

National Center for Complementary and Alternative Medicine. "Yoga: What You Need to Know." US Department of Health and Human Services, National Institutes of Health, 2015. https://nccih.nih.gov/health/yoga.

National Center for PTSD. "Late-Onset Stress Symptomatology (LOSS) Scale." US Department of Veterans Affairs, January 28, 2020. https://www.ptsd.va.gov/professional/assessment/adult-sr/loss_scale.asp.

National Institute on Drug Abuse. "Step by Step Guides to Finding Treatment for Drug Use Disorders: Treatment Information." US Department of Health and Human Services, National Institutes of Health, June 4, 2020. https://www.drugabuse.gov/publications/step-by-step-guides-to-finding-treatment-drug-use-disorders/if-your-adult-friend-or-loved-one-has-problem-drugs/treatment-information.

Niederkrotenthaler, T., M. Gould, G. Sonneck, S. Stack, and B. Till. "Predictors of Psychological Improvement on Non-Professional Suicide Message Boards: Content Analysis." *Psychological Medicine* 46, no. 16 (September 22, 2016): 3,429–3,442. https://doi.org/10.1017/s003329171600221x.

Nielson, E. C., R. S. Singh, K. L. Harper, and E. J. Teng. "Traditional Masculinity Ideology, Posttraumatic Stress Disorder (PTSD) Symptom Severity, and Treatment in Service Members and Veterans: A Systematic Review." *Psychology of Men & Masculinities* 21, no. 4 (January 27, 2020): 578–592. https://doi.org/10.1037/men0000257.

Nietzche, F. *Twilight of the Idols*. Translated by Duncan Large. Oxford: Oxford University Press, 1998. https://www.google.com/books/edition/_/iSNeybYAgNkC?hl=en&gbpv=0.

"Nine Strategies for Families Helping a Loved One in Recovery." Behavioral Health Evolution, Hazelden Foundation, 2016. http://www.bhevolution.org/public/family_support.page.

Nolan, C. R. "Bending Without Breaking: A Narrative Review of Trauma-Sensitive Yoga for Women with PTSD." *Complementary Therapies in Clinical Practice* 24 (August 2016): 32–40. https://doi.org/10.1016/j.ctcp.2016.05.006.

O'Hanlon, B. "Resolving Trauma Without Drama." *Possibilities*, October 2016. https://www.billohanlon.com.

Ojeda, V. D., and S. M. Bergstresser. "Gender, Race-Ethnicity, and Psychosocial Barriers to Mental Health Care: An Examination of Perceptions and Attitudes Among Adults Reporting Unmet Need." *Journal of Health and Social Behavior* 49, no. 3 (2008): 317–334. https://doi.org/10.1177/002214650804900306.

Olson, R. "Suicide Contagion and Suicide Clusters." Centre for Suicide Prevention, 2013. https://www.suicideinfo.ca/resource/suicidecontagion/.

Osório, C., T. Probert, E. Jones, A. H. Young, and I. Robbins. "Adapting to Stress: Understanding the Neurobiology of Resilience." *Behavioral Medicine* 43, no. 4 (2017): 307–322. https://doi-org.du.idm.oclc.org/10.1080/08964289.2016.1170661.

Owen, L., and B. Corfe. "The Role of Diet and Nutrition on Mental Health and Wellbeing." *Proceedings of the Nutrition Society* 76, no. 4 (July 14, 2017): 425–426. https://doi.org/10.1017/s0029665117001057.

Palmiter et al., "Building Your Resilience," American Psychological Association (2012), https://www.apa.org/topics/resilience.

Papazoglou, K., and B. Chopko. "The Role of Moral Suffering (Moral Distress and Moral Injury) in Police Compassion Fatigue and PTSD: An Unexplored Topic." *Frontiers in Psychology* 8, no. 1999 (November 15, 2017). https://doi.org/10.3389/fpsyg.2017.01999.

Parker, S., S. V. Bharati, and M. Fernandez. "Defining Yoga-Nidra: Traditional Accounts, Physiological Research, and Future Directions." *International Journal of Yoga Therapy* 23, no. 1 (January 1, 2013): 11–16. https://doi.org/10.17761/ijyt.23.1.t636651v22018148.

Patton, J. M. "5 Truths About Self-Care." Experience Life, August 19, 2019. https://experiencelife-com.cdn.ampproject.org/c/s/experiencelife.com/article/5-truths-about-self-care/amp/.

Pennebaker, J. W. "Writing About Emotional Experiences as a Therapeutic Process." *Psychological Science* 8, no. 3 (May 1, 1997): 162–166. https://doi.org/10.1111/j.1467-9280.1997.tb00403.x.

Perry, A., captain of the Peoria (IL) Fire Department and Illinois Firefighter Peer Support member. In discussion with the author at peer support team training session.

Polack, E. "New Cigna Study Reveals Loneliness at Epidemic Levels in America." *Cigna.com*, May 1, 2018. https://www.cigna.com/newsroom/news-releases/2018/new-cigna-study-reveals-loneliness-at-epidemic-levels-in-america.

Poll, H. "Celebrate Relationships, but Beware of Financial Infidelity." National Endowment for Financial Education, February 14, 2018. https://www.nefe.org/news/2018/02/celebrate-relationships-but-beware-of-financial-infideltiy.aspx.

Potter, C. M., A. P. Kaiser, L. A. King, D. W. King, E. H. Davison, A. V. Seligowski, C. B. Brady, and A. Spiro III. "Distinguishing Late-Onset Stress Symptomatology from Posttraumatic Stress Disorder in Older Combat Veterans." *Aging & Mental Health* 17, no. 2 (2013): 173–179. https://doi.org/10.1080/13607863.2012.717259.

Racine, M. "Chronic Pain and Suicide Risk: A Comprehensive Review." *Progress in Neuro-Psychopharmacology and Biological Psychiatry* 87 (2018): 269–280. https://doi.org/10.1016/j.pnpbp.2017.08.020.

Rando, T. A. *Grief, Dying, and Death Clinical Interventions for Caregivers*. Braille Jymico Inc., 2008.

"Recognizing Depression in Men." *Harvard Mental Health Letter*, June 2011. https://www.health.harvard.edu/newsletter_article/recognizing-depression-in-men.

Rice, V., and B. Liu. "Personal Resilience and Coping Part II: Identifying Resilience and Coping Among U.S. Military Service Members and Veterans with Implications for Work." *Work* 54, no. 2 (2016): 335–350. https://doi.org/10.3233/wor-162301.

Richards, D. "A Field Study of Critical Incident Stress Debriefing Versus Critical Incident Stress Management." *Journal of Mental Health* 10, no. 3 (2001): 351–362. https://doi.org/10.1080/09638230124190.

Riley, W. T., F. A. Treiber, and M. G. Woods. "Anger and Hostility in Depression." *Journal of Nervous and Mental Disease* 177, no. 11 (November 1989): 668–674. https://doi.org/10.1097/00005053-198911000-00002.

Robertson, I. "Stress Can Make Your Brain Stronger If You Know This." *Center for Brain Health*. University of Texas at Dallas, January 23, 2018. https://brainhealth.utdallas.edu/stress-can-make-your-brain-stronger-if-you-know-this/.

Robinson, L., J. Segal, and M. Smith, "Coping with a Life-Threatening Illness or Serious Health Event." HelpGuide, October 8, 2019. https://www.helpguide.org/articles/grief/coping-with-a-life-threatening-illness.htm.

Rodriguez-Raecke, R., A. Niemeier, K. Ihle, W. Ruether, and A. May. "Brain Gray Matter Decrease in Chronic Pain Is the Consequence and Not the Cause of Pain." *Journal of Neuroscience* 29, no. 44 (November 4, 2009): 13,746–13,750. https://doi.org/10.1523/jneurosci.3687-09.2009.

Rogers, M. L., M. A. Hom, T. E. Joiner. "Differentiating Acute Suicidal Affective Disturbance (ASAD) from Anxiety and Depression Symptoms: A Network Analysis." *Journal of Affective Disorders* 250 (May 1, 2019): 333–340. https://doi.org/10.1016/j.jad.2019.03.005.

Rovella, D., and A. McIntire. "Firefighters Are the Happiest Workers in America." Bloomberg, July 17, 2019. https://www.bloomberg.com/news/articles/2019-07-17/which-jobs-make-people-the-happiest-in-america.

Sanders, J. "Interview with Lieutenant Matt Olson." *The Fire Inside*, episode 3. Audio podcast. https://thefireinsidepodcast.com/003/.

Sapolsky, R. M. *Why Zebras Don't Get Ulcers: The Acclaimed Guide to Stress, Stress-Related Diseases, and Coping*. New York: Henry Holt and Co., 2004.

Satterle, C. "Closing the Communication Gap." PowerPoint presentation. n.d.

Schuster, C. R., and T. Thompson. "Self-Administration and Behavioral Dependence on Drugs." *Annual Review of Pharmacology and Toxicology* 9 (April 1969): 483–502. https://doi.org/10.1146/annurev.pa.09.040169.002411.

Seay, N. "How are Addiction, Depression and Suicide Linked?" *The Rehabs Journal*. American Addiction Centers, November 4, 2019. https://www.rehabs.com/blog/how-are-addiction-depression-and-suicide-linked/.

Shannahan, C., and L. Shannahan. *Deep Nutrition: Why Your Genes Need Traditional Food*. Flatiron Books, 2008.

Smith, C. "The Link Between Alcohol Use and Suicide." May 19, 2020. https://www.alcoholrehabguide.org/resources/dual-diagnosis/alcohol-and-suicide/.

Stankovic, L. "Transforming Trauma: A Qualitative Feasibility Study of Integrative Restoration (iRest) Yoga Nidra on Combat-Related Post-Traumatic Stress Disorder." *International Journal of Yoga Therapy* 21, no. 1 (October 1, 2011): 23–37. https://doi.org/10.17761/ijyt.21.1.v823454h5v57n160.

Stanley, I. H., M. A. Hom, and T. E. Joiner. "A Systematic Review of Suicidal Thoughts and Behaviors Among Police Officers, Firefighters, EMTs, and Paramedics." *Clinical Psychology Review* 44 (March 2016): 25–44. https://doi.org/10.1016/j.cpr.2015.12.002.

Stott, H., battalion chief, West Chicago Fire Protection District. In discussion with the author.

Streeter, C. C., P. L. Gerbarg, G. H. Nielsen, R. P. Brown, J. E. Jensen, and M. Silveri. "Effects of Yoga on Thalamic Gamma-Aminobutyric Acid, Mood and Depression: Analysis of Two Randomized Controlled Trials." *Neuropsychiatry (London)* 8, no. 6 (2018): 1,923–1,939. https://www.jneuropsychiatry.org/peer-review/effects-of-yoga-on-thalamic-gammaaminobutyric-acid-mood-and-depression-analysis-of-two-randomized-controlled-trials.pdf.

Streeter, C. C., P. L. Gerbarg, R. B. Saper, D. A. Ciraulo, and R. P. Brown. "Effects of Yoga on the Autonomic Nervous System, Gamma-Aminobutyric-Acid, and Allostasis in Epilepsy, Depression, and Post-Traumatic Stress Disorder." *Medical Hypotheses* 78, no. 5 (May 2012): 571–579. https://doi.org/10.1016/j.mehy.2012.01.021.

Substance Abuse and Mental Health Services Administration (SAMHSA). "First Responders: Behavioral Health Concerns, Emergency Response, and Trauma." *Disaster Technical Assistance Center Supplemental Research Bulletin*, May 2018. https://www.samhsa.gov/sites/default/files/dtac/supplementalresearchbulletin-firstresponders-may2018.pdf.

Substance Abuse and Mental Health Services Administration (SAMHSA). "Table 3.19, DSM-IV to DSM-5 Adjustment Disorders Comparison." June 2016. https://www.ncbi.nlm.nih.gov/books/NBK519704/table/ch3.t19/.

Surmeier, D. J., J. Plotkin, and W. Shen. "Dopamine and Synaptic Plasticity in Dorsal Striatal Circuits Controlling Action Selection." *Current Opinion in Neurobiology* 19, no. 6 (December 2009): 621–628. https://doi.org/10.1016/j.conb.2009.10.003.

Swaim, E. "10 Ways to Boost Dopamine and Serotonin Naturally." *Good Therapy*, December 12, 2017. https://www.goodtherapy.org/blog/10-ways-to-boost-dopamine-and-serotonin-naturally-1212177.

Tang, Y., Y. Ma, Y. Fan, H. Feng, J. Wang, S. Feng, Q. Lu, B. Hu, Y. Lin, J. Li, Y. Zhang, Y. Wang, L. Zhou, and M. Fan. "Central and Autonomic Nervous System Interaction Is Altered by Short-Term Meditation." *Proceedings of the National Academy of Sciences (US)* 106, no. 22 (June 2, 2009): 8,865–8,870. https://doi.org/10.1073/pnas.0904031106.

Tartakovsky, M. "4 Things Introverts Do That Makes Them Effective Leaders." *PsychCentral.com*, July 8, 2018. https://psychcentral.com/blog/archives/2013/09/28/4-things-introverts-do-that-makes-them-effective-leaders/.

Tull, M. "Exposure Therapy for Treating Post-Traumatic Stress Disorder Symptoms." VeryWellMind.com, February 25, 2020. https://www.verywellmind.com/exposure-therapy-for-ptsd-2797654.

Ungar, M. "Let Kids Be Bored (Occasionally)." *Psychology Today*, June 24, 2012. https://www.psychologytoday.com/blog/nurturing-resilience/201206/let-kids-be-bored-occasionally.

University of California–Riverside. "Wellness: Seven Dimensions of Wellness." June 4, 2014. https://www.wellness.ucr.edu/emotional_wellness.html.

University of Southern Florida College of Nursing. "USF College of Nursing Study Reveals Brief Therapy Treats PTSD Symptoms." *USF Nursing News*, July 2, 2014. https://hscweb3.hsc.usf.edu/nursingnews/usf-college-of-nursing-study-reveals-brief-therapy-treats-ptsd-symptoms/.

Ure, L. "Rethinking Anger." *Psychotherapy Networker*, October 1, 2020. https://www.psychotherapynetworker.org/blog/details/1824/rethinking-anger?utm_source=linkedin&utm_medium=social.

US Department of Health and Human Services (HHS). "Emotional Wellness Toolkit." National Institutes of Health, December 10, 2018. http://www.nih.gov/health-information/emotional-wellness-toolkit.

US Department of Health and Human Services (HHS), *2012 National Strategy for Suicide Prevention: Goals and Objectives for Action*. Washington, DC: Office of the Surgeon General and National Action Alliance for Suicide Prevention, HHS, September 2012. https://www.ncbi.nlm.nih.gov/books/NBK109917/.

van der Kolk, B. A. "Clinical Implications of Neuroscience Research in PTSD." *Annals of the New York Academy of Sciences* 1,071, no. 1 (July 26, 2006): 277–293. https://doi.org/10.1196/annals.1364.022.

van der Kolk, B. A., L. Stone, J. West, A. Rhodes, D. Emerson, M. Suvak, and J. Spinazzola. "Yoga as an Adjunctive Treatment for Posttraumatic Stress Disorder: A Randomized Controlled Trial." *Journal of Clinical Psychiatry* 75, no. 6 (June 2014): 559–565. https://doi.org/10.4088/jcp.13m08561.

Vincent, N. *Self-Made Man: One Woman's Year Disguised as a Man*. New York: Penguin Group, 2007.

Viney, L. L. "Loss of Life and Loss of Bodily Integrity: Two Different Sources of Threat for People Who Are Ill." *OMEGA—Journal of Death and Dying* 15, no. 3 (November 1, 1985): 207–222. https://doi.org/10.2190/x1b1-3c94-v2d5-amqj.

Viorst, J. *Necessary Losses: The Loves, Illusions, Dependencies, and Impossible Expectations That All of Us Have to Give Up in Order to Grow*. New York: Simon and Schuster, 1986.

Walker, P. "Emotional Flashback Management in the Treatment of Complex PTSD." Psychotherapy.net, September 2009. http://pete-walker.com/pdf/emotionalFlashbackManagement.pdf.

Wall, E., "Self-Care Practices and Attitudes Toward CISD and Seeking Mental Health Services Among Firefighters: A Close Look at a Mid-Sized Midwestern Urban City." UST Research Online, 2012. https://ir.stthomas.edu/ssw_mstrp/123.

Waters, B. "10 Traits of Emotionally Resilient People." *Psychology Today*, May 21, 2013. https://www.psychologytoday.com/us/blog/design-your-path/201305/10-traits-emotionally-resilient-people.

Webb, J. "Were You Raised in a Passive-Aggressive Family?" DrJoniceWebb.com, August 19, 2019. https://drjonicewebb.com/were-you-raised-in-a-passive-aggressive-family/.

Webb, J., and C. Musello. *Running on Empty: Overcome Your Childhood Emotional Neglect*. New York: Morgan James Publishing, 2019.

Weinberger, J. "How to Talk to Your Partner About Money (Without a Meltdown)." *Talk Space*, May 14, 2019. https://www.talkspace.com/blog/talk-to-partner-about-money/.

Weinstein, E. "Is It True: What Doesn't Kill You Makes You Stronger?" PsychCentral, October 28, 2017. https://psychcentral.com/lib/is-it-true-what-doesnt-kill-you-makes-you-stronger/.

Weiss, R. "Why Do People with Addictions Seek to Escape Rather Than Connect? A Look at the Approach to Addiction Treatment." *Consultant 360*, 56, no. 9 (September 2016), https://www.consultant360.com/articles/why-do-people-addictions-seek-escape-rather-connect-look-approach-addiction-treatment/.

Wilson, F. T., III., Illinois Firefighter Peer Support member. In discussion with the author.

Wilson, S. J., A. Woody, A. C. Padin, J. Lin, W. B. Malarkey, and J. K. Kiecolt-Glaser. "Loneliness and Telomere Length: Immune and Parasympathetic Function in Associations with Accelerated Aging." *Annals of Behavioral Medicine* 53, no. 6 (2018): 541–550. https://doi.org/10.1093/abm/kay064.

Wolin, S., and S. Wolin. "Resilience as Paradox." 1999. http://projectresilience.com/framesconcepts.htm.

Woźniak, K., and D. Iżycki. "Cancer: A Family at Risk." *Menopausal Review* 13, no. 4 (2014): 253–261. https://doi.org/10.5114/pm.2014.45002.

Yoga Therapy. Good Therapy, June 8, 2017. https://www.goodtherapy.org/learn-about-therapy/types/yoga-therapy.

Zaorsky, N. G., Y. Zhang, L. Tuanquin, S. M. Bluethmann, H. S. Park, and V. M. Chinchilli. "Suicide Among Cancer Patients." *Nature Communications* 10, no. 1 (2019). https://doi.org/10.1038/s41467-018-08170-1.

Zern, A., deputy chief, Sycamore (IL) Fire Department. In discussion with the author.

Index

A

accelerated resolution therapy (ART) 90, 173–174. *See also* therapy and treatment for PTSD
  effectiveness of 174–175
  late onset stress symptomatology (LOSS), overcoming 266
  personal stories on 175–176
  process of 174
  voluntary image replacement (VIR) and 174
ACEs (adverse childhood experiences). *See* adverse childhood experiences (ACEs)
acetylcholine (ACh) 163
active-constructive responding 188
active toughening 154, 198
acute stress disorder (ASD) 99. *See also* stress; trauma
acute suicidal affective disturbance (ASAD) 80–81
addiction. *See also* coping; recovery; relapse
  biology of 50–51
  connection and 54–57
  definition of 47–48
  depression and suicide, links to 83
  development of 49–50
  exercise and overcoming 156
  firefighting tactic analogy and 184
  identifying 54
  impulsive sex 58
  individuals and 48–49
  personal stories on 47, 54–55
  rat parks and 53
  reflection questions on 64
  retirement and 257
  risk factors for 51–52
  self-care and 57
  self-soothing and 48
  treatment for 64–65
  triggers and 57–58
adjustment disorder 136
adverse childhood experiences (ACEs) 34. *See also* trauma
  addiction and 51–52
  affective and cognitive empathy and 36

childhood emotional neglect (CEN) 52
  emotional neglect 29–30, 33
  inhibitory control and task-switching 35
  personal stories on 31–32
  physiological toughness and 197
affective empathy 36. *See also* adverse childhood experiences (ACEs)
aging, successful 271–272. *See also* retirement
alliances in fire and rescue 207–209
American Psychological Association (APA) 121
anger
  depression and 19–20
  exercise and 157
  redirecting 156–157
ANS (autonomic nervous system) 105
APA (American Psychological Association) 121
appraisal, challenge and threat 194–195
ART (accelerated resolution therapy). *See* accelerated resolution therapy (ART)
ASAD (acute suicidal affective disturbance) 80–81
ASD (acute stress disorder) 99. *See also* trauma
assessment, core values 189–190
autonomic nervous system (ANS) 105
avoidance
  imaginal/prolonged exposure therapy and 178
  money 222
  physiological toughness and 154
  post-traumatic stress disorder (PTSD) and 99

B

balanced people, actions of 3–4. *See also* connectedness
bipolar disorder misdiagnoses 107–109. *See also* stress
breakthroughs, firefighting tactics and 184
brotherhood 188–189
  community and xvii
  personal stories on 183–184
  reflection questions on 190
busyness
  first responders and 20–21
  self-neglect and 22–23

C

California Highway Patrol (CHP) 125
cancer
  children's questions about 243
  communication and 242–243
  emotional wellness and 240–241, 243
  emotions of 239–240
  family and 242
  finishing business 245
  help, offering 244
  marriage and 241
  nutrition and 240
  reflection questions on 245
  relationships and connectedness 241–242
  self-care and 240
  statistics for firefighters 239
  suicidal ideation and 243–244
  support groups 241
  truth statements on 243–244
CEN (childhood emotional neglect) 52. *See also* adverse childhood experiences (ACEs); emotional, neglect
Center for the Treatment and Study of Anxiety (CTSA) 180
challenge appraisal 194–195
change. *See also* retirement
  escaping 248–249
  processing 247–248
childhood emotional neglect (CEN) 52. *See also* adverse childhood experiences (ACEs); emotional, neglect
CHP (California Highway Patrol) 125
chronic stress 96–97, 155
CISD (critical incident stress debriefing). *See* critical incident stress debriefing (CISD)
clinical depression 18. *See also* depression
cluster suicide 76–80
cognitive empathy 36. *See also* adverse childhood experiences (ACEs)
communication 233–235
  cancer and 242–243
  deep listening 202
  in fires and at home 234
  intuition and senses 233–234
  paramilitary 214
  relationships and 214
compassion. *See* needing to be needed, relationship addiction
complex PTSD (CPTSD) 116–117. *See also* post-traumatic stress disorder (PTSD)
confirmation bias 127
connectedness 2–3. *See also* emotional wellness
  cancer and 241–242
  nature 7
  personal stories on 1–2
contagion suicide 76–80
coping. *See also* addiction
  definition of 122
  flexibility 132
  magical thinking 91
  mechanisms 91
  resiliency and 122
  self-medication and 16–17
core values assessment 189–190
CPTSD (complex PTSD) 116–117. *See also* post-traumatic stress disorder
creativity 130. *See also* resiliency
critical incident stress debriefing (CISD) xv
  effectiveness of 206–207
  potentially traumatic event (PTE) and 206–207
CTSA (Center for the Treatment and Study of Anxiety) 180

D

death
  escaping the idea of 248–249
  healthy fear of 250–251
  of situations 247–248
  philosophy of 249
  premortem for life 250–251
  reflection questions on 251
  relationship with 247
  transcending the self and 249
  view of, questions on 250
deep listening 202
delayed onset PTSD 116. *See also* post-traumatic stress disorder (PTSD)
depression
  addiction and suicide, links to 83
  anger and 19–20
  clinical 18
  exercise and 156
  first responders and 11, 13–14
  healing from 24
  hidden 18
  life change units (LCUs) and 253
  moral injury and 118
  moving meditation and 162
  nutrition and 158
  personal stories on 11–12
  reflection questions on 25
  rescuer's 13, 256
  retirement and 255–256
  situational 242
  sleep disturbances and 19–20
  vitamins and supplements for 159
dhyana. *See* meditation

Diagnostic and Statistical Manual of Mental Disorders, Fifth Edition (*DSM-5*) 114
Diagnostic and Statistical Manual of Mental Disorders, Sixth Edition (*DSM-6*) 81
differentiation 180
dissociation
  after trauma 105
  avoiding emotion 14–15
  personal stories on 31–32
DSM-5 (*Diagnostic and Statistical Manual of Mental Disorders*, Fifth Edition) 114
DSM-6 (*Diagnostic and Statistical Manual of Mental Disorders*, Sixth Edition) 81

E

emotional. *See also* adverse childhood experiences (ACEs); childhood emotional neglect (CEN); Feelings Wheel; trauma
  availability 29–30
  dangers 228
  neglect 29–30, 33
  processing 179
  vocabulary 17–18, 55–56
emotional wellness 2–4, 5–6, 155–156. *See also* connectedness; wellness, seven dimensions of
  cancer and 240–241
  exercise and 156
  first responder rhetoric and 24
  friendships and 131
  late onset stress symptomatology (LOSS) and 265
  reflection questions on 9
  retirement and 254–255
emotion, avoiding 14–15. *See also* dissociation
empathy, affective and cognitive 36. *See also* adverse childhood experiences (ACEs)
environmental wellness 7. *See also* wellness, seven dimensions of
Erikson's cognitive stages 263–264, 271
exercise
  emotional wellness and 152–153
  first responder performance improvements 153
  overcoming addiction and 156

F

family and first responders 29–30, 213
Feelings Wheel 17–18, 55–56
finances and relationships 217–218. *See also* money beliefs; relationships
  budgeting importance 223–225
  communicating goals 220–221
  money agreement topics, list of 225–226
  money scripts 218–220, 225
  money trauma and emotions 221
  personal stories on 218–219, 223–225
firefighters
  cancer statistics for 239
  hero mentality of 15–16
  traits of xvii, xviii
firefighter support web 4–5
first responder dilemma 16
first responder personality types 75
Five Cs of Successful Aging 271–272
flow 161
freeze response 101–104, 127. *See also* mind body intervention (MBX)
  personal stories on 101–103
  post-traumatic stress disorder as a result of 104
  trauma reactions 98–99
friendships
  emotional wellness and 131
  of the good 131
  of pleasure 131
  retirement and 254, 265
  of utility 131

G

gamma-aminobutyric acid (GABA) 166
gender pronouns xix–xx
gratitude 126. *See also* resiliency
  actions for happiness 161
  self-efficacy and 126
  the brain and 126–127
growth, personal xvii
gut microbiota 158

H

happiness
  self-care actions for 160–161
healing from depression 24
heart rate variability (HRV)
  narrative therapy and 177
  self-care actions and 163
hero mentality of firefighters 15–16
hidden depression 18. *See also* depression
high-intensity interval training (HIIT) 151
HRV (heart rate variability). *See* heart rate variability (HRV)
humor 130. *See also* resiliency

I

Illinois Firefighter Peer Support (ILFFPS) 141
imagery rescripting therapy 174

imaginal/prolonged exposure (reliving)
  therapy 174, 178–179. *See also* therapy
  and treatment for PTSD
  avoidance and 178
  differentiation and 180
  emotional processing 179
  late onset stress symptomatology (LOSS),
    overcoming 266
  process of 179–180
impulsive sex 58. *See also* addiction; relapse
incident commander 208
independence 129, 130. *See also* resiliency
inhibitory control 35. *See also* adverse
  childhood experiences (ACEs)
initiative 129–130. *See also* resiliency
insight 128–129, 130. *See also* resiliency
instrumental personality type 15
  business and 22
  emotional dangers and 228
  loneliness and 41
  retirement and 255
  self-medicating and 16
intellectual wellness 6. *See also* wellness,
  seven dimensions of
intentional breathing 162–163
  acetylcholine (ACh) and 163
  freeze response and 104
  heart rate variability and 163
  thought regulation and 127
  yoga and 166
interior chief 208
introverted leadership 186–187

J

Joiner's theories
  loneliness 40
  suicide 69–71

L

late-onset PTSD 259–261. *See also* post-
  traumatic stress disorder (PTSD)
  personal stories on 259–260
  research on 260
late onset stress symptomatology (LOSS) 98,
  262. *See also* trauma
  accelerated resolution therapy (ART) for 266
  imaginal/prolonged exposure (reliving)
    therapy for 266
  overcoming 265–266
  personal stories on 264–265
  reflection questions on 266
  retirement and 263
  rituals or ceremonies 265
LCU (life change unit) 253

leadership
  cutting-in-half technique 203
  deep listening and 202
  definition of 201
  five laws motivating 209
  maternalistic 202
  mitigating anxiety 202–204
  modeling vulnerability and coping
    flexibility 200–201
  paternalistic 79, 202
  after a potentially traumatic event
    (PTE) 205–206
  before a potentially traumatic event
    (PTE) 201–202
  during a potentially traumatic event
    (PTE) 204–205
  preventing PTSD with 196
  psychological body armor and 198
  psychological safety and 203
  reflection questions on 209–210
  training implications and 198–199
life change unit (LCU) 253
life review 254, 264
loneliness. *See also* needing to be needed
  broken attachments in childhood and 41–42
  emotional neglect and 29–30, 33
  epidemic of 39–40
  Joiner's theory on 40
  loop 43–44
  personal stories on 29–30, 39
  physical, mental, and behavioral impacts
    of 40–41, 42
  reflection questions on 44
  truth statements on 43
  work and 42–43
LOSS (late onset stress symptomatology).
  See late onset stress symptomatology
  (LOSS)

M

magical thinking 91. *See also* coping
manic episode 108–109. *See also* stress
mass-cluster suicides. *See* suicide
maternalistic leadership 202
MBX (mind body intervention). *See* mind body
  intervention (MBX)
meditation 151, 161
  cognitive occurrences during 167–168
  intrusive memories and 167
  mantra 163
  mindfulness 163
  moving 151, 161, 229, 266
  physiological occurrences during 166
  supporting science for 164–165
  yoga and 163

mental toughness xvi
mind body intervention (MBX) 104–105, 152–153, 165. *See also* freeze response
mindfulness 163
money beliefs 222–223. *See also* finances and relationships
money scripts 218–219
  establishing healthy 225
  examples of 220
moral injury 117–118
  personal stories on 118
  symptoms of 117
morality 130. *See also* resiliency
Myers-Briggs personality test 75, 255

N

National Center for Complementary and Integrative Health (NCCIH) 163
National Institute for Occupational Safety and Health (NIOSH) 239
National Institutes of Health (NIH) 3, 106, 165
National Registry of Evidence Based Programs and Practices (NREPP) 175
National Suicide Prevention Hotline 19, 83
nature connectedness 7
NCCIH (National Center for Complementary and Integrative Health) 163
needing to be needed 32–34. *See also* loneliness
  causes of 33–34
  life review and 254
  overcoming 38–39
  personal stories on 29–30, 39
  questions on 32
  reflection questions on 44
  relationship addiction and 37
  after retirement 256
  truth statements on 30, 38
  unhealthy behaviors of 37
neurogenesis 158
neuroplasticity 167
NIH (National Institutes of Health). *See* National Institutes of Health (NIH)
NIOSH (National Institute for Occupational Safety and Health) 239
NREPP (National Registry of Evidence Based Programs and Practices) 175
nutrition
  cancer and 240
  gut inflammation and 158
  hormonal imbalances and 158
  Mediterranean-style diet 158, 159
  mental health and 157–160
  vitamins and supplements for young women 159

O

occupational wellness 7–8. *See also* wellness, seven dimensions of

P

paramilitary communication style 214–215
parasympathetic nervous system (PNS) 105
  activating 166
  moving meditation and 162
  physical health and exercise and 155
passive toughening 154, 198
paternalistic leadership 79, 202
pediatric death
  difficulties of processing 88, 89
  personal stories on 87–88, 88–89, 89–90, 92
  processing and coping with 91
  reflection questions on 93
  truth statements on 91
peer support 187–188
  active-constructive responding and 188
  first responder brotherhood and 188–189
  self-disclosure and 188
personal growth xvii
personality types of first responders 75
physical wellness 5. *See also* wellness, seven dimensions of
physiological toughness 153–155
  active toughening 154, 198
  aging and 198
  avoidance and 154
  building, three factors for 154
  definition of 153
  early experiences 197
  exercise and 153
  mental and bodily experiences with 197
  passive toughening 154, 198
  training for 196–198
plans chief 208
PNS (parasympathetic nervous system). *See* parasympathetic nervous system (PNS)
point-cluster suicides. *See* suicide
post-traumatic growth (PTG) 125
  definition of 144
  pain, growing from 143–144
  personal stories on 141
  reflection questions on 145
post-traumatic stress disorder (PTSD) xvii, 12, 124. *See also* therapy and treatment for PTSD; trauma
  accelerated resolution therapy (ART) and 173
  assessment on 116
  avoidance and 99
  complex PTSD 116–117
  criteria for 114

post-traumatic stress disorder (PTSD) (*continued*)
  delayed onset PTSD 116
  emotion avoidance and 14
  late-onset PTSD 259–261
  leadership preventing 196
  mantra meditation and 163
  masculinity and 80
  meditation for 164–165
  mind body intervention (MBX) and 104–105
  moral injury and 118
  personal stories on 113, 115
  reflection questions on 119
  suicide and 74
  supporting science for yoga and meditation for 164–165
  symptoms of 114–115
  yoga for 163–164
potentially traumatic event (PTE) 193
  appraisal, factors influencing 194–195
  challenge appraisal and 194
  critical incident stress debriefing (CISD) and 206–207
  leadership after 205–206
  leadership before 201–202
  leadership during 204–205
  personal stories on 193
  preparing subordinates for 200–201
  preventing PTSD 196
  self-talk and 195
  threat appraisal and 194
  trajectories for, list of 194
premortem 250–251
prolonged exposure therapy. *See* imaginal/prolonged exposure (reliving) therapy
pronouns, gender xix–xx
psychological body armor
  components of 200
  leadership and 198
  relationships and 241
psychological injury. *See* trauma
psychologically safe work environment 203–204
PTE (potentially traumatic event). *See* potentially traumatic event (PTE)
PTG (post-traumatic growth). *See* post-traumatic growth (PTG)
PTSD (post-traumatic stress disorder). *See* post-traumatic stress disorder (PTSD)

Q

questions, reflection
  addiction 64
  brotherhood and strengths 190
  cancer 245
  death 251
  depression 25
  emotional wellness 9–10
  healthy relationships 237
  leadership 209–210
  loneliness 44
  needing to be needed 44
  pediatric death 93
  post-traumatic growth (PTG) 145
  post-traumatic stress disorder (PTSD) 119
  resiliency 137
  retirement 275
  retirement and late onset stress symptomatology (LOSS) 266
  self-care 168
  stress 110
  suicide 84
  therapy and treatment for PTSD 180
  trauma 110

R

rapid eye movement (REM) sleep 50
rapid intervention team (RIT) chief 208
recovery 61. *See also* addiction; relapse
  external 133
  helping someone in 62–63
  internal 133
  triggers and 57–58
reflection questions. *See* questions, reflection
relapse. *See also* addiction; recovery
  impulsive sex and 58
  risk factors for 57–58
  signs of 58–60
  triggers and 57–58
relationship addiction 37
relationships 129. *See also* finances and relationships; resiliency
  cancer and 241–242
  children 214
  communication 214–215
  conflict response 216–217
  first responder stress and intimacy 215–216
  improving 235–237
  listening well, basics of 217
  marriage, intimacy and longevity 231–232
  overworking 226–227
  positive interaction 229–230
  primary caretaker 214
  psychological body armor and 241
  reflection questions on 237
  relational excellence, basics of 236–237
  sex and intimacy 226
  sex and the emergency response 227–228
  spousal help, advice for 229–230

reliving therapy. *See* imaginal/prolonged exposure (reliving) therapy
REM (rapid eye movement) sleep 50
rescuer identity 31–32
rescuer's depression 13, 256
resiliency. *See also* gratitude
  advice on 123
  battery analogy for 133
  comfort with self and 132
  confirmation bias and 127
  control and 133–134
  coping and 122
  creativity 130
  definitions of 121–122
  humor 130
  independence 129, 130
  initiative 129–130
  insight 128–129, 130
  marks of 130–132
  mental recovery and 128
  morality 130
  optimism versus pessimism 125–127
  personal stories on 122–124, 136–137
  perspective and 131, 132
  physical activity and 132
  reflection questions on 137
  relationships 129
  relaxation exercises and 128
  self-awareness and 21–22
  self-doubt and 125
  self-efficacy and 126
  seven types of 128–130
  struggle versus growth 124–125
  tactical breathing 127–128
  thought regulation and 127–128
  through transition 134
retirement. *See also* change; late-onset PTSD; late onset stress symptomatology (LOSS)
  changing roles 271
  depression and 255–256
  emotional wellness and 254–255
  facing trauma after 259
  financing 272–273
  Five Cs of Successful Aging 271–272
  friendship and 254, 265
  late onset stress symptomatology (LOSS) and 262–263
  life review and 254
  male disconnection after 255
  meaningful productivity in 269
  personal stories on 254–255
  planning 270–271
  redefining 270
  reflection questions on 266, 275
  substance abuse and addictive behaviors 257
  successful, four pillars of a 275
  suicide and 258–259
  transitioning from full-time work 273–274
  working after 274
RIT (rapid intervention team) chief 208

S

safety chief 208
SAMHSA (Substance Abuse and Mental Health Services Administration) 21, 175
satisfaction, study on xvii
self-awareness 21–22. *See also* resiliency
self-care 38–39
  actions for treatment 161–164
  addiction and 57, 156
  anger, redirecting 156–157
  cancer and 240
  definition of 149
  exercise and 152–153
  how to begin 150
  intentional actions for happiness 160–161
  nature and 160
  nutrition and 157–160
  personal resiliency and 199
  personal stories on 147–149
  practices for first responders 151
  reflection questions on 168
  relationships and 228
self-disclosure 188
self-efficacy 126. *See also* gratitude
self-harm 75–76
self-medication 16–17. *See also* coping
  personal stories on 17
  retirement and 257
self-soothing
  addiction and 48
  emotional neglect and 33
  in rat parks 53
  substance abuse and 47
  unhealthy practices of 60
self-talk 195
sex
  addiction 33, 227
  compromising 227
  impulsive 58
  infidelity and 227
  intimacy and 226
  the emergency response and 227
Shneidman's theory on suicide 79–80, 82
SNS (sympathetic nervous system). *See* sympathetic nervous system (SNS)
Social Readjustment Rating Scale (SRRS) 253
social wellness 6. *See also* wellness, seven dimensions of

spiritual wellness 6–7. *See also* wellness, seven dimensions of
SRRS (Social Readjustment Rating Scale) 253
status, money 222
stories, personal
  abuse 31–32
  accelerated resolution therapy (ART) 175–176
  addiction 47, 54–55
  brotherhood and strengths 183–184
  connectedness 1–2
  depression 11–12
  dissociation 31–32
  finances and relationships 218–219, 223–225
  freeze response 101–103
  late-onset PTSD 259–260
  late onset stress symptomatology (LOSS) 264–265
  loneliness 29–30, 39
  moral injury 118
  needing to be needed 29–30, 39
  pediatric death 87–88, 88–89, 89–90, 92
  post-traumatic growth (PTG) 141
  post-traumatic stress disorder (PTSD) 113, 115
  potentially traumatic event (PTE) 193
  resiliency 122–124, 136–137
  retirement 254–255
  self-care 147–149
  self-medication 17
  substance abuse 17
  suicide 67–69, 71–72, 77–80
  training 199
  trauma 101–103
strengths, personal 185–186
  core values assessment 189–190
  introversion 186–187
  personal stories on 183–184
  reflection questions on 190
stress
  acute stress disorder 99
  areas affected by 107
  bipolar disorder misdiagnoses and 107–109
  chronic 96–97, 155
  common signs of 96
  definition of 95–96
  manic episode and 108–109
  reflection questions on 110
substance abuse
  personal stories on 17
  self-soothing and 47
Substance Abuse and Mental Health Services Administration (SAMHSA) 21, 175
suicide 19
  addiction and depression, links to 83
  cancer and 243–244
  contagion or cluster 76–77
  Joiner's theory on 69–71, 258
  moral injury and 118
  personal stories on 67–69, 71–72, 77–80
  reflection questions on 84
  retirement and 81–82, 258–259
  self-harm and 75–76
  Shneidman's theory on 79–80, 82
  situations that contribute to 73–74
  statistics on 68
  truth for those who consider 82–83
sympathetic nervous system (SNS) 105
  overactive SNS and relationships 216
  physical health and 155
  responses 106

T

tactical breathing 127–128
task-switching 35. *See also* adverse childhood experiences (ACEs)
therapy and treatment for PTSD. *See also* accelerated resolution therapy (ART); imaginal/prolonged exposure (reliving) therapy; post-traumatic stress disorder (PTSD)
  imagery rescripting 174
  narrative 148, 176–178
  reflection questions on 180
threat appraisal 194–195
tonic immobility. *See* freeze response
toughening
  active 154, 198
  passive 154, 198
training
  definition of 200
  implications for leaders 198
  personal stories on 199
  for physiological toughness 196–198
  psychological body armor and 198
transcending the self 249. *See also* death
transition
  processing 134–135
  resiliency through 134
trauma. *See also* adverse childhood experiences (ACEs); freeze response; late onset stress symptomatology (LOSS); post-traumatic stress disorder (PTSD)
  acute stress disorder (ASD) after 99
  attachment 34
  childhood 34
  definition of 97–98
  disconnecting after 105
  emotional neglect 29–30, 33
  experiencing 23–24
  human brain and 99–101
  life review and 254

mirror neurons and 101
money 221
nervous system and 105–107
personal stories on 101–103
reactions to 98–99
reflection questions on 110
after retirement, facing 259
traumatic events for first responders 97
truth statements on 261
trauma-sensitive yoga (TSY) 164
treatment, barriers to 173, 178
triggers 57–58. *See also* addiction; recovery; relapse
  identifying 59–60
truth statements
  adversity 35, 36
  cancer 243–244
  loneliness 43
  needing to be needed 30, 38
  pediatric death 91
  trauma 261
TSY (trauma-sensitive yoga) 164
twice-born individuals 141

V

veteran dilemma 16
vigilance, money 222
vocabulary, emotional 17–18, 55–56. *See also* Feelings Wheel
voluntary image replacement (VIR) 174. *See also* accelerated resolution therapy (ART)

W

wellness, seven dimensions of 4–8. *See also* emotional wellness
  emotional 5–6
  environmental 7
  intellectual 6
  occupational 7–8
  physical 5
  social 6
  spiritual 6–7
worship, money 222

Y

yoga
  assistance and support 135–136
  cognitive occurrences during 167–168
  definition of 163
  gray and white matter, effects on 167
  meditation and 163
  nidra 164, 165, 167
  physiological occurrences during 166
  PTSD and 163–164
  restorative 148, 163
  supporting science for 164–165
  transition and 134
  trauma-sensitive (TSY) 164